New Insights in Human Genomics

New Insights in Human Genomics

Editor: Joel White

Callisto Reference
www.callistoreference.com

Callisto Reference,
118-35 Queens Blvd., Suite 400,
Forest Hills, NY 11375, USA

Visit us on the World Wide Web at:
www.callistoreference.com

ISBN: 978-1-64116-797-0 (Hardback)

Cataloging-in-Publication Data

New insights in human genomics / edited by Joel White.
p. cm.
Includes bibliographical references and index.
ISBN 978-1-64116-797-0
1. Human genome. 2. Genomics. 3. Human genetics. 4. Medical genetics. I. White, Joel.
RB155.5 .E87 2023
616.042--dc23

Table of Contents

Preface

Human genomics is the area of study which focuses on human genes and chromosomes. The entire deoxyribonucleic acid (DNA) found in human beings is known as the human genome. It is composed of DNA found in mitochondria and chromosomes, which holds the genetic information. The human genome contains twenty three pairs of chromosomes, containing both protein-coding DNA sequences and different forms of non-coding DNA sequences. It is made up of long DNA polymers, similar to genomes of other living organisms. In each human cell, these polymers are kept in duplicate copies and encode the sequence of their constituent bases, which form the corresponding organism. The sequencing of the human genome has implications for a wide range of fields, including human evolution and molecular medicine. This book unfolds new insights in human genomics. It includes some of the vital pieces of work being conducted across the world, on various topics related to the human genome. This book is a resource guide for experts as well as students.

Various studies have approached the subject by analyzing it with a single perspective, but the present book provides diverse methodologies and techniques to address this field. This book contains theories and applications needed for understanding the subject from different perspectives. The aim is to keep the readers informed about the progresses in the field; therefore, the contributions were carefully examined to compile novel researches by specialists from across the globe.

Indeed, the job of the editor is the most crucial and challenging in compiling all chapters into a single book. In the end, I would extend my sincere thanks to the chapter authors for their profound work. I am also thankful for the support provided by my family and colleagues during the compilation of this book.

Editor

1

Chromatin Insulators and Topological Domains: Adding New Dimensions to 3D Genome Architecture

Navneet K. Matharu [1,*] and Sajad H. Ahanger [2,*]

1 Department of Bioengineering and Therapeutic Sciences, Institute for Human Genetics, University of California San Francisco, San Francisco, CA 94143, USA

2 Department of Ophthalmology, Lab for Retinal Cell Biology, University of Zurich, Wagistrasse 14, Zurich 8952, Switzerland

* Authors to whom correspondence should be addressed; E-Mails: navneet.matharu@ucsf.edu (N.K.M.); SajadHamid.Ahanger@usz.ch (S.H.A.).

Abstract: The spatial organization of metazoan genomes has a direct influence on fundamental nuclear processes that include transcription, replication, and DNA repair. It is imperative to understand the mechanisms that shape the 3D organization of the eukaryotic genomes. Chromatin insulators have emerged as one of the central components of the genome organization tool-kit across species. Recent advancements in chromatin conformation capture technologies have provided important insights into the architectural role of insulators in genomic structuring. Insulators are involved in 3D genome organization at multiple spatial scales and are important for dynamic reorganization of chromatin structure during reprogramming and differentiation. In this review, we will discuss the classical view and our renewed understanding of insulators as global genome organizers. We will also discuss the plasticity of chromatin structure and its re-organization during pluripotency and differentiation and in situations of cellular stress.

Keywords: insulators; architectural proteins; TAD boundaries; chromatin architecture

---

1. Introduction

To fit into the small volume of the nucleus, eukaryotes package their relatively large genome compactly into a nucleo-protein complex called chromatin. This packaging must ensure that specific regions in the genome remain accessible to large protein complexes that are responsible for important cellular processes such as transcription, replication, recombination, and DNA repair. At the same time, the effect of non-coding regulatory elements such as enhancers and repressors needs to be restricted to their cognate genes, preventing inappropriate activation or repression of neighboring expression domains. Additionally, developmental and tissue-specific expression demands dynamic reorganization of chromosomal domains at specific times and in specific subsets of cells. Eukaryotic genomes have evolved to organize chromatin fiber into a series of topologically and functionally independent domains. The structural organization of chromosomes at the interphase nucleus correlates with the transcription status of genomic regions. Highly transcribed regions are less dense and are placed towards the interior of the nucleus. DNA-FISH experiments have shown that different regions within the chromosome territory usually do not "intermingle" and therefore form separate compartments within a single chromosome [1]. The formation of such autonomous compartments in the chromosome can be achieved in a variety of ways, including establishment of a physical block to *cis* spreading of a chromatin state, recruitment of specific activities to a limited locus, or targeting to a sub-nuclear compartment associated with either silencing or activation. The study of factors and processes responsible for the formation and maintenance of such autonomous compartments is an active area of investigation. Specialized regulatory elements, termed "boundaries" or "insulators," have emerged as likely candidates to play this crucial role of chromatin compartmentalization. Such elements have been characterized by two experimentally defined properties involving altered gene expression. First, an insulator element acts as an enhancer-blocker by disrupting enhancer-promoter interactions, when positioned between an enhancer and a promoter, without rendering the enhancer inactive (as it is still capable of activating a "non-insulated" promoter) [2,3]. Second, when flanking a transgene, insulators are able to protect the transgene from position effects, particularly from the repressive effects of heterochromatin, allowing for position-independent gene expression [4]. This property of insulators is often referred to as barrier activity, since it involves blocking the spread of one chromatin state into another [5]. Some of the characterized boundaries have been shown to act primarily as barriers to heterochromatin in yeast, whereas others may possess both properties, *i.e.*, enhancer-blocking and barrier activity [6]. It is notable that, while barrier activity of boundaries prevents transcriptional repression, their enhancer-blocking property interferes with transcriptional activation. The two properties that define boundaries originate from the experimental assays used to identify and characterize them.

Insulator sequences and their associated proteins have been identified across species ranging from yeast to mammals. In yeast, TFIIIC is well characterized for its role in preventing spreading of repressive chromatin at RNA PolIII transcribed tRNA genes [7]. TFIIIC associates with some of the insulator elements in higher eukaryotes also. In mammals, a zinc-finger protein, CTCF, binds to most of the known insulator sequences. Several studies have shown that CTCF is involved in mediating chromatin interactions at several genomic loci in mice and humans that include but are not limited to beta-globin, H19/Igf2, MHC-II, HoxA, *etc.* [8]. It has also been found that both TFIIIC and CTCF require Cohesin for their insulator function in yeast and mammals, respectively. Studies have shown that tRNA genes can also function as enhancer-blockers in mammals, and that CTCF co-localizes with TFIIIC at several genomic loci, indicating a conserved mechanism of insulator function [9–11]. While many components of the insulator tool-kit are similar, there is considerable species-specific compositional variation. This is best reflected in *Drosophila*, which has evolved a diverse set of insulator factors and co-factors. These include the *Drosophila* homologue of vertebrate CTCF (dCTCF), Zest-white-5 (Zw5), Boundary Element Associated Factor-32 (BEAF-32), Suppressor of Hairy-wing [Su(Hw)], GAGA factor (GAF) and the recently identified proteins, Pita, and the zinc-finger protein interacting with CP190 (ZIPIC) [12–14]. These insulator binding proteins recruit co-factors such as Centrosomal Protein-190 (CP190) and Mod(mdg4), which are critical for their function [15,16]. This complexity of insulator tool-kit is well investigated in *Drosophila*; however, in mammals only CTCF has been studied as an insulator factor. It is possible that we are still far from our understanding of complex insulator functions in mammals. Recently, vertebrate GAF homologue has been investigated to function as an insulator component in mouse cells [17,18]. It has been also shown to tether chromatin loops to the nuclear periphery along with histone de-acetylase complex to repress gene expression [19]. Future studies are expected to enhance our understanding of genome insulation mechanisms in higher vertebrates.

2. Chromatin Insulators as a Genome Architectural Tool-Kit

The most widely accepted model of insulator function proposes that boundaries establish physical organization of the chromatin fiber into independent structural and functional domains. This model is based on the assumption that boundaries interact with each other or with components of "chromatin anchor points" such as nuclear matrix or nuclear envelope/lamina to loop out chromatin, providing steric and topological hindrance to enhancer-promoter interaction. The gypsy insulator in *Drosophila* provided one of the first illustrative examples of this model. A mutation in the *mod(mdg4)* gene that encodes one of the components of the gypsy insulator complex acts as an enhancer of position effect variegation (PEV) and has properties characteristic of the trithorax-group (trx-G) genes [20,21]. Additional support is presented by the fact that both Su(Hw) and Mod(mdg4) are found to be distributed at hundreds of sites on polytene chromosomes in salivary glands [22]. Given their distribution across the chromosome, one would expect to see a diffuse homogenous pattern of boundaries in the interphase chromosome of a diploid cell. Interestingly, this is not the case; instead, gypsy insulator proteins accumulate at a small number of nuclear locations. This observation has led to a suggestion that each of these sites where Su(Hw) and Mod(mdg4) proteins accumulate in the nucleus is actually several individual sites coming together, probably through their interaction with each other and/or with the components of other nuclear structure like lamina. However, the aggregation of these sites is not random; in fact 75% of them are present adjacent to nuclear lamina, suggesting that formation of such aggregates may require interaction between gypsy protein components and the nuclear lamina [23]. The tethering of gypsy insulators at the nuclear periphery, suggests that gypsy insulators might be component of MARs/SARs [22]. This idea is supported by the finding that sequences containing the gypsy insulators possess MARs activity [24]. This arrangement of gypsy insulators might create physically and topologically independent domains that interfere with the transmission of the signal from an enhancer located in one domain to a promoter located in the adjacent domain. Other insulator proteins in *Drosophila* such as BEAF 32B (Boundary element associated factor, isoform 32B) binds to scs and scs' boundary elements and tether these regions to nuclear lamina [25]. Loss of BEAF leads to abrogation of the boundary activity of scs and scs' regions and hence dissociation from nuclear lamina. Although boundaries primarily organize chromatin by physically restricting the legitimate enhancer-promoter communication, it is to be noted that not all insulators work via tethering chromatin to nuclear matrix or lamina proteins to bring about this organization.

All known vertebrate boundaries require CTCF for their insulator function. CTCF has been shown to target boundaries of the β-globin locus in chicken, mice, and humans to nucleolar periphery [26,27]. Co-purification studies have shown that CTCF co-purifies with nucleolar proteins present at the nuclear periphery. It was shown that CTCF interacts with nucleophosmin in nuclear matrix [26]. Nucleophosmin localizes at CHS4 insulators *in vivo*, along with other CTCF binding sites in the genome [26]. In addition, many CTCF-dependent boundary elements are known to tether to nuclear periphery. It was proposed that, like in *Drosophila*, CTCF-associated insulator complex tethers chromatin to nuclear matrix and thus subdivides chromosomes into active and inactive looped domains in mammals. Using 3C (chromosome conformation capture) assay, long-range interactions were shown to occur between murine β-globin locus and enhancer elements in the LCR, 40 to 60 kb away [28]. This particular interaction was only seen in erythroid cells, where β-globin locus was transcriptionally active [29,30]. At an earlier developmental time point in mice, the same LCR shifts its interaction with other globin genes, switching its transcriptional program in a developmental stage-specific manner [29]. Similar results were observed using RNA-TRAP to demonstrate that the active β-globin locus and enhancers of the LCR come in close contact *in vivo* [31]. The interaction of LCR with globin gene promoter is mediated by CTCF [30]. However, there is no direct evidence that this interaction involves nuclear matrix components.

CTCF-dependent chromatin boundaries of the imprinted Igf2/H19 locus have also been shown to function through the formation of chromatin loops. The parent-specific expression of Igf2/H19 was shown to be mediated through the establishment of chromatin loop domains formed by specific interactions between two differentially methylated regions (DMR) of H19 and Igf2 loci [32]. The differential interaction of H19 DMR and Igf2 DMRs generates a simple epigenetic switch for Igf2 transcription, through which it is localized to either inactive or active nuclear compartments. While in active state, the Igf2 gene promoter lies close to the downstream enhancers of the H19 locus, but not in its inactive state. These results suggest that boundaries may have an ability to relocate specific domains to the new nuclear environment as part of their mechanism of action.

It is emerging from the above studies that insulator elements function by organizing chromatin fiber into independent domains of gene activity. However, some results do not fit well with the structural model for insulation. For example, as with the gypsy insulator elements, interaction of the scs and scs' boundaries could explain their insulator function; however, the interaction between scs and scs' also depends upon the adjacent sequences, which may have insulation properties [33]. For example, homologous pairing of gypsy insulators or heterologous pairing of gypsy and binding sites for GAGA factor have been shown to neutralize insulator activity [34]. When 11 homologous and heterologous combination of boundaries were tested, it was seen that heterologous combination of gypsy with other boundaries or homologous pairing of other insulator elements such as scs or SF1 do not always reduce their enhancer-blocking activity [35]. In fact, some paired insulator elements exhibit increased enhancer-blocking activity, suggesting that they can function additively or independently. These early observations developed the "promoter-decoy model" for insulation. According to this model, an insulator sequence is essentially capable of recruiting the components of transcription machinery and hence can compete with the promoter to interact with the enhancer [36]. Co-localization of looped H19ICR (H19 Imprinting control region) insulator with its promoter and the trapping of HS2 enhancer complex of the human ε-globin gene by the inserted insulator substantiate the proposition of a promoter decoy model [37,38]. In the case of ε-globin gene, the HS2 enhancer delivers the transcription machinery to its promoter by tracking along intervening DNA sequence. An insulator inserted between the HS2 enhancer and the globin gene promoter stalls the transcription machinery at insulator site, thereby blocking long-range enhancer function [38]. Nevertheless, not all observed promoter-insulator interactions are explained by the promoter decoy model alone and there is no single inclusive model of insulator function [37]. Recently, the chromatin loop domain model has gained attention owing to its technological advancements, and hence it remains the most plausible model to explain the majority of enhancer-promoter interactions.

3. Insulators Organize Topological Domains of the Genome

Advances in genome technologies such as ChIP-seq, Chromosome Conformation assays (4C, 5C, Hi-C), and ChIA-PET have rapidly enhanced our understanding of the 3D architecture of the genome [39–45]. Chromatin conformation capture studies have revealed that the genomes of both vertebrates and invertebrates are divided into megabase to sub-megabase size chromatin domains (Figure 1a), within which smaller sub-domains are formed by local interactions at kilobase scale (Figure 1b). The large Mb scale domains are also termed "Topologically associated domains" or TADs. In *Drosophila* TADs are smaller than those of vertebrates, ranging from tens of kilobases to less than a megabase in structure [46]. They are largely conserved not only in multiple cell types but also across different species and are thought to serve an architectural role in the nucleus. It has been observed that the frequency of intra-TAD interactions is higher, while inter-TAD interactions are generally low. This has been attributed to flanking regions of the TADs termed as TAD-borders, which have low to very low interaction frequencies. Depending on the scale of DNA interaction frequency, these TAD borders can be strong or weak (Figure 1c). The TAD border strength is calculated as the ratio of intra/inter TAD interactions [47]. Data obtained using Hi-C have shown that there is a negative correlation between the strength of the TAD boundary and the frequency of the inter-TAD interaction. A stronger TAD boundary ensures lesser interaction frequency between neighboring TADs. Some of the TAD borders or "TAD boundaries" are rich in tRNA genes, SINE elements (both known to possess insulator activity), and housekeeping genes [48]. They also harbor DNase hypersensitive sites and have high occupancy of insulator proteins [47].

In *Drosophila*, several proteins such as BEAF-32, CTCF, Su(Hw), DREF, Rad21 (cohesin), TFIIIC, Cap-H2 (condesin II), Mod(mdg4), CP190, Z4, Chromator, L(3)mbt, and Fs1h-L were investigated for their binding on TAD borders [47]. While CTCF is known to mediate long-range enhancer-promoter interactions as well as acting as RNA PolII stalling site [49], it was also found to frequently associated with TFIIIC, Cohesin, and PRDM5 at TAD boundaries. Barring few sites, Su(Hw), on the other hand, was generally absent from TAD borders and mostly localized within TADs. Interestingly, strong TAD borders correlated with high occupancy of insulator factors. TAD borders that have high occupancy of insulator factors showed robust enhancer-blocking activity, while those with fewer insulator-binding sites behaved as weak enhancer-blockers [47]. These results suggest that genomic regions that have a high concentration of bound insulator factors (such as TAD boundaries) may serve genome architectural roles. However, we still do not know if TAD borders associate with nuclear lamina to loop out chromatin, as discussed in the previous section.

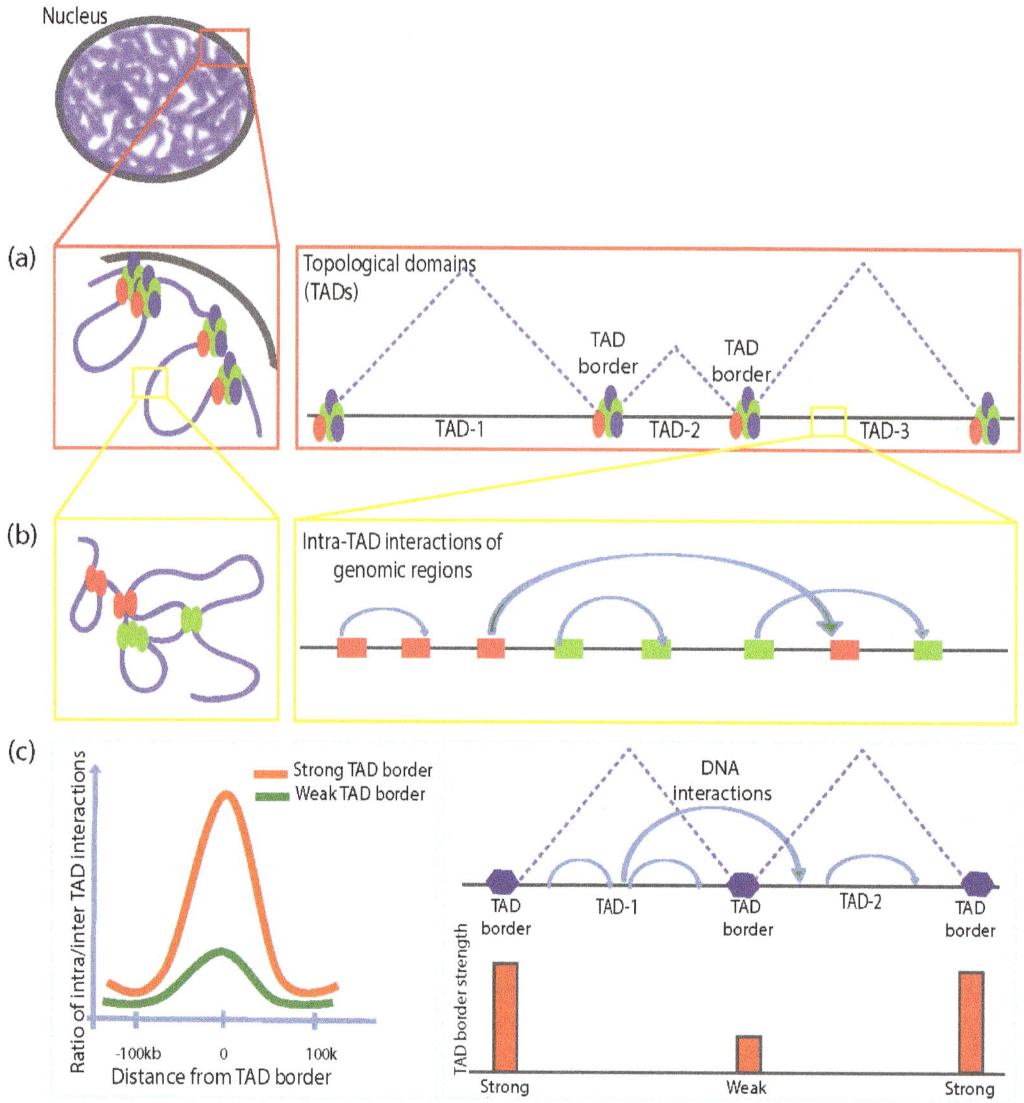

Figure 1. Chromatin nuclear organization with perspective of TAD structuring. (**a**) Megabase-scale chromatin looping and associated topological domain structuring. Insulator factors (shown in colored ovals) are decorated over chromatin mediating the Mb level chromatin looping; (**b**) Kilobase-scale chromatin looping mediated by insulator factors and associated co-factors facilitates specific interactions of gene regulatory regions. Kb-scale looping defines the transcriptional network on the genome; (**c**) TAD border strength and the DNA interaction frequency. A stronger TAD border is defined by the lesser frequency of inter-TAD interactions.

We are still in the early stages of our understanding of the composition of mammalian TAD boundaries. Only a fraction of CTCF sites are present at the mammalian TAD boundaries, while majority of the CTCF sites are found within TADs [48]. Several studies have shown that CTCF and Cohesin can differentially modulate inter and intra-TAD interactions. For example, loss of Cohesin in HEK289T and mouse thymocytes was shown to result in loss of intra-TAD interactions without affecting TAD architecture [50,51]. However, depletion of CTCF not only affected intra-TAD interactions but also led to loss of TAD boundaries and resulted in an increase in inter-TAD interaction [50]. Other studies have shown that depletion of Cohesin resulted in overall loss of both inter- as well as intra-TAD interactions by affecting Cohesin/CTCF-anchored sites [52]. CTCF and Cohesin are also involved in mediating enhancer-promoter looping; therefore, the loss of either or both results in the abrogation of enhancer-promoter interactions [53]. Future studies are expected to shed more light on the components of TAD structuring in mammals. Taken together, these results suggest that eukaryotic genome is organized by twofold architectural looping; (1) at the Mb-scale, it is organized into topological domains (TADs); and (2) each TAD is further organized at the Kb-scale, to facilitate more local interactions between gene regulatory elements necessary for generating unique transcriptional outputs [54,55]. The role of insulator factors at both levels of genome organization is beginning to emerge. Insulator factors binding to TAD borders at high concentrations may act as "super-insulators" to maintain Mb-level TAD structures. However, at the kilobase scale, insulator factors along with other transcription factors may facilitate more local looping interactions within the TAD structures. Therefore, at least in *Drosophila*, most of the typical "insulator factors" are emerging to be "genome architectural proteins", thus deviating from their classical definition.

Given the critical role of TAD boundaries in genome organization, it is not surprising that genomic defects that disrupt TAD boundaries can cause debilitating diseases. A number of human diseases are associated with chromosomal aberrations such as deletion, inversion, translocation, and duplication. To what extent such genomic variations affect chromatin topology and their subsequent contribution to the disease phenotype is nearly untouched. A recent study demonstrated how structural anomalies in the genome could disrupt TAD organization and result in at least three related human genetic disorders [56]. Three different types of limb malformations, namely brachydactyly (short digits), F-syndrome syndactyly (fused axial digits), and polysyndactyly (duplicated and fused digits), identified in three different families, were investigated. By performing comparative genomic hybridization (CGH), the above mentioned malformations were shown to be associated with genomic aberrations in the q arm of chromosome 2, having four important coding genes, WNT6, IHH, EPHA4, and PAX3. Investigating the TAD organization of the locus revealed that it is structurally divided into three independent TADs, PAX3-TAD, EPHA4-TAD, and WNT6/IHH-TAD (Figure 2a). The brachydactyly family has a deletion that encompasses portions of EPHA4-TAD as well as the boundary separating it from PAX3-TAD (Figure 2b). The F-syndrome family has inversions or duplication having breakpoints within WNT6/IHH TAD and EPHA4-TAD, encompassing the TAD boundary between these two (Figure 2c). The polysyndactyly family has duplications and deletions within WNT6/IHH TAD, also disturbing its TAD boundary (Figure 2d). All these chromosomal aberrations were re-engineered in a mouse model using the CRSIPR/Cas9 system as well as in hESC (human embryonic stem cells) to map genomic interactions using 4C. The gene expression profile of the locus revealed non-cognate association of the gene promoter with the enhancers. These severe limb malformations clearly resulted from perturbations in the TAD structure and its boundaries, which relocate enhancers with gene promoters. These TAD boundaries are associated with CTCF-loop domains in mouse limbs. This study provides strong evidence that disruption of TADs and TAD boundaries could cause severe developmental disorders in humans. Deciphering the structural basis of X-inactivation in *Caenorhabditis elegans* also revealed the importance of TAD boundaries. A dosage-compensated X-chromosome was observed to have discrete self-interacting 1Mb domains similar to TADs. The boundaries of these TADs are much stronger and are decorated by dosage-compensation condensin complex (DCC) [57]. Loss of DCC complex or deletion of its binding sites results in weaker TAD boundaries and defective dosage compensation. However, these TAD borders were not tested for classical boundary function. These studies indicate that, as in *Drosophila*, TAD borders in other organisms may also harbor insulator function and TAD borders may represent another class of insulator elements. Future studies are expected to delineate their discrete nature from classical insulator elements.

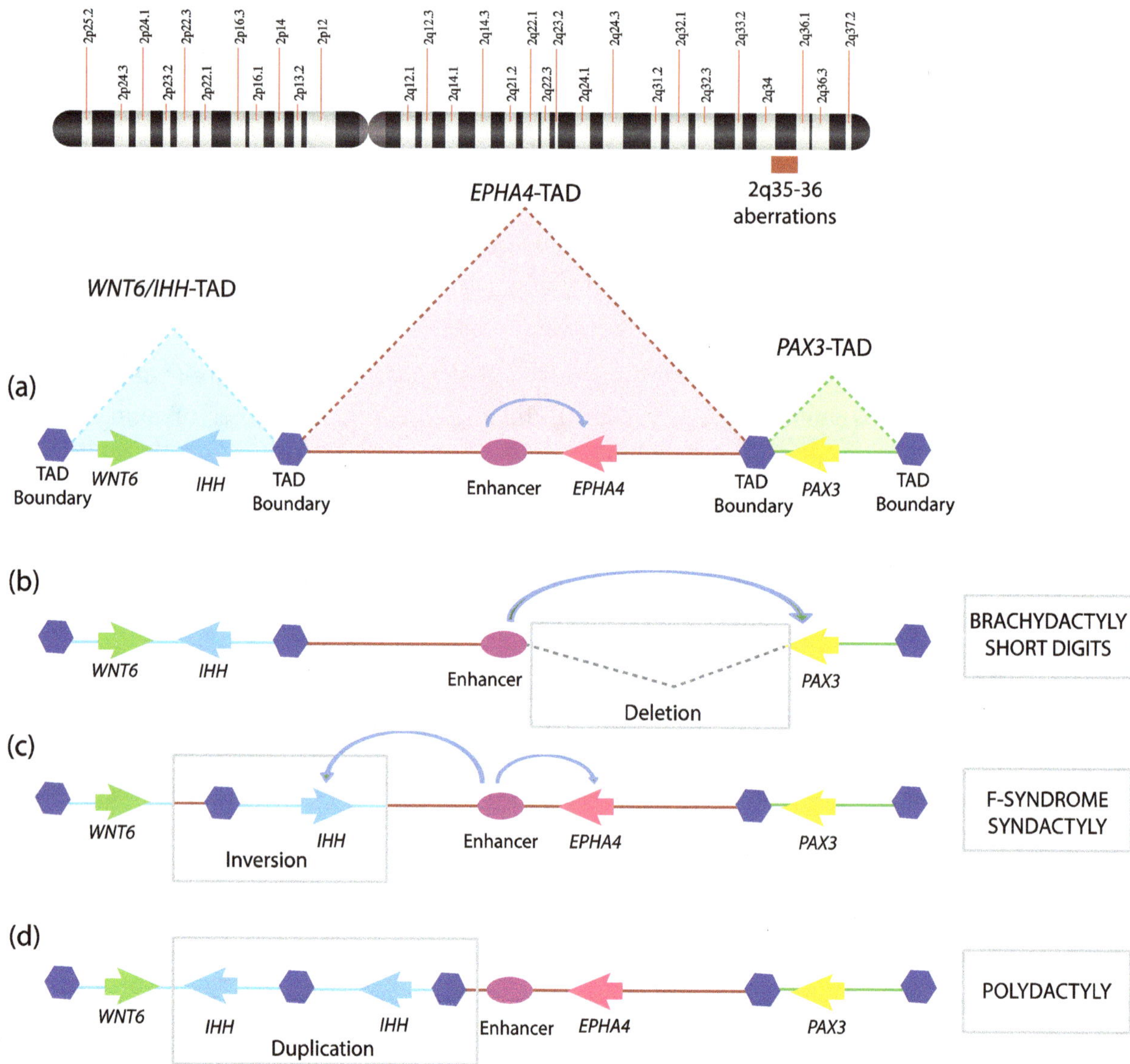

Figure 2. Topological domain disruption linked to limb malformations. Mutations encompassing the *WNT6-IHH/EPHA4/PAX3* locus on human chromosome 2q arm disturb TAD structuring and are thus involved in limb phenotypes; (**a**) The *WNT6-IHH/EPHA4/PAX3* locus is organized into three TADs with intra-TAD specific interactions playing a key role in setting up the transcriptional program during limb development; (**b**) A deletion encompassing the TAD boundary between *EPHA4/PAX3* leads to mis-assignment of *EPHA4* enhancer to *PAX3* promoter and hence short digits; (**c**) Inversion surrounding *IHH* and its nearest TAD boundary allows *EPHA4* enhancer to drive the IHH promoter, which was prevented by structural hindrance posed by the TAD boundary in a wild-type scenario, hence leading to F-syndrome; (**d**) Duplication of *IHH*/TAD boundary results in extra TAD, which causes the polydactyly phenotype.

4. Stress Response and Insulator-Mediated Chromatin Dynamics

Cellular stress response is an evolutionarily conserved mechanism that enables organisms to adapt to several environmental challenges such as elevated temperatures, environmental genotoxic agents, and mechanical injury. During conditions of stress, cells undergo a wide range of molecular changes that include the shutting down of many actively transcribed genes and transcriptional activation of several previously silenced genes. Thus cellular response to stress presents a particularly intriguing paradigm to study chromatin dynamics. Recently, insulator factors have been implicated in chromatin re-structuring during temperature stress response in *Drosophila* [58]. It was observed that elevated temperature resulted in disassociation of chromatin-bound CP190, the common co-factor of many *Drosophila* insulator proteins, resulting in global chromatin changes and looping interactions. Indeed, this was shown in a recent report in which heat shock induced re-localization of insulator proteins to ectopic sites, resulting in inter-TAD interactions and large-scale chromatin reconfiguration [59]. Using Hi-C methods, the chromatin dynamics of long-range interactions were analyzed in *Drosophila* Kc167 cells before and after heat shock [59]. Surprisingly, heat shock conditions resulted in more inter-TAD interactions with a concomitant decrease in TAD border strength, whereas before heat shock intra-TAD interactions were predominant. Before heat stress, in Kc167 cells, ChIP-Seq studies using antibodies against a variety of insulator/architectural proteins like BEAF-32, CTCF, Su(Hw), DREF, Rad21 (cohesin), TFIIIC, Cap-H2 (condesin II), Mod(mdg4), CP190, Z4, chromator, L(3)mbt, and Fs1h-L showed that except for Rad21 and Cap-H2, all other architectural/insulator proteins show higher distribution of DNA-bound peaks at TAD boundaries (Figure 3a). Upon heat shock. the distribution of these factors was found to be abundant in interior regions of TAD, at enhancers and promoters away from TAD boundaries, which results in a weakening of TAD boundary strength (Figure 3b). Additionally, the distribution of histone modification marks before and after heat shock in *Drosophila* presented drastic changes in H3K27me3 marks along with ectopic Polycomb (Pc) occupancy at enhancer regions that were previously marked with H3K27ac [59]. This indicates that regulatory regions are repressed upon heat shock and Pc is implicated in the mediated silencing of enhancers upon heat shock. However, it is still not clear if Pc silencing is the direct cause of reorganized insulator factor binding or a secondary response to temperature stress. From these studies, it emerges that chromatin-bound insulators relocate themselves to ectopic sites to re-organize the chromatin upon temperature stress, which leads to changes in transcriptional output. However, in mammals, similar findings are yet to be investigated.

Other studies that have linked insulator factors to stress response include the formation of insulator bodies under conditions of osmotic stress [60]. Insulator bodies are nuclear foci consisting of multiple insulator factors such as Su(Hw), Mod (mdg4), dCTCF, and CP190 that were previously thought to represent functional looping contacts between insulator-bound proteins [61], although this was later refuted in other studies [62]. However, recent studies have indicated that these structures are not formed by chromatin-bound insulator contacts but rather as a result of insulator factors disassociated from chromatin that dramatically alter the chromatin configuration in response to osmotic stress [60]. This osmotic stress response was shown to be specific to insulator factors, as other speckle-forming proteins like Polycomb factors and HP1 foci did not change under osmolarity stress. Finally, it was also shown that the loss of chromatin-bound Su(Hw) protein is reversed once cells were shifted to isotonicity [60]. Although the mechanisms of temperature and osmolarity stress response differ markedly, both of them bring about changes in chromatin dynamics. Taken together, these findings suggest that there is a dramatic disassociation and relocation of chromatin-bound insulator factors during cellular stress. It remains to be seen whether insulator-bound-factors other than Su(Hw) are also relocated in response to osmolarity stress and, like heat stress, if these changes are related to TAD structuring.

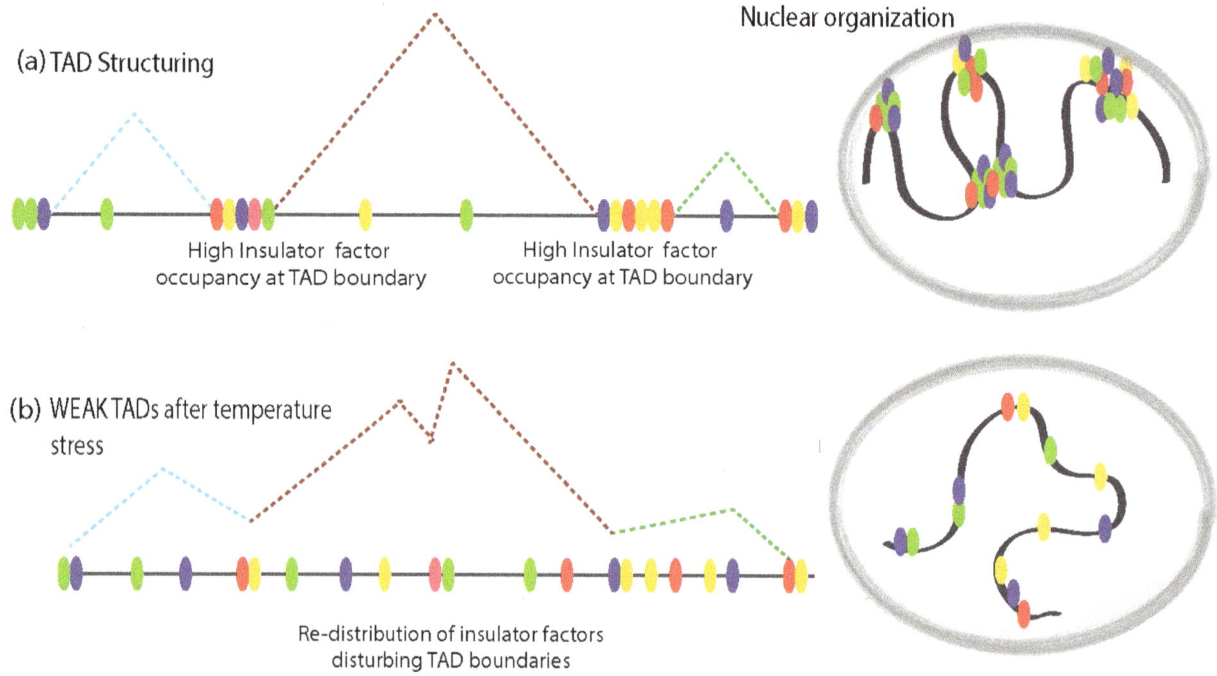

Figure 3. Stress induces redistribution of insulator factors and reconfiguration of chromatin structure. (**a**) Normal TAD is structured by architectural/insulator proteins clustered around their boundaries, maintaining the Megabase scale chromatin structuring, while intra-TAD looping in mediated by discrete motifs for insulator factor binding; (**b**) Under temperature stress, TAD boundaries get weakened due to re-distribution of insulator factors to intra-TAD regions. This may cause promiscuous looping interactions resulting in aberrant transcriptional program.

5. Chromatin Dynamics in Pluripotency and Differentiation

Stem cells have the ability to self-renew and differentiate into multiple lineages of functionally specialized cells in response to appropriate differentiation cues. This cellular plasticity is attributed to several unique structural and functional features of the stem cell genomes. These include hyper-dynamic association of chromatin-interacting factors, disorganized heterochromatin, enhanced global transcriptional activity in coding and non-coding regions, and activation of the second X-chromosome in the female cells [63,64]. The process of differentiation is accompanied by repression in transcription of pluripotency-associated genes, silencing of repeat elements, more stable interaction of chromatin-binding factors, and inactivation of the X-chromosome and formation of distinct heterochromatin foci [64]. Conversely, during reprogramming of differentiated cells into pluripotent state (iPSCs), cells must re-acquire the structural and functional features associated with pluripotency. The mechanism(s) and factors responsible for this genomic plasticity remain elusive. Findings over the past five years have strongly supported a role of insulator factors such as CTCF in shaping the 3D genome architecture in pluripotent cells and its re-organization during differentiation (and *vice versa*). Using ChIA-PET (chromatin immunoprecipitation interaction assay with paired end tagging), Handako *et al.* presented a global and high-resolution CTCF associated interaction map in mouse Embryonic Stem (ES) cells [65]. Their study revealed that CTCF configures the genome in ES cells into distinct chromatin domains and sub-nuclear compartments that bear unique epigenetic features and transcriptional activities. Other studies have shown that CTCF mediated TADs are invariant to stem cells and are conserved not only among diverse cell types but also across species [48]. This indicates that insulator proteins such as CTCF may act as general architectural proteins and that cell-type-specific interactions are mediated by other factors. Indeed, it was recently shown that Cohesin and Mediator but not CTCF play a key role in mediating such cell-type-specific interaction [66]. For example, Cohesin was shown to co-bind genomic sites of CEBPA and estrogen receptor (ER) in HepG2 and MC7 cells respectively, to generate cell specific transcription program [67]. It was also shown that Cohesin is important for mediating chromatin interaction involving ER in MC7 cells. Similarly, Cohesin and Mediator along with the Cohesin loading factor Nipbl (Nipped-B like) were shown to specifically mediate ES cell specific enhancer–promoter interactions [53]. In contrast to CTCF/Cohesin co-bound sites, Cohesin and Mediator co-occupied sites are cell-type-specific and often overlap with pluripotency factors such as Nanog, Oct4, and Sox2 in ES cells. In the same study it was also shown that Mediator and Cohesin occupy different promoters in different cell types and mediate chromatin looping that generates a cell-type-specific expression program (Figure 4a,b).

Since large, megabase-scale TADs are largely conserved between cell types, it is hypothesized that chromatin interactions at sub-megabase level could account for the lineage specific expression programs (Figure 4a,b). To test this, Phillips-Cremins *et al.* used 5C to map interactions in ESCs and Neural progenitor cells (NPCs). They found that large, megabase-scale interactions were enriched in CTCF/Cohesin and were conserved between ESCc and NPCs, a result consistent with previous reports of conservation of TADs across cell types [66]. Cohesin and Mediator co-bound sites were involved in sub-megabase level promoter–enhancer interactions within TADs that were specific to ESCs and NPCs. Depletion of Cohesin or Mediator components resulted in disruption of this cell-type-specific interaction and downregulation of genes within these interactions. It is to be noted that these cell-type-specific interactions are defined by lineage-specific transcription factors during differentiation and are reinforced/maintained by Cohesin/Mediator complex. During T-cell differentiation, multipotent naïve cells, upon receiving antigenic signals, make the lineage choice between Th1 and Th2 cells. These two differential lineages maintain overall similar chromosome interactions, except at the minor looping level, especially at the cytokine locus. Variable genomic contacts between Th1 and Th2 are maintained by DNA-binding, cell-specific transcription factor STAT [68]. In the absence of STAT, the cytokine locus is unable to lose its promiscuous contact with naïve T cells, thus indicating the importance of cell-type-specific factors in chromatin looping rearrangement [68]. Recently, it has been shown that TAD structures are relatively stable across cell types; nevertheless, there are considerable variations accounted for during differentiation [69]. More studies in future are expected to develop a comprehensive and better understanding of what factors and cellular demands lead to TAD re-structuring.

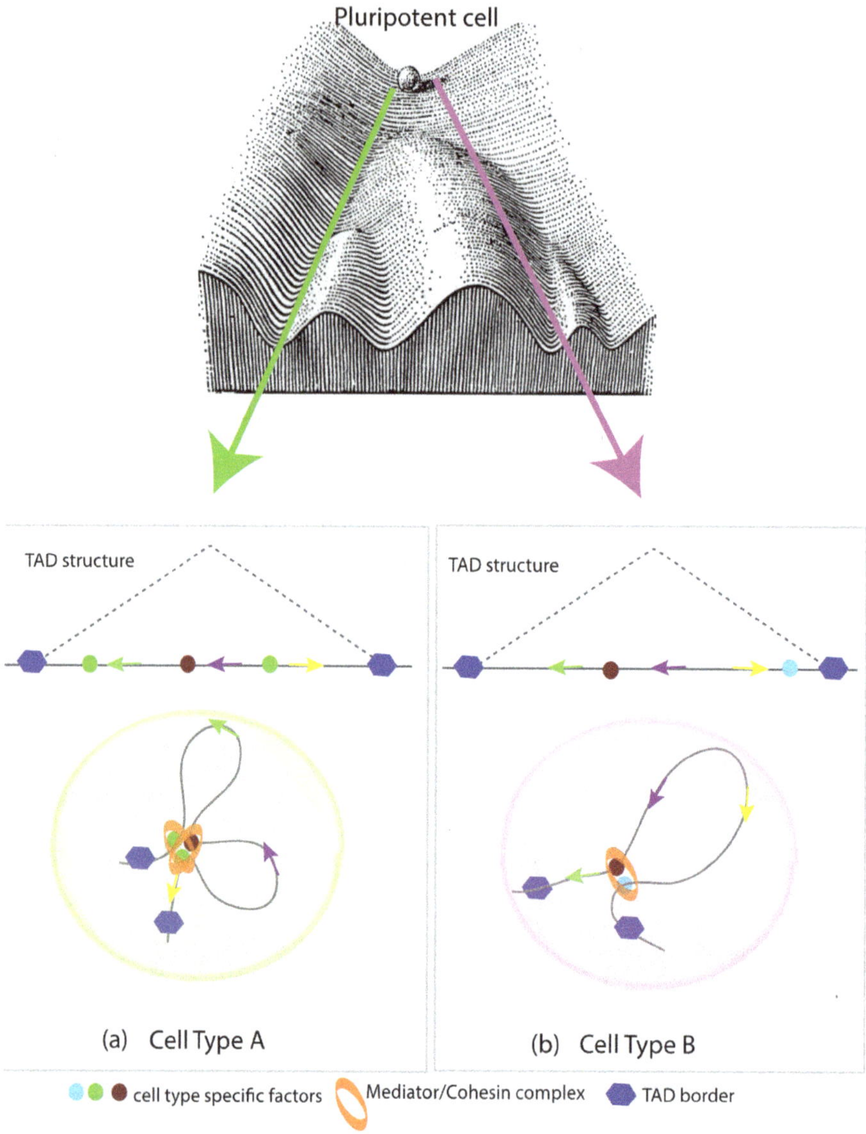

Figure 4. Cell-type-specific genome wiring at the kilobase level. This schematic shows chromatin organization defining the cell type specificity for pluripotent cell along the Waddington's epigenetic landscape. (**a**) Intra-TAD looping in Cell type A, mediated by the Mediator/Cohesin complex, is instrumental in specifying cell type chromatin folding; (**b**) Despite the overall TAD structure remaining consistent, it is the kilobase-scale interactions in Cell type B that assign the transcriptional program of the cell leading to its lineage specification.

The reprogramming of somatic cells into pluripotent state is a very inefficient process *in vitro*, and our understanding of the barriers to this process is limited. Only 1%–3% of somatic cells undergo reprogramming to attain pluripotency. Several recent studies have analyzed the interactions in ESCs, iPSCs, and their differentiated progenies and have uncovered chromatin interactions that are unique to successfully reprogrammed cells. One of these studies showed that during reprogramming, overexpressed pluripotency factors (OCT4 and NANOG) are not only bound at endogenous gene loci in successfully reprogrammed cells (iPSCs) but also in un-reprogammed cells [70]. However, specific interaction between pluripotency genes, OCT4, NANOG, and SOX2 was only observed in iPSCs that resulted in transcription from these gene loci [70]. These interactions were shown to be mediated by the Cohesin complex. Using 5C in ESCs, it was shown that Nanog promoter interactions are rearranged upon differentiation and re-established in induced pluripotent cells [71]. A large fraction of Nanog-interacting loci were bound by Mediator or Cohesin and the depletion of either in ECSs resulted in disruption of contacts and re-acquisition of a differentiation-specific interaction pattern. Either loss of chromatin interaction in ESCs or their re-acquisition during induced reprogramming often preceded transcriptional changes, suggesting a functional link between chromatin interactions and transcriptional change. Interaction studies of Oct4 locus using 4C and FISH also suggested a causative role for chromatin interaction and transcriptional activity. Several regions that frequently interacted with the Oct4 locus were identified, which showed that these interactions precede the transcriptional activation of endogenous Oct4 during reprogramming [72]. Klf4 was shown to be critical for these interactions as depletion of Klf4 lead to Cohesin unloading from the Oct4 promoter and loss of transcription of the endogenous Oct4. Studying the long-range interactions in ESC, iPSCs, and differentiated cells revealed that genomic contacts made by pluripotent factors like NANOG are more independent than other classes of contacts such as those mediated by Polycomb proteins. Loss of the Polycomb protein EED resulted in the loss of Polycomb-specific contacts only, preserving the interactions made by pluripotent factors necessary for differentiation [73].

Taken together, the above findings suggest that there are key interactions between pluripotency genes and architectural proteins that are specific to ESCc and iPSCs. During differentiation, looping contacts are specified by cell-type-specific transcription factors within TADs and are mainly maintained by Cohesin/Mediator. However, major loops defining TAD structures coincide with CTCF and Cohesin binding, which are proposed to play a more fundamental role in chromosomal folding.

6. Regulation of Insulator Function

Given the idea that insulators compartmentalize the genome into structurally and functionally independent domains of gene activity, it is conceivable that as cells differentiate, insulator-mediated changes in chromatin organization precede or accompany cell differentiation. Such changes may be crucial for the establishment or maintenance of specific patterns of gene expression. If this is true, then cells must possess mechanisms to regulate insulator activity to establish distinct patterns of chromatin organization that are cell type specific. Indeed, several examples exist that demonstrate that insulator function can be modulated at several stages. These mechanisms may involve interfering with protein-protein and/or protein-DNA interactions via protein modification, DNA modification, or competition for DNA binding. Analysis of BEAF-32 binding sites in *Drosophila* provided the first evidence for regulation of insulator activity by competition [74]. Two regions that bind to BEAF-32 act as insulators and are also bound by DREF. BEAF-32 and DREF occupy the sites independently, which leads to a model where these two proteins compete for DNA binding [74]. This observation suggested that the same boundary sequence, when bound to BEAF-32, forms a site of insulation, interfering with enhancer-promoter interaction while DREF binding blocks BEAF-32 and allows enhancer-promoter communication. In a similar manner, covalent modification of the insulator binding sites has been shown to regulate insulator function at the Igf2/H19 locus. DNA methylation of the CTCF binding site on the paternal chromosome blocks its binding, which renders the insulator inactive, resulting in only Igf2 expression in mammals. On the maternal chromosome, the CTCF binding site is not methylated and therefore the insulator is functional, resulting in H19 expression only [75,76]. This mark is imposed during male germ cell development and stably maintained thereafter. This is an example of the epigenetic regulation of insulator activity using a mechanism that involves inhibition of binding of the insulator protein to its DNA sequence.

Covalent modification of insulator proteins may serve as the most predominant means of regulating insulator function. For example, in *Drosophila*, ubiquitination of Su(Hw) by the E3 ubiquitin ligase dTopors increases its insulator activity [77]. A mutation in the dTopors gene reduces the insulator function of Su(Hw), while its over-expression can rescue insulator activity. Since one of the models proposes that insulators function via loop formation by interacting with each other or the components of nuclear structure, it is likely that their function is also regulated through modulation of loop formation (Figure 5a). Poly(ADP-ribosyl)ation or PARylation of CTCF has been shown to regulate CTCF-dependent insulator function in *Drosophila*. Using ChIP-on-chip, it was found that Poly (ADP-ribose) (PAR) co-localizes with around 78% of the CTCF genome-wide binding sites in mouse fetal liver cells. PARylation is necessary for the insulator activity of CTCF at H19-ICR locus, as treatment with 3-aminobenzamide (an inhibitor of PARylation) results in loss of insulator function [78]. Since PARylation has been found to promote protein-protein interactions, it is likely that PARylation regulates insulator activity by promoting protein-protein interaction between insulator sites [79] (Figure 5b). Indeed, PARylation has been shown to be important for the attachment of Nuclear Matrix-associated regions to the nuclear periphery. Inhibition of PARylation results in loss of CP190, Su(Hw), CTCF, and Mod(mdg4) in the nuclear matrix fraction and a two-fold reduction in their binding peaks in ChIP-Seq analysis. Inhibiting the process of PARylation results in loss of intra-chromosomal interactions [80]. When TAD structures were investigated in *Drosophila*, the loss of PARylation did not disrupt the binding of insulator factors on TAD boundaries; however, it is not known if these insulator factors are PARylated while bound to TAD boundaries [80].

Modification of Mod(mdg4)2.2 and CP190, the components of the Su(Hw) insulator complex, by SUMO (small ubiquitin-like modification) is another example of regulation of insulator activity by modulating protein–protein interactions [81]. SUMO co-localizes with a number of insulator sites throughout the genome, as shown by immunostaining of polytene chromosomes in *Drosophila*. However, in contrast to PAR, SUMO appears to negatively regulate insulator function as mutations in the SUMOylation pathway result in enhanced insulator activity (Figure 5c). Furthermore, over-expression of SUMOylation pathway components has been shown to disrupt the formation of insulator bodies in diploid nuclei from third instar larvae of *Drosophila*. This observation suggests that SUMOylation regulates insulator function by interfering with the ability of insulators from different sites to interact with each other (Figure 5c).

A few years ago, it was shown that RNAi machinery also plays a role in regulating interaction between insulator-associated proteins. The Su(Hw) insulator requires RNAi machinery to make RNAs that mediate the interactions between individual insulator sites to form insulator bodies. Mutations that affect the components of RNAi machinery and presumably impair the production of these RNAs affect insulator function. For example, loss of function and over-expression studies showed that the helicase Rm62 has a negative effect on insulator function, whereas Argonaute (Ago) proteins facilitate their activity [82]. Moshkovich and colleagues showed an RNAi-independent role of Argonaute-2 in dCTCF/CP190-dependent insulator function [83]. Ago2 was shown to physically interact with dCTCF and CP190 and mutation in Ago2 results in a reduction of chromosomal looping interaction and loss of insulator activity.

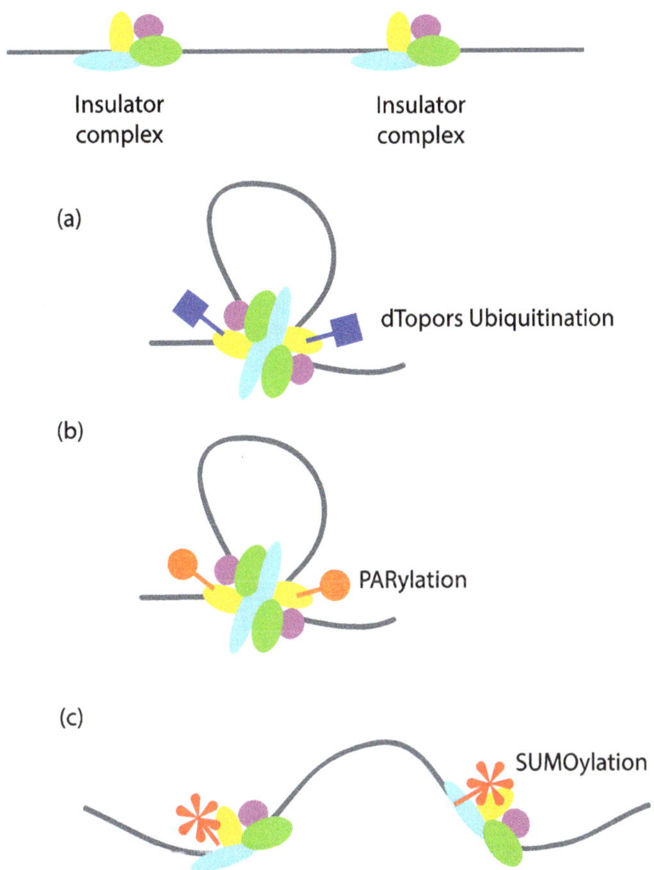

Figure 5. Regulation of insulator activity by covalent modification of insulator factors. Covalent modifications of insulator factors can have both positive and negative effects on insulator-mediated looping interactions. (**a**) Ubiquitination of Su(Hw) by dTopors potentiates its insulator activity by mediating looping interactions; (**b**) PARylation of CTCF regulates its intrachromosomal interactions and hence its insulator activity; (**c**) SUMOylation of Mod(mdg4) and CP190 negatively regulates their activity in *Drosophila* by interfering with the chromatin loop formation.

Apart from the aforementioned ways of regulating insulator activity, there are other modes through which insulator function can be modulated. For example, the *Abdominal-B (Abd-B)* locus in *Drosophila* consists of a series of developmentally regulated enhancers, silencers, and insulators. The question of how a correct enhancer targets the *Abd-B* gene in a given cell type is intriguing. Several insulator elements in the *Abd-B* region have been shown to exhibit insulator bypass (pairing of insulators), resulting in loss of insulator function [84–87]. Utilizing this mechanism, specific enhancers can be targeted to their cognate promoters in a particular cell type and at the correct developmental stage. Another way to achieve developmentally regulated expression of the *Abd-B* locus is through Promoter Targeting Sequences (PTS). Such sequences have been found at the 3' region of two of the insulator elements, Fab-7 and Fab-8, and allow enhancers to overcome dCTCF and Su(Hw) enhancer-blocking activity in transgenic assays [88,89]. Nearby sequences affecting insulator function have also been found in vertebrates. Thyroid response elements (TRE) found upstream of the chicken *lysozyme* gene and TRE upstream of the *c-myc* gene have been found to reduce CTCF enhancer-blocking activity [90]. CTCF remains bound to DNA in the presence of the hormone, indicating that TRE must inactivate CTCF.

An additional layer of regulation of insulation comes from various interacting partners [91]. CHD8 enhances insulator activity, whereas Kaiso has a negative effect on the CTCF-mediated enhancer blocking [92,93].

Given the vast number of strategies for regulating insulator function both at the level of DNA binding and in modulating protein–protein interactions that mediate loop formation, insulator activity is dynamic. Interfering at the level of DNA binding could be a more permanent form of regulation, whereas protein competition and modification are reversible modes of regulation that could be adapted by cells to respond to changing needs during development and differentiation or during cellular stress.

7. Conclusions and Future Directions

Recent studies have extensively mapped chromatin interactions at multiple spatial scales and greatly expanded our understanding of the organizational role of insulators/architectural proteins in 3D genome organization. Several questions remain unanswered that need to be addressed in future studies. For example, it remains to be established whether gene expression is a cause or effect of such interactions, although recent studies have provided some evidence in favor of the latter [72]. Moreover, chromatin interactions have been predominantly analyzed in cases of gene activation; equally important will be to map looping interactions that result in gene silencing, which may be important not only for maintaining pluripotency (by silencing differentiation genes) but also for generating/maintaining cell type specificity. It is also important to keep in mind that interactions maps are generated from a large pool of fixed cells and therefore do not provide the dynamics of those interactions. Future studies aiming to image the dynamic clustering of insulator factors might throw more light on this proposition. Perhaps the most important challenge in the field will be to link chromatin interactions to their functional outcomes. Recent advancements in gene editing technologies such as the CRISPR-Cas9 system will allow for perturbing or manipulating these interactions to understand their underlying functions. Finally, the perturbation of TADs and TAD boundaries as the underlying cause of human genetic defects is a powerful example of how chromatin organization is relevant to human disease. Therefore, it is essential to consider the impact of genetic defects on chromatin topology in order to fully understand the molecular mechanism of their pathogenesis.

Acknowledgments

Navneet K. Matharu acknowledges Nadav Ahituv's group at the Department of Bioengineering and Therapeutics, University of California, San Francisco for their help. Sajad H. Ahanger thanks his current group, led by Christian Grimm of the Department of Ophthalmology, University of Zurich, and his past lab led by Rainer Renkawitz at the Institute of Genetics, Justus-Liebig University, Germany, for their support. The authors also thank Rakesh Mishra's team at the Center for Cellular and Molecular Biology, CSIR, India for their support. We apologize to those colleagues whose work was not cited due to space limitations.

Author Contributions

Both authors contributed equally. Sajad H. Ahanger and Navneet K. Matharu reviewed literature and wrote the manuscript. Navneet K. Matharu and Sajad H. Ahanger made figures.

References

1. Goetze, S.; Mateos-Langerak, J.; Gierman, H.J.; de Leeuw, W.; Giromus, O.; Indemans, M.H.; Koster, J.; Ondrej, V.; Versteeg, R.; van Driel, R. The three-dimensional structure of human interphase chromosomes is related to the transcriptome map. *Mol. Cell. Biol.* **2007**, *27*, 4475–4487.
2. Geyer, P.K.; Corces, V.G. DNA position-specific repression of transcription by a *Drosophila* zinc finger protein. *Genes Dev.* **1992**, *6*, 1865–1873.
3. Kellum, R.; Schedl, P. A group of scs elements function as domain boundaries in an enhancer-blocking assay. *Mol. Cell. Biol.* **1992**, *12*, 2424–2431.
4. Kellum, R.; Schedl, P. A position-effect assay for boundaries of higher order chromosomal domains. *Cell* **1991**, *64*, 941–950.
5. Sun, F.L.; Elgin, S.C. Putting boundaries on silence. *Cell* **1999**, *99*, 459–462.
6. West, A.G.; Gaszner, M.; Felsenfeld, G. Insulators: Many functions, many mechanisms. *Genes Dev.* **2002**, *16*, 271–288.
7. Simms, T.A.; Dugas, S.L.; Gremillion, J.C.; Ibos, M.E.; Dandurand, M.N.; Toliver, T.T.; Edwards, D.J.; Donze, D. TFIIIC binding sites function as both heterochromatin barriers and chromatin insulators in *Saccharomyces cerevisiae*. *Eukaryot. Cell* **2008**, *7*, 2078–2086.
8. Ohlsson, R.; Bartkuhn, M.; Renkawitz, R. CTCF shapes chromatin by multiple mechanisms: The impact of 20 years of CTCF research on understanding the workings of chromatin. *Chromosoma* **2010**, *119*, 351–360.
9. Hiraga, S.; Botsios, S.; Donze, D.; Donaldson, A.D. TFIIIC localizes budding yeast ETC sites to the nuclear periphery. *Mol. Biol. Cell* **2012**, *23*, 2741–2754.
10. Noma, K.; Cam, H.P.; Maraia, R.J.; Grewal, S.I. A role for TFIIIC transcription factor complex in genome organization. *Cell* **2006**, *125*, 859–872.
11. Valenzuela, L.; Dhillon, N.; Kamakaka, R.T. Transcription independent insulation at TFIIIC-dependent insulators. *Genetics* **2009**, *183*, 131–148.
12. Dai, Q.; Ren, A.; Westholm, J.O.; Duan, H.; Patel, D.J.; Lai, E.C. Common and distinct DNA-binding and regulatory activities of the BEN-solo transcription factor family. *Genes Dev.* **2015**, *29*, 48–62.
13. Maksimenko, O.; Bartkuhn, M.; Stakhov, V.; Herold, M.; Zolotarev, N.; Jox, T.; Buxa, M.K.; Kirsch, R.; Bonchuk, A.; Fedotova, A.; *et al.* Two new insulator proteins, Pita and ZIPIC, target CP190 to chromatin. *Genome Res.* **2015**, *25*, 89–99.
14. Vogelmann, J.; le Gall, A.; Dejardin, S.; Allemand, F.; Gamot, A.; Labesse, G.; Cuvier, O.; Negre, N.; Cohen-Gonsaud, M.; Margeat, E.; Nollmann, M. Chromatin insulator factors involved in long-range DNA interactions and their role in the folding of the Drosophila genome. *PLoS Genet.* **2014**, *10*, e1004544.
15. Ahanger, H.S.; Shouche, Y.S.; Mishra, R.K. Functional sub-division of the *Drosophila* genome via chromatin looping: The emerging importance of CP190. *Nucleus* **2013**, *4*, 115–122.
16. Pai, C.Y.; Lei, E.P.; Ghosh, D.; Corces, V.G. The centrosomal protein CP190 is a component of the gypsy chromatin insulator. *Mol. Cell* **2004**, *16*, 737–748.

17. Matharu, N.K.; Hussain, T.; Sankaranarayanan, R.; Mishra, R.K. Vertebrate homologue of *Drosophila* GAGA factor. *J. Mol. Biol.* **2010**, *400*, 434–447.
18. Srivastava, S.; Puri, D.; Garapati, H.S.; Dhawan, J.; Mishra, R.K. Vertebrate GAGA factor associated insulator elements demarcate homeotic genes in the HOX clusters. *Epigenet. Chromatin* **2013**, doi:10.1186/1756-8935-6-8.
19. Zullo, J.M.; Demarco, I.A.; Pique-Regi, R.; Gaffney, D.J.; Epstein, C.B.; Spooner, C.J.; Luperchio, T.R.; Bernstein, B.E.; Pritchard, J.K.; Reddy, K.L. DNA sequence-dependent compartmentalization and silencing of chromatin at the nuclear lamina. *Cell* **2012**, *149*, 1474–1487
20. Dorn, R.; Krauss, V.; Reuter, G.; Saumweber, H. The enhancer of position-effect variegation of Drosophila, E(var)3-93D, codes for a chromatin protein containing a conserved domain common to several transcriptional regulators. *Proc. Natl. Acad. Sci. USA* **1993**, *90*, 11376–11380.
21. Gerasimova, T.I.; Gdula, D.A.; Gerasimov, D.V.; Simonova, O.; Corces, V.G. A *Drosophila* protein that imparts directionality on a chromatin insulator is an enhancer of position-effect variegation. *Cell* **1995**, *82*, 587–597.
22. Gerasimova, T.I.; Corces, V.G. Polycomb and trithorax group proteins mediate the function of a chromatin insulator. *Cell* **1998**, *92*, 511–521.
23. Gerasimova, T.I.; Byrd, K.; Corces, V.G. A chromatin insulator determines the nuclear localization of DNA. *Mol. Cell* **2000**, *6*, 1025–1035.
24. Nabirochkin, S.; Ossokina, M.; Heidmann, T. A nuclear matrix/scaffold attachment region co-localizes with the gypsy retrotransposon insulator sequence. *J. Biol. Chem.* **1998**, *273*, 2473–2479.
25. Pathak, R.U.; Rangaraj, N.; Kallappagoudar, S.; Mishra, K.; Mishra, R.K. Boundary element-associated factor 32B connects chromatin domains to the nuclear matrix. *Mol. Cell. Biol.* **2007**, *27*, 4796–4806.
26. Yusufzai, T.M.; Felsenfeld, G. The 5'-HS4 chicken beta-globin insulator is a CTCF-dependent nuclear matrix-associated element. *Proc. Natl. Acad. Sci. USA* **2004**, *101*, 8620–8624.
27. Yusufzai, T.M.; Tagami, H.; Nakatani, Y.; Felsenfeld, G. CTCF tethers an insulator to subnuclear sites, suggesting shared insulator mechanisms across species. *Mol. Cell* **2004**, *13*, 291–298.
28. Tolhuis, B.; Palstra, R.J.; Splinter, E.; Grosveld, F.; de Laat, W. Looping and interaction between hypersensitive sites in the active beta-globin locus. *Mol. Cell* **2002**, *10*, 1453–1465.
29. Palstra, R.J.; Tolhuis, B.; Splinter, E.; Nijmeijer, R.; Grosveld, F.; de Laat, W. The beta-globin nuclear compartment in development and erythroid differentiation. *Nat. Genet.* **2003**, *35*, 190–194.
30. Splinter, E.; Heath, H.; Kooren, J.; Palstra, R.J.; Klous, P.; Grosveld, F.; Galjart, N.; de Laat, W. CTCF mediates long-range chromatin looping and local histone modification in the beta-globin locus. *Genes Dev.* **2006**, *20*, 2349–2354.
31. Carter, D.; Chakalova, L.; Osborne, C.S.; Dai, Y.F.; Fraser, P. Long-range chromatin regulatory interactions *in vivo*. *Nat. Genet.* **2002**, *32*, 623–626.
32. Murrell, A.; Heeson, S.; Reik, W. Interaction between differentially methylated regions partitions the imprinted genes Igf2 and H19 into parent-specific chromatin loops. *Nat. Genet.* **2004**, *36*, 889–893.
33. Kuhn, E.J.; Hart, C.M.; Geyer, P.K. Studies of the role of the *Drosophila* scs and scs' insulators in defining boundaries of a chromosome puff. *Mol. Cell. Biol.* **2004**, *24*, 1470–1480.

34. Melnikova, L.; Juge, F.; Gruzdeva, N.; Mazur, A.; Cavalli, G.; Georgiev, P. Interaction between the GAGA factor and Mod(mdg4) proteins promotes insulator bypass in *Drosophila. Proc. Natl. Acad. Sci. USA* **2004** *101*, 14806–14811.
35. Majumder, P.; Cai, H.N. The functional analysis of insulator interactions in the *Drosophila* embryo. *Proc. Natl. Acad. Sci. USA* **2003**, *100*, 5223–5228.
36. Geyer, P.K. The role of insulator elements in defining domains of gene expression. *Curr. Opin. Genet. Dev.* **1997**, *7*, 242–248.
37. Yoon, Y.S.; Jeong, S.; Rong, Q.; Park, K.Y.; Chung, J.H.; Pfeifer, K. Analysis of the H19ICR insulator. *Mol. Cell. Biol.* **2007**, *27*, 3499–3510.
38. Zhu, X.; Ling, J.; Zhang, L.; Pi, W.; Wu, M.; Tuan, D. A facilitated tracking and transcription mechanism of long-range enhancer function. *Nucleic Acids Res.* **2007**, *35*, 5532–5544.
39. Dekker, J.; Rippe, K.; Dekker, M.; Kleckner, N. Capturing chromosome conformation. *Science* **2002**, *295*, 1306–1311.
40. Dostie, J.; Richmond, T.A.; Arnaout, R.A.; Selzer, R.R.; Lee, W.L.; Honan, T.A.; Rubio, E.D.; Krumm, A.; Lamb, J.; Nusbaum, C.; *et al.* Chromosome Conformation Capture Carbon Copy (5C): A massively parallel solution for mapping interactions between genomic elements. *Genome Res.* **2006**, *16*, 1299–1309.
41. Fullwood, M.J.; Liu, M.H.; Pan, Y.F.; Liu, J.; Xu, H.; Mohamed, Y.B.; Orlov, Y.L.; Velkov, S.; Ho, A.; Mei, P.H.; *et al.* An oestrogen-receptor-alpha-bound human chromatin interactome. *Nature* **2009**, *462*, 58–64.
42. Kieffer-Kwon, K.R.; Tang, Z.; Mathe, E.; Qian, J.; Sung, M.H.; Li, G.; Resch, W.; Baek, S.; Pruett, N.; Grontved, L.; *et al.* Interactome maps of mouse gene regulatory domains reveal basic principles of transcriptional regulation. *Cell* **2013**, *155*, 1507–1520.
43. Van Berkum, N.L.; Lieberman-Aiden, E.; Williams, L.; Imakaev, M.; Gnirke, A.; Mirny, L.A.; Dekker, J.; Lander, E.S. Hi-C: A method to study the three-dimensional architecture of genomes. *J. Vis. Exp.* **2010**, doi:10.3791/1869.
44. Zhang, Y.; Wong, C.H.; Birnbaum, R.Y.; Li, G.; Favaro, R.; Ngan, C.Y.; Lim, J.; Tai, E.; Poh, H.M.; Wong, E.; *et al.* Chromatin connectivity maps reveal dynamic promoter-enhancer long-range associations. *Nature* **2013**, *504*, 306–310.
45. Zhao, Z.; Tavoosidana, G.; Sjolinder, M.; Gondor, A.; Mariano, P.; Wang, S.; Kanduri, C.; Lezcano, M.; Sandhu, K.S.; Singh, U.; *et al.* Circular chromosome conformation capture (4C) uncovers extensive networks of epigenetically regulated intra- and interchromosomal interactions. *Nat. Genet.* **2006**, *38*, 1341–1347.
46. Sexton, T.; Yaffe, E.; Kenigsberg, E.; Bantignies, F.; Leblanc, B.; Hoichman, M.; Parrinello, H.; Tanay, A.; Cavalli, G. Three-dimensional folding and functional organization principles of the *Drosophila* genome. *Cell* **2012**, *148*, 458–472.
47. Van Bortle, K.; Nichols, M.H.; Li, L.; Ong, C.T.; Takenaka, N.; Qin, Z.S.; Corces, V.G. Insulator function and topological domain border strength scale with architectural protein occupancy. *Genome Biol.* **2014**, doi:10.1186/gb-2014-15-5-r82.
48. Dixon, J.R.; Selvaraj, S.; Yue, F.; Kim, A.; Li, Y.; Shen, Y.; Hu, M.; Liu, J.S.; Ren, B. Topological domains in mammalian genomes identified by analysis of chromatin interactions. *Nature* **2012**, *485*, 376–380.

49. Ghavi-Helm, Y.; Klein, F.A.; Pakozdi, T.; Ciglar, L.; Noordermeer, D.; Huber, W.; Furlong, E.E. Enhancer loops appear stable during development and are associated with paused polymerase. *Nature* **2014**, *512*, 96–100.
50. Zuin, J.; Dixon, J.R.; van der Reijden, M.I.; Ye, Z.; Kolovos, P.; Brouwer, R.W.; van de Corput, M.P.; van de Werken, H.J.; Knoch, T.A.; van, I.W.F.; *et al.* Cohesin and CTCF differentially affect chromatin architecture and gene expression in human cells. *Proc. Natl. Acad. Sci. USA* **2014**, *111*, 996–1001.
51. Seitan, V.C.; Faure, A.J.; Zhan, Y.; McCord, R.P.; Lajoie, B.R.; Ing-Simmons, E.; Lenhard, B.; Giorgetti, L.; Heard, E.; Fisher, A.G.; *et al.* Cohesin-based chromatin interactions enable regulated gene expression within preexisting architectural compartments. *Genome Res.* **2013**, *23*, 2066–2077.
52. Sofueva, S.; Yaffe, E.; Chan, W.C.; Georgopoulou, D.; Rudan, M.V.; Mira-Bontenbal, H.; Pollard, S.M.; Schroth, G.P.; Tanay, A.; Hadjur, S. Cohesin-mediated interactions organize chromosomal domain architecture. *EMBO J.* **2013**, *32*, 3119–3129.
53. Kagey, M.H.; Newman, J.J.; Bilodeau, S.; Zhan, Y.; Orlando, D.A.; van Berkum, N.L.; Ebmeier, C.C.; Goossens, J.; Rahl, P.B.; Levine, S.S.; *et al.* Mediator and cohesin connect gene expression and chromatin architecture. *Nature* **2010**, *467*, 430–435.
54. Rao, S.S.; Huntley, M.H.; Durand, N.C.; Stamenova, E.K.; Bochkov, I.D.; Robinson, J.T.; Sanborn, A.L.; Machol, I.; Omer, A.D.; Lander, E.S.; *et al.* A 3D map of the human genome at kilobase resolution reveals principles of chromatin looping. *Cell* **2014**, *159*, 1665–1680.
55. Tark-Dame, M.; Jerabek, H.; Manders, E.M.; Heermann, D.W.; van Driel, R. Depletion of the chromatin looping proteins CTCF and cohesin causes chromatin compaction: Insight into chromatin folding by polymer modelling. *PLoS Comput. Biol.* **2014**, *10*, e1003877.
56. Lupianez, D.G.; Kraft, K.; Heinrich, V.; Krawitz, P.; Brancati, F.; Klopocki, E.; Horn, D.; Kayserili, H.; Opitz, J.M.; Laxova, R.; *et al.* Disruptions of topological chromatin domains cause pathogenic rewiring of gene-enhancer interactions. *Cell* **2015**, *161*, 1012–1025.
57. Crane, E.; Bian, Q.; McCord, R.P.; Lajoie, B.R.; Wheeler, B.S.; Ralston, E.J.; Uzawa, S.; Dekker, J.; Meyer, B.J. Condensin-driven remodelling of X chromosome topology during dosage compensation. *Nature* **2015**, *523*, 240–244.
58. Wood, A.M.; van Bortle, K.; Ramos, E.; Takenaka, N.; Rohrbaugh, M.; Jones, B.C.; Jones, K.C.; Corces, V.G. Regulation of chromatin organization and inducible gene expression by a *Drosophila* insulator. *Mol. Cell* **2011**, *44*, 29–38.
59. Li, L.; Lyu, X.; Hou, C.; Takenaka, N.; Nguyen, H.Q.; Ong, C.T.; Cubenas-Potts, C.; Hu, M.; Lei, E.P.; Bosco, G.; *et al.* Widespread rearrangement of 3D chromatin organization underlies polycomb-mediated stress-induced silencing. *Mol. Cell* **2015**, *58*, 216–231.
60. Schoborg, T.; Rickels, R.; Barrios, J.; Labrador, M. Chromatin insulator bodies are nuclear structures that form in response to osmotic stress and cell death. *J. Cell Biol.* **2013**, *202*, 261–276.
61. Capelson, M.; Corces, V.G. Boundary elements and nuclear organization. *Biol. Cell* **2004**, *96*, 617–629.
62. Golovnin, A.; Melnikova, L.; Volkov, I.; Kostuchenko, M.; Galkin, A.V.; Georgiev, P. "Insulator bodies" are aggregates of proteins but not of insulators. *EMBO Rep.* **2008**, *9*, 440–445.
63. Efroni, S.; Duttagupta, R.; Cheng, J.; Dehghani, H.; Hoeppner, D.J.; Dash, C.; Bazett-Jones, D.P.; le Grice, S.; McKay, R.D.; Buetow, K.H.; *et al.* Global transcription in pluripotent embryonic stem cells. *Cell Stem Cell* **2008**, *2*, 437–447.

64. Denholtz, M.; Plath, K. Pluripotency in 3D: Genome organization in pluripotent cells. *Curr. Opin. Cell. Biol.* **2012**, *24*, 793–801.
65. Handoko, L.; Xu, H.; Li, G.; Ngan, C.Y.; Chew, E.; Schnapp, M.; Lee, C.W.; Ye, C.; Ping, J.L.; Mulawadi, F.; *et al.* CTCF-mediated functional chromatin interactome in pluripotent cells. *Nat. Genet.* **2011**, *43*, 630–638.
66. Phillips-Cremins, J.E.; Sauria, M.E.; Sanyal, A.; Gerasimova, T.I.; Lajoie, B.R.; Bell, J.S.; Ong, C.T.; Hookway, T.A.; Guo, C.; Sun, Y.; *et al.* Architectural protein subclasses shape 3D organization of genomes during lineage commitment. *Cell* **2013**, *153*, 1281–1295.
67. Schmidt, D.; Wilson, M.D.; Ballester, B.; Schwalie, P.C.; Brown, G.D.; Marshall, A.; Kutter, C.; Watt, S.; Martinez-Jimenez, C.P.; Mackay, S.; *et al.* Five-vertebrate ChIP-seq reveals the evolutionary dynamics of transcription factor binding. *Science* **2010**, *328*, 1036–1040.
68. Hakim, O.; Sung, M.H.; Nakayamada, S.; Voss, T.C.; Baek, S.; Hager, G.L. Spatial congregation of STAT binding directs selective nuclear architecture during T-cell functional differentiation. *Genome Res.* **2013**, *23*, 462–472.
69. Dixon, J.R.; Jung, I.; Selvaraj, S.; Shen, Y.; Antosiewicz-Bourget, J.E.; Lee, A.Y.; Ye, Z.; Kim, A.; Rajagopal, N.; Xie, W.; *et al.* Chromatin architecture reorganization during stem cell differentiation. *Nature* **2015**, *518*, 331–336.
70. Zhang, H.; Jiao, W.; Sun, L.; Fan, J.; Chen, M.; Wang, H.; Xu, X.; Shen, A.; Li, T.; Niu, B.; Ge, S.; *et al.* Intrachromosomal looping is required for activation of endogenous pluripotency genes during reprogramming. *Cell Stem Cell* **2013**, *13*, 30–35.
71. Apostolou, E.; Ferrari, F.; Walsh, R.M.; Bar-Nur, O.; Stadtfeld, M.; Cheloufi, S.; Stuart, H.T.; Polo, J.M.; Ohsumi, T.K.; Borowsky, M.L.; *et al.* Genome-wide chromatin interactions of the Nanog locus in pluripotency, differentiation, and reprogramming. *Cell Stem Cell* **2013**, *12*, 699–712.
72. Wei, Z.; Gao, F.; Kim, S.; Yang, H.; Lyu, J.; An, W.; Wang, K.; Lu, W. Klf4 organizes long-range chromosomal interactions with the oct4 locus in reprogramming and pluripotency. *Cell Stem Cell* **2013**, *13*, 36–47.
73. Denholtz, M.; Bonora, G.; Chronis, C.; Splinter, E.; de Laat, W.; Ernst, J.; Pellegrini, M.; Plath, K. Long-range chromatin contacts in embryonic stem cells reveal a role for pluripotency factors and polycomb proteins in genome organization. *Cell Stem Cell* **2013**, *13*, 602–616.
74. Hart, C.M.; Cuvier, O.; Laemmli, U.K. Evidence for an antagonistic relationship between the boundary element-associated factor BEAF and the transcription factor DREF. *Chromosoma* **1999**, *108*, 375–383.
75. Bell, C.A.; Felsenfeld, G. Methylation of a CTCF-dependent boundary controls imprinted expression of the Igf2 gene. *Nature* **2000**, *405*, 482–485.
76. Hark, A.T.; Schoenherr, C.J.; Katz, D.J.; Ingram, R.S.; Levorse, J.M.; Tilghman, S.M. CTCF mediates methylation-sensitive enhancer-blocking activity at the H19/Igf2 locus. *Nature* **2000**, *405*, 486–489.
77. Capelson, M.; Corces, V.G. The ubiquitin ligase dTopors directs the nuclear organization of a chromatin insulator. *Mol. Cell* **2005**, *20*, 105–116.
78. Yu, W.; Ginjala, V.; Pant, V.; Chernukhin, I.; Whitehead, J.; Docquier, F.; Farrar, D.; Tavoosidana, G.; Mukhopadhyay, R.; Kanduri, C.; Oshimura, M.; *et al.* Poly(ADP-ribosyl)ation regulates CTCF-dependent chromatin insulation. *Nat. Genet.* **2004**, *36*, 1105–1110.

79. Klenova, E.; Ohlsson, R. Poly(ADP-ribosyl)ation and epigenetics. Is CTCF PARt of the plot? *Cell Cycle* **2005**, *4*, 96–101.
80. Ong, C.T.; van Bortle, K.; Ramos, E.; Corces, V.G. Poly(ADP-ribosyl)ation regulates insulator function and intrachromosomal interactions in *Drosophila*. *Cell* **2013**, *155*, 148–159.
81. Capelson, M.; Corces, V.G. SUMO conjugation attenuates the activity of the gypsy chromatin insulator. *EMBO J.* **2006**, *25*, 1906–1914.
82. Lei, E.P.; Corces, V.G. RNA interference machinery influences the nuclear organization of a chromatin insulator. *Nat. Genet.* **2006**, *38*, 936–941.
83. Moshkovich, N.; Nisha, P.; Boyle, P.J.; Thompson, B.A.; Dale, R.K.; Lei, E.P. RNAi-independent role for Argonaute2 in CTCF/CP190 chromatin insulator function. *Genes Dev.* **2011**, *25*, 1686–1701.
84. Gruzdeva, N.; Kyrchanova, O.; Parshikov, A.; Kullyev, A.; Georgiev, P. The Mcp element from the bithorax complex contains an insulator that is capable of pairwise interactions and can facilitate enhancer-promoter communication. *Mol. Cell. Biol.* **2005**, *25*, 3682–3689.
85. Kyrchanova, O.; Toshchakov, S.; Parshikov, A.; Georgiev, P. Study of the functional interaction between Mcp insulators from the Drosophila bithorax complex: Effects of insulator pairing on enhancer-promoter communication. *Mol. Cell. Biol.* **2007**, *27*, 3035–3043.
86. Kyrchanova, O.; Toshchakov, S.; Podstreshnaya, Y.; Parshikov, A.; Georgiev, P. Functional interaction between the Fab-7 and Fab-8 boundaries and the upstream promoter region in the *Drosophila* Abd-B gene. *Mol. Cell. Biol.* **2008**, *28*, 4188–4195.
87. Rodin, S.; Kyrchanova, O.; Pomerantseva, E.; Parshikov, A.; Georgiev, P. New properties of *Drosophila* fab-7 insulator. *Genetics* **2007**, *177*, 113–121.
88. Chen, Q.; Lin, L.; Smith, S.; Lin, Q.; Zhou, J. Multiple promoter targeting sequences exist in abdominal-B to regulate long-range gene activation. *Dev. Biol.* **2005**, *286*, 629–636.
89. Zhou, J.; Levine, M. A novel cis-regulatory element, the PTS, mediates an anti-insulator activity in the *Drosophila* embryo. *Cell* **1999**, *99*, 567–575.
90. Lutz, M.; Burke, L.J.; LeFevre, P.; Myers, F.A.; Thorne, A.W.; Crane-Robinson, C.; Bonifer, C.; Filippova, G.N.; Lobanenkov, V.; Renkawitz, R. Thyroid hormone-regulated enhancer blocking: Cooperation of CTCF and thyroid hormone receptor. *EMBO J.* **2003**, *22*, 1579–1587.
91. Wallace, J.A.; Felsenfeld, G. We gather together: Insulators and genome organization. *Curr. Opin. Genet. Dev.* **2007**, *17*, 400–407.
92. Defossez, P.A.; Kelly, K.F.; Filion, G.J.; Perez-Torrado, R.; Magdinier, F.; Menoni, H.; Nordgaard, C.L.; Daniel, J.M.; Gilson, E. The human enhancer blocker CTC-binding factor interacts with the transcription factor Kaiso. *J. Biol. Chem.* **2005**, *280*, 43017–43023.
93. Ishihara, K.; Oshimura, M.; Nakao, M. CTCF-dependent chromatin insulator is linked to epigenetic remodeling. *Mol. Cell* **2006**, *23*, 733–742.

2

The Revolution in Human Monogenic Disease Mapping

Emma Duncan [1,2], **Matthew Brown** [1] **and Eileen M. Shore** [3,4,*]

[1] University of Queensland Diamantina Institute, Translational Research Institute, Princess Alexandra Hospital, 37 Kent Road, Woolloongabba, Brisbane 4102, Queensland, Australia; E-Mails: emma.duncan@uq.edu.au (E.D.); matt.brown@uq.edu.au (M.B.)
[2] Department of Endocrinology and Diabetes, Royal Brisbane and Women's Hospital, Herston 4029, Queensland, Australia
[3] Center for Research in FOP and Related Disorders, Department of Orthopaedic Surgery, Perelman School of Medicine, University of Pennsylvania, 3450 Hamilton Walk, Philadelphia, PA 19104, USA
[4] Department of Genetics, Perelman School of Medicine, University of Pennsylvania, Curie Boulevard, Philadelphia, PA 19104, USA

* Author to whom correspondence should be addressed; E-Mail: shore@mail.med.upenn.edu;

Abstract: The successful completion of the Human Genome Project (HGP) was an unprecedented scientific advance that has become an invaluable resource in the search for genes that cause monogenic and common (polygenic) diseases. Prior to the HGP, linkage analysis had successfully mapped many disease genes for monogenic disorders; however, the limitations of this approach were particularly evident for identifying causative genes in rare genetic disorders affecting lifespan and/or reproductive fitness, such as skeletal dysplasias. In this review, we illustrate the challenges of mapping disease genes in such conditions through the ultra-rare disorder fibrodysplasia ossificans progressiva (FOP) and we discuss the advances that are being made through current massively parallel ("next generation") sequencing (MPS) technologies.

Keywords: Human Genome Project; fibrodysplasia ossificans progressiva; monogenic diseases; disease gene discovery; NGS

---

1. Introduction

Until very recently, mapping the causative gene for monogenic diseases depended on finding families with demonstrable Mendelian inheritance of the disease, preferably in multiple generations. Linkage approaches in such families were successful in identifying mutations responsible for many of the more frequent monogenic diseases and traits [1]. However, identifying the underlying mutations in rare genetic diseases, especially those associated with low reproductive fitness, late onset diseases, or diseases with early lethality, proved much more challenging.

The Human Genome Project (HGP) set the stage for success in meeting these challenges. The ultra-rare disorder fibrodysplasia ossificans progressiva (FOP), with a frequency of one in two million, is an example of such a success. Linkage in FOP families identified chromosomal regions of interest; Human Genome Project databases then facilitated the identification of candidate genes within the linkage region and permitted the efficient identification of altered DNA sequences. FOP was found to be caused by a recurrent single nucleotide substitution occurring in >95% of patients. Despite the ultimate success in mapping the FOP gene, this discovery took decades of effort to identify families with inheritance of the disease (linkage analysis was eventually accomplished with just five two-generation families) followed by the time-consuming tasks of conducting genome-wide linkage analysis, and subsequent identification and re-sequencing of candidate genes within the linkage intervals. Knowing the genetic mutation in FOP has led quickly to better understanding of the underlying pathology and directed strategies for treatments, neither of which were possible before molecular aetiology was determined.

Building on the technology, computation, and scientific information generated through the HGP, the continued advances in mapping disease genes have been extraordinarily rapid. Faced today with the challenge of identifying a rare gene mutation in a disease like FOP, high-throughput exome and/or whole genome sequencing approaches would identify the genetic mutation rapidly. Moreover, gene identification may require sequencing of only a small number of unrelated cases, small families, or, in some cases, even a single proband. These breakthroughs are leading to an upsurge in disease gene discoveries with their associated benefits [2].

After mapping a disease-causing gene, many challenges remain in understanding additional genetic contributions to disease onset and progression. In FOP, for example, although the main disease characteristics are unique and readily recognized there is significant variability in the age of disease onset, the rate of disease progression, and the severity of disease, likely to arise from as yet un-identified genetic causes. Similar issues exist with many common heritable diseases such as osteogenesis imperfecta and Marfan's syndrome. With further development of genome technologies, the ability to understand phenotypic variability and the participation of genetic modifiers is becoming a reality.

2. Linkage Mapping

2.1. History of Linkage Mapping

At the turn of the twentieth century, Archibald Garrod coined the term "inborn errors of metabolism" to explain the increased incidence of alkaptonuria (and, subsequently, also cystinuria, pentosuria, and albinism) in consanguineous families compared with the general population, and suggested that these conditions were caused by transmissible elements within families [3]. Since this time, the challenge has been to identify these errors.

For most of the twentieth century, linkage mapping was the standard means of identifying the gene/s underlying an inherited disorder. Linkage is the co-segregation of a genetic region with disease phenotype within a family. Markers, such as restriction fragment length polymorphisms, microsatellites, or single nucleotide polymorphisms (SNPs), are genotyped in family members. Markers close to a disease-causing mutation will be co-inherited with the disease-causing mutation, unless separated by a meiotic event—the closer the marker to the disease-causing gene, the less likely it will be separated at meiosis. An area of linkage within a family may thus extend a considerable genetic distance. Whole genome linkage scans, in which approximately 300–400 microsatellite markers are genotyped in family members, usually result in identification of regions of linkage spanning 10–20 cM (on average ~10–20 million DNA bases). Such a region may harbor many hundreds of genes, and fine mapping (by further marker genotyping and/or sequencing of candidate genes) will usually be necessary to identify the exact causative gene.

2.2. Weaknesses of the Linkage Approach

Traditional "parametric" linkage analysis compares the likelihood of the observed transmission of genetic markers in relation to the trait or disease, in the context of a specified model of inheritance. Non-parametric methods not requiring knowledge of mode of inheritance can be used, though are less powerful when the correct mode of inheritance is known. Diseases with late onset of clinical features or with incomplete penetrance are harder to map by linkage due to possible incorrect attribution of disease status among family members. Diseases with significant gene/environment interaction present similar issues, unless all family members are exposed to the same environmental stressors. The ability to map a gene also depends on the number of informative meioses within a pedigree. Thus, large families with many affected individuals—especially distantly-related individuals who will have a high number of meioses and recombination events between them—are the most useful for linkage mapping. Diseases that affect reproductive fitness (such as skeletal dysplasias) are less likely to have such large informative pedigrees. One solution to this problem is to use many families affected by the same condition in order to identify a common linkage region shared by affected persons within each family. This strategy requires that the disease be caused by mutations in the same single common locus in all families, although the causative mutation within this common locus may be unique in each individual family. The approach will, therefore, not be successful for diseases with "phenocopies", in which a clinical phenotype may arise from mutations in more than one gene. (For example, phenocopies might result if mutations in various genes along a biological pathway resulted in a common phenotypic endpoint.) "Pooling" linkage information from families with disparate underlying causes would not be a successful strategy. Extremely rare diseases are, by definition, extremely rare; obtaining sufficient number of families to pool their linkage information will usually require international collaboration. Lastly, novel/spontaneous mutations (those newly occurring within an individual) cannot be mapped by linkage.

2.3. Successes in Linkage Mapping: Monogenic vs. Complex Diseases

Despite these limitations, mapping monogenic diseases by linkage has been quite successful, even for rare diseases, with well over 1000 monogenic (Mendelian) disorders mapped by the turn of the century [1]. Approximately two-thirds of the 400 or so recognized skeletal dysplasias were mapped by linkage or similar approaches by 2010 [4]. In contrast, mapping complex diseases by linkage was much less successful. Complex genetic diseases are typically polygenic in nature: affected individuals have different variants in multiple, but overlapping, sets of genes, with each variant contributing only a small part to the final overall phenotype. Before the era of high throughput microarray genotyping and the advent of genome-wide association studies (GWAS), only a handful of genes had been identified for complex diseases using linkage [5].

3. Mapping the FOP Gene Highlights the Challenges

Fibrodysplasia ossificans progressiva (FOP; MIM 135100) is a severe disorder of progressive and extensive extra-skeletal ossification. Heterotopic ossification in FOP begins in childhood within connective tissues, such as skeletal muscle, tendon, and ligament. Onset of bone formation can be induced by trauma, or may occur spontaneously. Bone formation is episodic, leading to cumulative disability and shortened lifespan [6].

Reproductive fitness is low in FOP, resulting in infrequent inheritance and a population frequency (about one per two million) that reflects the rate of new mutations. When the search for the FOP gene began, only very few cases of familial inheritance of FOP had been reported, with most known cases occurring *de novo* in families (reviewed in [7,8]). Although these few family pedigrees suggested an autosomal dominant mode of inheritance, the paucity of families with transmission of FOP made genome-wide linkage analysis, the state-of-the-art approach at the time, an impractical strategy for gene identification.

Eventually, four small pedigrees with autosomal dominant inheritance of FOP were identified, although some had ambiguous clinical features. An initial genome-wide linkage analysis using 240 microsatellite markers (spaced $\leq$ 25 cM) was conducted, although it was recognized that there was incomplete/uneven marker coverage across the genome and many markers lacked sufficient informativity [9]. Whilst the initial linkage analysis focused attention on a chromosome 4 locus [9], further analysis revealed additional linked loci on chromosomes 2 and 6. However, subsequent sequencing analysis of numerous candidate genes in the linkage regions revealed no identifiable mutations. The limited information available at this time about the human genome, including gene locations and sequence, made this process much more challenging.

As genome analysis technologies continued to develop and additional families with autosomal dominant transmission of FOP were identified, further genome-wide linkage studies were performed. These used both a higher density SNP marker panel as well as more dense and informative microsatellite panels, which were combined in a single analysis. The international team involved focused their studies on five two-generation families with stringent and unambiguous phenotypic features of FOP in all affected family members (congenital malformation of the great toes and progressive heterotopic ossification in characteristic anatomic patterns). Consistent linkage was then demonstrated with the chromosome 2q23–24 interval in all five families (LOD score 2.3) [8], with no other locus segregating with the disease in all pedigrees. Using better characterized and annotated human DNA sequences generated through the Human Genome Project (HGP), we selected candidate genes within the 16.5 Mb linkage interval for sequencing and mutation identification. The interval contained more than 40 known genes, however, concurrent investigations had identified the BMP signaling pathway as altered in FOP [10–13], and genes in this pathway were given high priority. One such gene was the *Activin A type I receptor* gene (*ACVR1*; OMIM 102576; also known as *Alk2* or *ActRI*), encoding a receptor for bone morphogenetic protein (BMP).

DNA sequence analysis of all *ACVR1* protein-coding exons and splice junctions identified a heterozygous c.617G>A (Arg206His; CGC ≥ CAC) mutation present in all affected members in these FOP families, with the same mutation present in multiple sporadic cases of FOP [8].

Cumulative data over the past eight years shows that FOP is caused by this recurrent single nucleotide substitution in >95% of patients ([14–16]). In exceptional cases of FOP, mainly those whose phenotype varies slightly from the description above, affected individuals have mutations at other amino acid positions in *ACVR1* in the glycine-serine (GS) or protein kinase domains [14,17]. Thus far, all patients clinically diagnosed with the “classical” FOP phenotype have *ACVR1* mutations, and these mutations are fully penetrant.

4. Massive Parallel Sequencing: A New Era

The mapping of rare monogenic disorders has been completely transformed by the advent of massive parallel sequencing (MPS), also known as “next-generation” sequencing. In the last few years, the causative genes for dozens of monogenic disorders have been identified using MPS, and the rate of discovery has been exponential. To illustrate this latter point, we recently published a review of MPS in skeletal dysplasias; at the time of paper submission (April 2013) 26 papers had been published using MPS to identify the causative gene for 22 skeletal dysplasias; by the time of paper acceptance (July 2013) a further ten papers had added another six novel skeletal dysplasia genes to the list [2]. Since then, further causative genes for skeletal dysplasias, as well as a host of other Mendelian disorders, have been identified, and it is likely that many of the remaining unmapped monogenic diseases will prove tractable to mapping by MPS.

4.1. MPS Technologies

Disease gene identification by MPS became possible because of three main developments: the technology of sequencing multiple genetic regions simultaneously; the success of the Human Genome Project in providing a complete and reliable reference genome for comparison with the test sequence data; and the availability of large databases of genomic information from healthy individuals, and increasingly in patients with disease, to assist in assessment of variants observed.

The pivotal technological breakthrough for MPS was the development of technologies and platforms for simultaneous sequencing of multiple regions of fragmented DNA or RNA in a single experiment. It is beyond the scope of this paper to provide a comprehensive discussion of these technologies. However, briefly: DNA is fragmented and common adapters are ligated to the fragment ends. The fragments are subsequently amplified by PCR, followed by sequencing-by-synthesis, the common adapters providing uniform starting templates for both the amplification and sequencing reactions for all fragments (more recent technological developments, so-called "third-generation" technologies, involve sequencing without this step of any preceding amplification, improving both accuracy and speed of MPS). Sequencing-by-synthesis involves addition of bases to a growing strand: as each base is added, a signal is generated and "read" by the software, thus generating the sequence of each fragment. The sequence of each fragment is then mapped against the human genome, allowing identification of genetic variants present in the sequenced individual(s).

Large databases of genetic variation (such as The HapMap project, UK10K, 1000genomes, Human Variome Project, and NCBI dbSNP) are used in interpreting sequence data obtained through MPS: identified variants can be characterized as part of the "normal" variability seen in the population, or as novel or rare variants and thus more likely to be pathogenic. Of note, population genetic variability differs among populations of different ethnicities; the sequence data of an individual should be compared against an ethnically-matched reference genome sequence. Once the sequence data have been generated and compared with the appropriate reference human genome, the data can be analyzed empirically based on the observed inheritance and population prevalence of the condition under examination.

Although MPS was developed for whole-genome sequencing (WGS), targeted sequencing proved more cost-effective and efficient initially. Thus, prior to amplification, a library of fragments containing regions of particular interest may be selected (for example, by using probes that anneal to specific regions of interest, allowing their subsequent identification and isolation for PCR amplification and sequencing). Whole exome sequencing (WES) with capture and sequencing limited to gene exons may be particularly suited for rare Mendelian disorders since prior to the advent of MPS 85% of monogenic diseases were predicted to arise from protein-coding mutations [18]—whether this will hold true as more causative mutations are identified is as of yet unknown.

The power of MPS methodology is illustrated by the many causative genetic mutations identified since its advent, especially since they are frequently mapped through sequencing of remarkably few individuals. Indeed, some disease genes have been identified through sequencing of a single proband [2], although confirmation of pathogenicity requires subsequent validation, such as genetic evidence in other affected individuals and/or functional studies of the candidate gene.

4.2. Mapping Strategies for Monogenic Diseases Using MPS

The experimental design for mapping a monogenic disease by WES does not necessarily require any prior linkage or association data. Rather it depends on the population frequency of disease, the mode of inheritance (including penetrance), and the presence or absence of consanguinity of the affected individuals. These parameters then determine an appropriately parsimonious experimental design—how many and which individuals should be sequenced and what empirical approach should be adopted for analysis of the sequence data. For example, a rare autosomal recessive condition in a non-consanguineous family is likely to arise from compound heterozygosity; identification of two novel (or very rare) damaging variants in a single gene provides strong evidence of likely causality even if only a single affected individual from the family is sequenced. In contrast, mapping an autosomal recessive condition in a consanguineous family is more difficult. In this circumstance, the disease is more likely to arise from homozygous carriage of a novel (or rare) variant carried by both parents. However, a high number of homozygous rare variants would be expected due to consanguinity anyway; determining which of these is most likely pathogenic can be difficult. For an autosomal dominant condition, the most parsimonious experimental design is to sequence distantly-related affected individuals with a large number of meioses (and, by implication, recombination events) separating the affected cases—with n meioses between the individuals, the chance of a dominant variant segregating with affection status by chance is $1/2^n$. It is also possible to map *de novo* dominant conditions by sequencing unrelated individuals and analyzing the data either for a single variant carried by all affected individuals or for unique variants carried in a common gene by each individual [19]. These last analysis strategies depend crucially upon correct clinical phenotyping of the unrelated affected individuals. The inclusion of phenocopies in the analysis would decrease the success of mapping the causative gene—unless the stringent parameters of analysis are relaxed to allow for their possible presence. For example, one can search for a common shared gene amongst only a proportion of affected individuals rather than requiring a variant to be present in the same gene in every sequenced case. An alternative approach includes pathway analysis (searching for variants in a common pathway amongst individuals with a common phenotype). From a clinical viewpoint, large online databases cataloguing observed variants/mutations in patients with common conditions are also useful in identifying likely disease-causing mutation(s) (as examples, the Leiden Open Variation Database and the Genome Medicine Database of Japan).

4.3. MPS Limitations

Like all technologies, MPS has limitations. Good sequencing data depend upon sufficient capture of the causative gene by coverage of sufficient depth of sequencing to call homozygous or heterozygous variants (typically, 10-fold coverage is required for calling a homozygous variant and 15-fold for a heterozygous variant). WES off-the-shelf target platforms vary in their coverage of the "whole exome" [20,21], which may result in a gene of interest failing to be sequenced. For example, we (and others) failed to identify the disease-causing mutation in OI type V despite sequencing several families with the condition; the causative mutation was identified in the 5' UTR of *IFITM5* [22,23], a region not captured with the whole exome capture platform we had employed. Less-than-complete whole exome capture can arise for several additional reasons, including new gene annotation subsequent to platform development and production [24]; a manufacturing decision to target only the main transcript for a gene rather than all known transcripts of a gene; and the technical challenges of capturing GC-rich sequences (which are common in the first exons of many genes [25]). Conversely, there are some regions that amplify excessively: if duplicates are not removed, strand-specific PCR-introduced errors may result, skewing variant allele frequencies with consequent effects on variant detection sensitivity and specificity [26]. Regions of genomic sequence similarity may result in non-specificity of target fragment selection—for example, a probe may anneal not only to the desired target exon but also to an unwanted region of high homology which, when incidentally captured and sequenced, results in apparently novel heterozygous variants at those points of difference between the two selected regions (a phenomenon known as multi-mapping [27]).

However, despite these limitations, faced today with the challenge of identifying a rare gene mutation in a disease like FOP, WES of a small number of FOP patients would likely rapidly reveal the recurrent c.617G>A (R206H) ACVR1 mutation, leading to quick recognition of *ACVR1* as the disease-causing gene. If MPS technologies had been available when the search for the FOP gene began, the answer could have been found in 15 weeks, not 15 years. Certainly, this has proved to be the case in many other skeletal dysplasias where researchers are faced with similar issues of small families afflicted with diseases having a detrimental effect upon reproduction and lifespan [2,4].

4.4. Rare Variants as a Cause of Complex Diseases?

Although we focused in this paper on the use of new mapping approaches for Mendelian disorders, MPS has also been used for mapping rare variants that contribute to the heritability of complex diseases. The contributions of rare variation in loci that also harbor common susceptibility alleles, or in genomic regions without common susceptibility alleles, are still the subject of active research. Whilst many examples exist of rare variants contributing to loci harboring common variant associations, these are few by comparison with the total number of common variant associations identified to date. Indeed, targeted sequencing of 25 loci associated with autoimmune disorders in nearly 25,000 individuals with six autoimmune phenotypes and just over 17,000 controls failed to identify any rare variants contributing significantly to immune-mediated disease susceptibility [28]. Styrkarsdottir *et al.* used whole genome scanning in the Icelandic population to identify a rare variant in *LGR4* associated with both bone mineral density and fracture risk [29]. WES identified mutations in *WNT1* as the cause of autosomal dominant early-onset osteoporosis in some families; however, as mutations in *WNT1* were also identified in consanguineous families with autosomal recessive osteogenesis imperfecta it would perhaps be more correct to regard the families diagnosed with AD osteoporosis as having a subtle form of OI and/or a monogenic skeletal dysplasia rather than the common polygenic disease of osteoporosis [30,31]. Studies that conducted dense rare-variant genotyping, such as Immunochip-based analyses of immune-mediated diseases, have had little success in identifying novel rare variant associations, despite large sample sizes [29,32,33].

5. Conclusions and Challenges

After mapping a disease-causing gene, many challenges remain, and many of these challenges are likely to be met through the resources and technologies that continue to build on the foundation of the Human Genome Project.

One important consideration is in understanding the multiple genetic contributions to disease onset and progression in addition to the primary causative gene in monogenic disorders. In FOP, for example, although the main disease characteristics are unique and readily recognized, variability in the age of disease onset, the rate of disease progression, and the severity of disease can be high, even in the context of the same ACVR1 c.617G>A mutation. Such phenotypic variability is likely influenced by underlying genetic causes. Identification of genetic modifiers that "protect" an individual with an FOP mutation, for example by directing a late onset or less aggressive disease progression, would provide new therapeutic targets and strategies for treating the disease. Similar issues exist for many common heritable diseases, such as osteogenesis imperfecta and Marfan's syndrome. With further development of genome technologies, the ability to understand phenotypic variability and the participation of genetic modifiers is becoming a reality.

The ultimate challenge is to elucidate the functions of the target gene and the consequences of its mutant forms, and, most importantly, the translation of this knowledge to treatments. Identification of the specific mutation in *ACVR1* has clear and important diagnostic value, providing a means to confirm suspicion of FOP based on toe malformations or, in cases of potential inheritance (and when sequencing in early life becomes more commonplace), a means to diagnose the condition before irreversible clinical manifestations occur allowing for early intervention. Identification of the target pathway and the specific mutation mechanism in FOP has opened up therapeutic strategies for this disease. Additionally, although the roles of BMP signaling in a wide range of tissue development and homeostasis functions had been well established and the signaling pathway and its components were well defined prior to identifying the FOP mutation, the roles of ACVR1/ALK2 in these processes were unrecognized and poorly understood. FOP identified ACVR1 as a key regulator of skeletal development and bone formation, providing an important new focus for skeletal biology and regenerative medicine. This has been the case for many skeletal dysplasias mapped to date, in which the causative gene was often not known to be involved in bone prior to its identification [34]. There are many examples of such findings providing important insights into musculoskeletal development and pathology, and many current treatments have been developed based on genetic discoveries—for example the development of anti-sclerostin antibodies based on the discovery that the high bone mass conditions of sclerosteosis and van Buchem's disease arise from mutations in the gene for sclerostin [35,36]. The power of MPS to map disease-associated mutations will thus benefit not only affected individuals and families, but also lead to a dramatic expansion in our understanding of human diseases. This will inform development of new therapies not only for rare monogenic disorders but also for diseases common in the general population.

Acknowledgments

The authors acknowledge the help of Murray Hargrave in editing the manuscript. This work was supported in part by the International Fibrodysplasia Ossificans Progressiva Association, the Center for Research in FOP and Related Disorders, the Cali-Weldon Professorship in FOP (to EMS), and the National Institutes of Health (NIH R01-AR41916).

Author Contributions

Emma Duncan, Matthew Brown and Eileen M. Shore all wrote the manuscript.

References

1. Peltonen, L.; McKusick, V.A. Genomics and medicine. Dissecting human disease in the postgenomic era. *Science* **2001**, *291*, 1224–1229.
2. Lazarus, S.; Zankl, A.; Duncan, E.L. Next-generation sequencing: A frameshift in skeletal dysplasia gene discovery. *Osteoporos. Int.* **2014**, *25*, 407–422.

3. Garrod, A.E. The Croonian Lectures on inborn errors of metabolism. *Lancet* **1908**, *172*, 214–220.
4. Warman, M.L.; Cormier-Daire, V.; Hall, C.; Krakow, D.; Lachman, R.; LeMerrer, M.; Mortier, G.; Mundlos, S.; Nishimura, G.; Rimoin, D.L.; *et al.* Nosology and classification of genetic skeletal disorders: 2010 revision. *Am. J. Med. Genet. A* **2011**, *155A*, 943–968.
5. Glazier, A.M.; Nadeau, J.H.; Aitman, T.J. Finding genes that underlie complex traits. *Science* **2002**, *298*, 2345–2349.
6. Shore, E.M.; Kaplan, F.S. Inherited human diseases of heterotopic bone formation. *Nat. Rev. Rheumatol.* **2010**, *6*, 518–527.
7. Shore, E.M.; Feldman, G.J.; Xu, M.; Kaplan, F.S. The genetics of fibrodysplasia ossificans progressiva. *Clin. Rev. Bone Miner. Metab.* **2005**, *3*, 201–204.
8. Shore, E.M.; Xu, M.; Feldman, G.J.; Fenstermacher, D.A.; Cho, T.-J.; Choi, I.H.; Connor, J.M.; Delai, P.; Glaser, D.L.; LeMerrer, M.; *et al.* A recurrent mutation in the BMP type I receptor ACVR1 causes inherited and sporadic fibrodysplasia ossificans progressiva. *Nat. Genet.* **2006**, *38*, 525–527.
9. Feldman, G.; Li, M.; Martin, S.; Urbanek, M.; Urtizberea, J.A.; Fardeau, M.; LeMerrer, M.; Connor, J.M.; Triffitt, J.; Smith, R.; *et al.* Fibrodysplasia ossificans progressiva, a heritable disorder of severe heterotopic ossification, maps to human chromosome 4q27–31. *Am. J. Hum. Genet.* **2000**, *66*, 128–135.
10. Gannon, F.H.; Kaplan, F.S.; Olmsted, E.; Finkel, G.C.; Zasloff, M.A.; Shore, E. Bone morphogenetic protein 2/4 in early fibromatous lesions of fibrodysplasia ossificans progressiva. *Hum. Pathol.* **1997**, *28*, 339–343.
11. Virdi, A.S.; Shore, E.M.; Oreffo, R.O.; Li, M.; Connor, J.M.; Smith, R.; Kaplan, F.S.; Triffitt, J.T. Phenotypic and molecular heterogeneity in fibrodysplasia ossificans progressiva. *Calcif. Tissue Int.* **1999**, *65*, 250–255.
12. Fiori, J.L.; Billings, P.C.; Serrano de la Pena, L.S.; Kaplan, F.S.; Shore, E.M. Dysregulation of the BMP-p38 MAPK signaling pathway in cells from patients with fibrodysplasia ossificans progressiva (FOP). *J. Bone Miner. Res.* **2006**, *21*, 902–909.
13. Serrano de la Pena, L.S.; Billings, P.C.; Fiori, J.L.; Ahn, J.; Kaplan, F.S.; Shore, E.M. Fibrodysplasia ossificans progressiva (FOP), a disorder of ectopic osteogenesis, misregulates cell surface expression and trafficking of BMPRIA. *J. Bone Miner. Res.* **2005**, *20*, 1168–1176.
14. Kaplan, F.S.; Xu, M.; Seemann, P.; Connor, J.M.; Glaser, D.L.; Carroll, L.; Delai, P.; Fastnacht-Urban, E.; Forman, S.J.; Gillessen-Kaesbach, G.; *et al.* Classic and atypical fibrodysplasia ossificans progressiva (FOP) phenotypes are caused by mutations in the bone morphogenetic protein (BMP) type I receptor ACVR1. *Hum. Mutat.* **2009**, *30*, 379–390.
15. Zhang, W.; Zhang, K.; Song, L.; Pang, J.; Ma, H.; Shore, E.M.; Kaplan, F.S.; Wang, P. The phenotype and genotype of fibrodysplasia ossificans progressiva in China: A report of 72 cases. *Bone* **2013**, *57*, 386–391.
16. Shore, E.M. Department of Orthopaedic Surgery, Perelman School of Medicine, University of Pennsylvania, Philadelphia, PA, USA. Unpublished work, 2014.

17. Culbert, A.L.; Chakkalakal, S.A.; Convente, M.R.; Lounev, V.Y.; Kaplan, F.S.; Shore, E.M. Fibrodysplasia (myositis) ossificans progressiva. In *Genetics of Bone Biology and Skeletal Disease*; Thakker, R.V., Whyte, M.P., Eisman, J.A., Igarashi, T., Eds.; Academic Press: New York, NY, USA, 2013; pp. 375–393.
18. Cooper, D.N.; Krawczak, M.; Antonorakis, S.E. The nature and mechanisms of human gene mutation. In *The Metabolic and Molecular Bases of Inherited Disease*; Scriver, C., Beaudet, A.L., Sly, W.S., Valle, D., Eds.; McGraw-Hill: New York, NY, USA, 1995; pp. 259–291.
19. Zankl, A.; Duncan, E.L.; Leo, P.J.; Clark, G.R.; Glazov, E.A.; Addor, M.C.; Herlin, T.; Kim, C.A.; Leheup, B.P.; McGill, J.; *et al.* Multicentric carpotarsal osteolysis is caused by mutations clustering in the amino-terminal transcriptional activation domain of MAFB. *Am. J. Hum. Genet.* **2012**, *90*, 494–501.
20. McInerney-Leo, A.M.; Marshall, M.S.; Gardiner, B.; Benn, D.E.; McFarlane, J.; Robinson, B.G.; Brown, M.A.; Leo, P.J.; Clifton-Bligh, R.J.; Duncan, E.L. Whole exome sequencing is an efficient and sensitive method for detection of germline mutations in patients with phaeochromcytomas and paragangliomas. *Clin. Endocrinol.* **2014**, *80*, 25–33.
21. McInerney-Leo, A.M.; Marshall, M.S.; Gardiner, B.; Coucke, P.J.; van Laer, L.; Loeys, B.L.; Summers, K.M.; Symoens, S.; West, J.A.; West, M.J.; *et al.* Whole exome sequencing is an efficient, sensitive and specific method of mutation detection in osteogenesis imperfecta and Marfan syndrome. *Bonekey Rep.* **2013**, *2*, 456.
22. Cho, T.J.; Lee, K.E.; Lee, S.K.; Song, S.J.; Kim, K.J.; Jeon, D.; Lee, G.; Kim, H.N.; Lee, H.R.; Eom, H.H.; *et al.* A single recurrent mutation in the 5'-UTR of IFITM5 causes osteogenesis imperfecta type V. *Am. J. Hum. Genet.* **2012**, *91*, 343–348.
23. Semler, O.; Garbes, L.; Keupp, K.; Swan, D.; Zimmermann, K.; Becker, J.; Iden, S.; Wirth, B.; Eysel, P.; Koerber, F.; *et al.* A mutation in the 5'-UTR of IFITM5 creates an in-frame start codon and causes autosomal-dominant osteogenesis imperfecta type V with hyperplastic callus. *Am. J. Hum. Genet.* **2012**, *91*, 349–357.
24. Brunham, L.R.; Hayden, M.R. Hunting human disease genes: Lessons from the past, challenges for the future. *Hum. Genet.* **2013**, *132*, 603–617.
25. Hoischen, A.; Gilissen, C.; Arts, P.; Wieskamp, N.; van der Vliet, W.; Vermeer, S.; Steehouwer, M.; de Vries, P.; Meijer, R.; Seiqueros, J.; *et al.* Massively parallel sequencing of ataxia genes after array-based enrichment. *Hum. Mutat.* **2010**, *31*, 494–499.
26. Koboldt, D.C.; Ding, L.; Mardis, E.R.; Wilson, R.K. Challenges of sequencing human genomes. *Brief Bioinform.* **2010**, *11*, 484–498.
27. Hashimoto, T.; de Hoon, M.J.; Grimmond, S.M.; Daub, C.O.; Hayashizaki, Y.; Faulkner, G.J. Probabilistic resolution of multi-mapping reads in massively parallel sequencing data using MuMRescueLite. *Bioinformatics* **2009**, *25*, 2613–2614.
28. Hunt, K.A.; Mistry, V.; Bockett, N.A.; Ahmad, T.; Ban, M.; Barker, J.N.; Barrett, J.C.; Blackburn, H.; Brand, O.; Burren, O.; *et al.* Negligible impact of rare autoimmune-locus coding-region variants on missing heritability. *Nature* **2013**, *498*, 232–235.

29. Styrkarsdottir, U.; Thorleifsson, G.; Sulem, P.; Gudbjartsson, D.F.; Sigurdsson, A.; Jonasdottir, A.; Jonasdottir, A.; Oddsson, A.; Helgason, A.; Magnusson, O.Y.; *et al.* Nonsense mutation in the LGR4 gene is associated with several human diseases and other traits. *Nature* **2013**, *497*, 517–520.

30. Laine, C.M.; Joeng, K.S.; Campeau, P.M.; Kiviranta, R.; Tarkkonen, K.; Grover, M.; Lu, J.T.; Pekkinen, M.; Wessman, M.; Heino, T.J.; *et al.* WNT1 mutations in early-onset osteoporosis and osteogenesis imperfecta. *N. Engl. J. Med.* **2013**, *368*, 1809–1816.

31. Pyott, S.M.; Tran, T.T.; Leistritz, D.F.; Pepin, M.G.; Mendelsohn, N.J.; Temme, R.T.; Fernandez, B.A.; Elsayed, S.M.; Elsobky, E.; Verma, I.; *et al.* WNT1 mutations in families affected by moderately severe and progressive recessive osteogenesis imperfecta. *Am. J. Hum. Genet.* **2013**, *92*, 590–597.

32. Parkes, M.; Cortes, A.; van Heel, D.A.; Brown, M.A. Genetic insights into common pathways and complex relationships among immune-mediated diseases. *Nat. Rev. Genet.* **2013**, *14*, 661–673.

33. Keupp, K.; Beleggia, F.; Kayserili, H.; Barnes, A.M.; Steiner, M.; Semler, O.; Fischer, B.; Yigit, G.; Janda, C.Y.; Becker, J.; *et al.* Mutations in WNT1 cause different forms of bone fragility. *Am. J. Hum. Genet.* **2013**, *92*, 565–574.

34. Glazov, E.A.; Beleggia, F.; Kayserili, H.; Barnes, A.M.; Steiner, M.; Semler, O.; Fischer, B.; Yigit, G.; Janda, C.Y.; Becker, J.; *et al.* Whole-exome re-sequencing in a family quartet identifies POP1 mutations as the cause of a novel skeletal dysplasia. *PLoS Genet.* **2011**, *7*, e1002027.

35. Balemans, W.; Patel, N.; Ebeling, M.; van Hul, E.; Wuyts, W.; Lacza, C.; Dioszegi, M.; Dikkers, F.G.; Hildering, P.; Willems, P.J.; *et al.* Identification of a 52 kb deletion downstream of the SOST gene in patients with van Buchem disease. *J. Med. Genet.* **2002**, *39*, 91–97.

36. Van Hul, W.; Balemans, W.; van Hul, E.; Dikkers, F.G.; Obee, H.; Stokroos, R.J.; Hildering, P.; Vanhoenacker, F.; van Camp, G.; Willems, P.J. Van Buchem disease (hyperostosis corticalis generalisata) maps to chromosome 17q12–q21. *Am. J. Hum. Genet.* **1998**, *62*, 391–399.

3

The Impact of the Human Genome Project on Complex Disease

Jessica N. Cooke Bailey [1,*], Margaret A. Pericak-Vance [2] and Jonathan L. Haines [1,*]

[1] Department of Epidemiology and Biostatistics, Case Western Reserve University Medical Center, Cleveland, OH 44106, USA

[2] Hussman Institute for Human Genomics, Miller School of Medicine, University of Miami, Miami, FL 33136, USA; E-Mail: mpericak@med.miami.edu

* Author to whom correspondence should be addressed; E-Mails: jnc43@case.edu (J.N.C.B.); jonathan.haines@case.edu (J.L.H.).

Abstract: In the decade that has passed since the initial release of the Human Genome, numerous advancements in science and technology within and beyond genetics and genomics have been encouraged and enhanced by the availability of this vast and remarkable data resource. Progress in understanding three common, complex diseases: age-related macular degeneration (AMD), Alzheimer's disease (AD), and multiple sclerosis (MS), are three exemplars of the incredible impact on the elucidation of the genetic architecture of disease. The approaches used in these diseases have been successfully applied to numerous other complex diseases. For example, the heritability of AMD was confirmed upon the release of the first genome-wide association study (GWAS) along with confirmatory reports that supported the findings of that state-of-the art method, thus setting the foundation for future GWAS in other heritable diseases. Following this seminal discovery and applying it to other diseases including AD and MS, the genetic knowledge of AD expanded far beyond the well-known *APOE* locus and now includes more than 20 loci. MS genetics saw a similar increase beyond the *HLA* loci and now has more than 100 known risk loci. Ongoing and future efforts will seek to define the remaining heritability of these diseases; the next decade could very well hold the key to attaining this goal.

Keywords: human genome project; age-related macular degeneration; Alzheimer's disease; multiple sclerosis; genetics; genomics; genome-wide association study

---

1. Introduction

In celebration of the 10th anniversary of the completion of the Human Genome Project, it is pertinent to take a step back and reflect on the progress that has been made in genetic and genomic research over the past decade by exploring the knowledge gleaned from the extensive wealth of information provided by the Human Genome Project (HGP). Herein we provide a concise historical overview of three signature human diseases that have strong but complex genetic etiologies: age-related macular degeneration (AMD), multiple sclerosis (MS), and Alzheimer's disease (AD). The significant progress in defining the genetic architecture of these diseases, beginning with the pre-genome-wide association study (GWAS)-era and concluding with the current state of each, and what lies ahead for these complex diseases reflects the great progress that has been made in general in the study of multifactorial diseases, and provides a brief glimpse at what we can hope the next decade of genomic research will provide.

2. Age-Related Macular Degeneration (AMD) and the First Genome-Wide Association Study

Age-related macular degeneration an ocular neurodegenerative disease that results primarily in loss of central vision, is a major cause of visual impairment and blindness in elderly populations worldwide. Although there was at one time substantial controversy over the strength of the genetic effects in AMD, genetic and epidemiological research, established that there is a significant genetic component to AMD, estimated to be 45%–70% [1]. This was supported by twin studies that reported higher incidence of disease in monozygotic *versus* dizygotic twins [1–5] and family studies in which risk for developing AMD between first degree relatives ranges from 2–3 [3,6,7]. This knowledge encouraged the application of increasingly sophisticated genomic techniques to elucidate the genetic etiology of AMD susceptibility and pathogenesis. Prior to major genetic breakthroughs such as the completion of the HGP, it was well established that inflammatory and immunologic mediators contribute to AMD (e.g., [8–13]). However, this knowledge did not lead to identification of any confirmed genetic loci for AMD. Following the trends at that time in applying the available statistical genetic techniques, numerous genetic linkage studies using multiplex families and (primarily) affected sibships were attempted [14–22]. Notably, The *ABCA4* (*ABCR*) locus on chromosome 1p21, identified for its involvement in autosomal recessive Stargardt disease retinopathy [23–27], was one of the first loci identified as involved in AMD, though not all reports have been consistent [23–31]. While linkage studies continued to provide suggestive evidence of a role of genetics, they did not find any definitive locus for AMD. In a large meta-analysis of most of these genetic linkage studies, several chromosomal regions were identified as highly likely to harbor AMD genes, most convincingly including chromosome 1q23.3–q32 and 10q26 [32].

With the continuing evolution of HGP resources, in particular the identification of very large numbers of single nucleotide polymorphisms (SNPs) [33,34] multiple new experimental designs for identifying AMD loci were employed. SNPs provided several advantages over the then prevalent microsatellite markers; the two most important were the high density of SNPs across the genome, and their much higher fidelity in genotyping. The culmination of the efforts of four independent studies using four complementary study designs was a convergence on the discovery of the association between AMD and the gene encoding the complement factor H protein (*CFH*), located on chromosome 1q32. In one of the first reported genome-wide association studies (GWAS), Klein *et al.* [35]

screened 96 AMD cases and 50 non-AMD controls to evaluate variants associated with AMD. The GWAS method implements a hypothesis-free approach in which a large number of SNPs are genotyped across the genome and evaluated for association with disease. This particular study evaluated 116,204 successfully genotyped SNPs and detected association between AMD and an intronic SNP in *CFH* ($p < 10^{-7}$). Linkage disequilibrium analysis and localization using resequencing in this region led to the discovery of a nonsynonymous SNP in exon 9 of *CFH*; this SNP, rs1061170, causes the substitution of a histidine for a tyrosine at amino acid 402 (Y402H).

Independently and concurrently with the Klein *et al.* study [35], Haines *et al.* [36] also identified the association with the Y402H variant, but by implementing a purely locational genomic approach. By focusing on and extending slightly beyond the 24 Mb region implicated by linkage studies of AMD [14,15,22] they identified a five-SNP haplotype spanning a 261 kb region surrounding the Regulators of Complement Activation (RCA) gene cluster by genotyping only 61 SNPs in two independent datasets. Both affected and unaffected individuals homozygous for the risk haplotype were sequenced for the genes residing in this haplotype. Their hypothesis was that having controlled for the locus specific genetic background (e.g., the haplotype), frequency differences for variants between cases and controls would identify the causal variation. Scanning the coding region of *CFH* in those individuals, Y402H was by far the most significantly different of the 11 detected variants. Follow-up genotyping in the original datasets confirmed that the Y402H variant was significantly associated with risk for AMD and that a surprisingly high proportion of the genetic variation in AMD could be attributed to the Y402H variant.

Implementing yet another independent, concurrent, and complimentary approach to localize AMD-causing variants, Edwards *et al.* also identified the Y402H variant using a fine-mapping approach focused on this same general region on chromosome 1 [37]. This study centered efforts on 86 SNPs located in coding sequences encompassing the RCA locus in a case-control sample. The most significant of the 29 associated variants located in the RCA was again rs1061170 (Y402H) in *CFH*. Replication analysis evaluating this and 13 additional SNPs typed in a second case-control sample confirmed the association of Y402H with AMD. Further analysis established that that C (risk)-allele carrying individuals accounted for approximately half of cases.

Hageman and colleagues also confirmed the Y402H variant, applying yet a fourth genetic analysis method [38]. They applied prior biological knowledge of the involvement of *CFH* (also called *HF1*) in membranoproliferative glomerulonephritis type II (MPGNII), a disease in which patients develop ocular drusen nearly identical to those found in AMD patients. The genetic lesion for MPGNII resides in the same chromosome 1q31–32 region that was also implicated in linkage studies of AMD [14,15,22]. Evaluating two samples of unrelated individuals for AMD-associated variation in *CFH*, this group also detected evidence for association between AMD and the Y402H variant.

These four studies simultaneously reported the role of variation in a chromosome 1 region that had previously been highlighted in AMD linkage studies [14,15,22]; identifying this major genetic determinant of AMD, something that even a year earlier was thought not to exist, was a major landmark in genetics of complex disease. These results, while obviously important for AMD research, provided the first validation of the GWAS approach. Up to that point, hundreds of papers had been written about the potential of the GWAS study design, but very little had been published on actual implementation. Dr. Elias Zherhouni, then Director of NIH, highlighted these studies as a major breakthrough in health research [39]. This very strong validation imparted the necessary confidence in GWAS to invigorate its application to numerous other diseases. Over the past nine years more than 2000 GWAS studies that have been published [40].

Since the initial discovery of the Y402H *CFH* variant, substantial progress has been made in understanding the genetics of AMD. This includes the localization of the strongest single genetic effect in AMD on chromosome 10q26 (through positional localization approach) to the region containing *ARMS2* and *HTRA1* (e.g., [41–47]), though there is still controversy whether either one or both of these genes contains the causal variant (e.g. [46–49]).

In the *CFH* region, in addition to the high-effect Y402H variant, a deletion of *CFHR1* and *CFHR3* was detected and determine to be protective for AMD [50] (e.g., [51–53]). Various additional studies focused on the potential role of additional complement mediators in AMD. Gold *et al.* explored additional alternative complement pathway activators beyond CFH and determined that variants in complement factor b (*BF/CFB*) and *C2* are highly protective against AMD [54]. A coding variant in *C3* was also determined to be associated with AMD [55–57]. A variant upstream of the *CFI* gene was also determined to influence AMD risk [58]. A variant in the *CFD* gene was associated with AMD but replicated almost solely in females [59]. More recently discovered AMD-associated loci have been detect in/near genes *ADAMTS9*, *B3GALTL*, *CETP*, *COL8A1-FILIP1L*, *IER3-DDR1*, *LIPC*, *RAD51B*, *SLC16A8*, *TGFBR1* and *TIMP3* [60–62].

GWAS studies in AMD now include over 1 million markers [61,63–65]. Though the traditional GWAS approach has been incredibly informative for many diseases, a great deal of the genetic proportion of many of these diseases remains to be fully elucidated even after applying straightforward and complex GWAS methods [66]. Approaches to enhance the detection of genetic variation associated with disease have necessarily expanded beyond the traditional GWAS to broaden the range of discovery and increase the power of detection; one technique that aids in this process is imputing variants—using known genetic information from a reference sample. Obviously increasing the sample size of genetic studies of complex diseases is crucial to accelerate the identification of disease-specific variants. Expanding the number of testable variants is now a more attainable goal using imputation, a technique that can significantly increase the number of tested variants beyond those interrogated by a GWAS through informing genotypes of untyped SNPs [67,68]. Combining known genotypes at GWAS-interrogated SNPs with available sequence data from a reference panel and inferring untyped SNPs in the dataset based on haplotype frequencies allows for the inference of numerous SNPs with varying degrees of confidence and accuracy. This method increases the power of GWAS by increasing the number of SNPs that can be tested, it can also lead to more efficient identification of causal variants and/or SNPs in high linkage disequilibrium with a causal variant [67,68]. Imputation has been implemented in several studies of AMD to enhance the ability to detect associated variants [61,63,69]. For example, the most recent publication from the AMD Gene Consortium reports seven novel variants that were detected using imputed data in addition to confirming 12 previously identified variants [61].

An additional method to utilize genome-wide data beyond the traditional association analysis is to perform pathway enrichment analyses. The goal of such analyses is identify biological relationships between associated genetic signals and pathways of interest in a particular disease. Pathway analysis can be performed by a comprehensive review of GWAS results to assess overrepresentation of SNPs meeting a specific threshold that occur within biological pathways [70]. These enhance GWAS by evaluating potentially biologically relevant signals that might otherwise be overlooked because of the numerous false-positive results that occur in large GWAS studies [70]. These have the potential to highlight otherwise undetected small and/or interactive effects that are important to evaluate in addition to and in the context of the overall genome-wide results. Using the INRICH (Interval-based Enrichment Analysis Tool for Genome Wide Association Studies) pathway analysis tool [71] to evaluate overall results, the AMD Gene Consortium not only confirmed previously implicated AMD pathways, but also determined additional pathways of interest in the most recent publication which detected enrichment of complement and atherosclerotic pathway-encoding genes as well as genes involved in pathways of collagen and extracellular region, complement and coagulation cascades, lipoprotein metabolism, and regulation of apoptosis [61].

The impressive impact that genetic information can have on our understanding of disease pathophysiology is highlighted in the recent publication by Yang *et al.* in which they report that *ARMS2/HTRA1* risk alleles contribute to AMD pathogenesis by decreasing the defense capabilities of superoxide dismutase 2 (SOD2) and thereby cause the retinal pigment epithelium to be more susceptibility to oxidative damage [72]. Having an explanation for the role the variants have in the disease is crucial to further elucidating disease mechanisms both genetically and physiologically. Additionally, genetic studies implicate VEGF as having a role in AMD and current AMD treatment and clinical trials utilize this information for treatment of neovascular AMD (reviewed in [73]), thus highlighting the utility of genetic data for clinical impact.

3. Alzheimer's Disease

Alzheimer's disease (AD) is a genetically heterogeneous neurologic disorder that is the leading cause of dementia among the elderly. It is characterized by the progressive loss of cognitive ability beyond what is normally associated with aging. AD is a complex disease that is influenced by both environmental and genetic mediators, the most significant of which is age [74,75]. The heritability of AD is estimated between 60%–80%. Before 1985, there was very significant debate about whether or not genetics played any role in AD (e.g., [76–80]). However, in 1987, using some of the earliest technologies employing genomic markers, a locus for the rare early onset AD (EOAD) was identified [81], and in 1991 the responsible variation in the *APP* gene was located [41].

Expansions of genomic marker sets, developed through early HGP efforts, were used to further identify two additional early onset genes in the early 1990's [82–86]. Simultaneously and independently, the emerging technologies of genomic markers and genetic linkage analysis were applied to the far more common late onset Alzheimer's disease (LOAD), which accounts for 99% of AD cases [87]. Using these techniques, Pericak-Vance *et al.* identified a locus on chromosome 19 near the gene encoding apolipoprotein E (*APOE*) [88,89], which was at that time thought to only be involved in cardiovascular disease. This locus has three distinct alleles: ∈ 2, ∈ 3, and ∈ 4. Corder *et al.* characterized a dose-dependent association between the *APOE*-∈ 4 allele and an increased risk of LOAD [90]. Mutations in the EOAD genes are causal, with very high penetrance, and opened avenues for exploring the pathophysiology of AD. However, in aggregate they explain less than 1% of AD. In contrast, *APOE* explains at least 25% of AD. A year after determining the role of the ∈ 4 variant in

LOAD susceptibility, it became apparent that the ∈ 2 allele provided an independently protective effect on LOAD [91].

The *APOE* finding was pivotal for two reasons. Within the AD research community, it provided a new avenue and a completely different view of the genetic etiology of AD. More generally, however, this was one of the very first examples of how the emerging technologies of the HGP could be successfully applied to diseases lacking a simple Mendelian inheritance pattern, *i.e.*, what are commonly called complex diseases. The finding of the *APOE*-∈ 2 allele protective effect was also one of the first examples of different alleles carrying different effects on a complex disease, a pivotal moment in AD research and broadly in the field of genetics.

Innumerable attempts to identify additional genomic variations modulating the risk of LOAD followed these groundbreaking *APOE* discoveries, using the increasingly dense set of known variations and emerging sequencing techniques (cataloged in Alzgene.org). These efforts were primarily applied to specific genes of interest; that is, employing a focused candidate gene approach. Although there were numerous reports of significant associations, no consensus arose that any of these were true effects. It was not until GWAS became a viable approach [92–94], and multiple datasets were combined, that additional LOAD loci become visible and confirmed [95–97]. The most recent efforts by the Alzheimer Disease Genetics Consortium (ADGC) and the International Genomics of Alzheimer's Project (IGAP) have greatly increased the number of known loci associated with LOAD. In the 2011 Naj *et al.* report, a three-stage design (discovery stage 1, replication stages 2–3) was utilized; this analysis evaluated >18,000 cases and >29,000 controls using both joint- and meta-analysis approaches and novel genome-wide significant hits were detected at SNPs in *MS4A4A*, *CD2AP*, *EPHA1* and *CD33* [96]. In Lambert *et al.* 2013, the IGAP reported an additional eleven novel LOAD susceptibility loci after analyzing genotyped and imputed data in a two-stage meta-analysis of >25,000 cases and >48,000 controls [95]. There are now over 20 loci identified that influence LOAD [95]. Importantly, using the pathway approach, the amyloid precursor protein and tau pathways are confirmed by this most recent large GWAS in addition to the newly implicated hippocampal synaptic function, cytoskeletal function and axonal transport, regulation of gene expression and post-translational modification of proteins, and microglial and myeloid cell function pathways [95].

4. Multiple Sclerosis

Multiple sclerosis (MS) is a common cause of neurological disability involving inflammatory demyelination of the central nervous system [98–101]. There is ample evidence that MS has a strong genetic component, but like so many other complex diseases, non-genetic influences are also important (e.g., [99,102–104]). MS is also a complex, heterogeneous disease in which significant efforts to unravel the role of genetics have been made. Unlike both AMD and LOAD, the first and strongest genetic effect in MS was identified well before the HGP was undertaken. Because MS is an autoimmune disease, it was strongly suspected that the major histocompatibility locus (*MHC*) would be involved. More specifically, there was a focus on the human leukocyte antigen loci on chromosome 6. In the early 1970's the *HLA* loci could be genotyped using blood antigen reactions, allowing assignment of genotypes without directly examining the DNA. Through a number of efforts (e.g., [99,103–109]) a strong risk association with the *HLA-DR* locus, and specifically the 15*01 allele was identified.

Despite this auspicious beginning, identifying additional MS loci languished. As with the other complex diseases, genetic linkage analysis was applied to multiplex MS families, with varying results. Some early genetic linkage studies confirmed the role of HLA [110], while others did not [111]. Additional studies, using the increasingly dense DNA marker sets and larger datasets, ultimately demonstrated and confirmed that the *HLA* locus was the single largest genetic effect, and that any other MS loci would have at most modest individual effects [109,110,112–116]. These studies did highlight several other possible loci, but did not have the resolution to identify specific associated genetic variations [115,116].

Finally, in 2007, nearly 30 years after the initial association finding, a second locus for MS was identified [117,118]. Gregory *et al.* employed a genomic convergence approach that integrated data from genetic linkage studies, genetic association studies, model system gene expression data, and *in vitro* functional data to narrow in on a specific locus and a functional polymorphism in the interleukin-7 receptor α chain (*IL7R*) [117]. Independently, the International Multiple Sclerosis Genetics Consortium (IMSGC) published results from one of the first large-scale GWAS studies, using 334,923 GWAS SNPs. The IMSGC used a hybrid study design that included a family-based study of 931 family trios and an independent dataset set of cases and controls [118]. These analyses confirmed the role of genetic variation in *IL7RA* and also highlighted variations in *IL2RA*.

These results also had broad implications for the field of MS resarch. The IMSGC GWAS was still one of the first such studies done with a well-powered dataset and demonstrated that family-based and case-control GWAS approaches were both useful methods for exploring genetic information. In addition, like AMD, the convergence of independent approaches (GWAS and gene-targeted methods) further validated that GWAS could identify relevant associated loci. Subsequent studies with much larger datasets [96,119,120] have now identified over 100 total loci associated with MS.

Efforts in MS have shown substantial increases in the number of independent loci associated with this disease. The most recent IMSGC study evaluated in two stages more than 80,000 individuals of European ancestry [119]. This analysis expanded the known MS loci by 48, raising the total number of discrete MS-associated loci to 103. In addition, the IMSGC interrogated specific genomic regions and hypotheses using a custom array, the Immunochip. The group efficiently utilized extensive amounts of data by assembling multiple studies and utilizing imputation methods and then applying conditional and joint analysis methods [119]. Such methods are becoming more common in the efforts to expand the power to detect common variation in many multifactorial diseases. The strongest of the novel hits from this analysis implicates a SNP in the region between *BCL10* and *DDAH1* [119]; *BCL10* is an activator of nuclear factor (NF)-κB signaling which is involved in gene expression control of inflammation, immunity, cell proliferation and apoptosis and has been explored as a clinical target for MS [73,119,121].

Pathway analysis in MS has also proven useful. The IMSGC study additionally sought to evaluate the Gene Ontology (GO) processes of the associated variants using the MetaCore ([122]); their results indicated, as expected, that most variants fall in or near genes with immune function [119]. Another recent endeavor to evaluate pathways involved in MS utilized results from eight MS GWAS datasets and prioritized genes in the cell adhesion molecule (CAM) biological pathway with the Cytoscape software [120,123]. Their findings highlighted five networks that were associated with susceptibility to MS—again supporting the utility of expanding beyond traditional case-control association analyses of GWAS data and encouraging the use of multiple datasets to determine enrichment of signals that might otherwise not have been detected using a traditional GWAS approach [120].

5. Conclusions and Future Directions

Since the completion of the HGP and shortly after the first GWAS, thousands more GWAS have been reported [40]; these have brought forth great progress in numerous diseases that previously were only hypothesized to have a genetic component. Large-scale collaborative efforts have raised the number of known AMD, AD, and MS loci to 19 [61], 20 [95], and 103 [119], respectively. Efforts to increase sample size have been successful, as evidenced by the largest and most recently reported analyses of AMD [61], AD [95] and MS [92], which each evaluated >74,000 individuals; however, other techniques are necessary to evaluate and explore as data becomes increasingly large and complex. Whole exome and whole genome sequencing are more recent approaches to generating genetic data that allow investigation far beyond the capabilities of the GWAS, and their utility is just starting to take shape in studies of many complex common diseases, including those mentioned herein. These will enable the study of rare and low frequency variants, which have been implicated as a potential source of missing heritability in many genetic diseases [66]. Analysis of data from exome arrays, designed to jointly interrogate data relevant to association studies of common variants and sequencing studies of rare variants, will improve genetic analysis of disease by providing greater coverage of known susceptibility loci and enhancing the likelihood for discovery of novel disease loci.

For each of these diseases, the fact that there is a genetic component is irrefutable; this knowledge has, over the past decade, most certainly been confirmed and expounded upon with the completion of the human genome sequence. As genetic knowledge continues to grow, and as clinical phenotyping techniques improve, further genetic variation influencing AMD, AD, and MS will likely be detectable and, hopefully, their roles in these diseases will be more clearly defined [124]. We can anticipate that as our understanding of genetic etiology of these diseases grows, future studies will further explore rare variations contributing to disease, the role of copy number variants, and the genetics of these diseases in non-European populations. Additionally, the role of currently undetermined environmental factors and their interactions with genetic variants must continue to be elucidated. The global objective of prior and ongoing studies is certainly to improve the current comprehension of new and existing disease loci in order that the biology of these diseases can be fully explicated in the hope of attaining improved strategies for disease treatment and prevention in the future.

The last decade has doubtlessly ushered in dramatic advances in the amount of shared data available to genetic researchers. Resources such as the NHLBI Grand Opportunity Exome Sequencing Project [125], the 1000Genomes [126], ENCODE [127], and the International HapMap Project [128,129] provide seemingly limitless amounts of data—all geared toward further understanding the intricacies of the human genome and how alterations of it influence human variation. We have provided a brief genetic history of three diseases that are exemplars of developing approaches to apply the incredible resources of the HGP. Such progress will most certainly continue to improve and exponentially increase in the next decade to facilitate a greater understanding of these and other complex diseases, as well as usher in the realization of personalized medicine.

Acknowledgments

This work was supported by NIH T32 EY007157 (Jessica N. Cooke Bailey), Alzheimer's Disease Genetics Consortium, funded by NIA grant U01AG032984 (Margaret A. Pericak-Vance, Jonathan L. Haines), BrightFocus Foundation grant A2011048 (Margaret A. Pericak-Vance), DOD grant W81XWH-12-1-0013 (Margaret A. Pericak-Vance, Jonathan L. Haines), NEI grants 7R01EY012118 (Margaret A. Pericak-Vance, Jonathan L. Haines), 1R01EY022310 (Jonathan L. Haines, Margaret A. Pericak-Vance), 1R01EY023164 (Margaret A. Pericak-Vance), and 1R01EY020928 (Margaret A. Pericak-Vance), NIA grants 1R01AG027944 (Margaret A. Pericak-Vance, Jonathan L. Haines) and R01AG19085 (Margaret A. Pericak-Vance, Jonathan L. Haines) and NINDS grant R01NS032830 (Margaret A. Pericak-Vance, Jonathan L. Haines).

Author Contributions

Jessica N. Cooke Bailey wrote and edited the manuscript. Margaret A. Pericak-Vance conceived the idea for the manuscript and reviewed and edited the manuscript. Jonathan L. Haines conceived the idea for the manuscript, wrote, reviewed and edited the manuscript.

References

1. Seddon, J.M.; Cote, J.; Page, W.F.; Aggen, S.H.; Neale, M.C. The US twin study of age-related macular degeneration: Relative roles of genetic and environmental influences. *Arch. Ophthalmol.* **2005**, *123*, 321–327.
2. Hammond, C.J.; Webster, A.R.; Snieder, H.; Bird, A.C.; Gilbert, C.E.; Spector, T.D. Genetic influence on early age-related maculopathy: A twin study. *Ophthalmology* **2002**, *109*, 730–736.
3. Klaver, C.C.; Wolfs, R.C.; Assink, J.J.; van Duijn, C.M.; Hofman, A.; de Jong, P.T. Genetic risk of age-related maculopathy. Population-based familial aggregation study. *Arch. Ophthalmol.* **1998**, *116*, 1646–1651.
4. Klein, M.L.; Mauldin, W.M.; Stoumbos, V.D. Heredity and age-related macular degeneration. Observations in monozygotic twins. *Arch. Ophthalmol.* **1994**, *112*, 932–937.
5. Meyers, S.M. A twin study on age-related macular degeneration. *Trans. Am. Ophthalmol. Soc.* **1994**, *92*, 775–843.
6. Heiba, I.M.; Elston, R.C.; Klein, B.E.; Klein, R. Sibling correlations and segregation analysis of age-related maculopathy: The beaver dam eye study. *Genet. Epidemiol.* **1994**, *11*, 51–67.

7. Seddon, J.M.; Cote, J.; Davis, N.; Rosner, B. Progression of age-related macular degeneration: Association with body mass index, waist circumference, and waist-hip ratio. *Arch. Ophthalmol.* **2003**, *121*, 785–792.
8. Anderson, D.H.; Mullins, R.F.; Hageman, G.S.; Johnson, L.V. A role for local inflammation in the formation of drusen in the aging eye. *Am. J. Ophthalmol.* **2002**, *134*, 411–431.
9. Anderson, D.H.; Radeke, M.J.; Gallo, N.B.; Chapin, E.A.; Johnson, P.T.; Curletti, C.R.; Hancox, L.S.; Hu, J.; Ebright, J.N.; Malek, G.; *et al.* The pivotal role of the complement system in aging and age-related macular degeneration: Hypothesis re-visited. *Prog. Retin. Eye Res.* **2010**, *29*, 95–112.
10. Ding, X.; Patel, M.; Chan, C.C. Molecular pathology of age-related macular degeneration. *Prog. Retin. Eye Res.* **2009**, *28*, 1–18.
11. Patel, M.; Chan, C.C. Immunopathological aspects of age-related macular degeneration. *Semin. Immunopathol.* **2008**, *30*, 97–110.
12. Penfold, P.L.; Killingsworth, M.C.; Sarks, S.H. Senile macular degeneration: The involvement of immunocompetent cells. *Graefes Arch. Clin. Exp. Ophthalmol.* **1985**, *223*, 69–76.
13. Tuo, J.; Grob, S.; Zhang, K.; Chan, C.C. Genetics of immunological and inflammatory components in age-related macular degeneration. *Ocul. Immunol. Inflamm.* **2012**, *20*, 27–36.
14. Abecasis, G.R.; Yashar, B.M.; Zhao, Y.; Ghiasvand, N.M.; Zareparsi, S.; Branham, K.E.; Reddick, A.C.; Trager, E.H.; Yoshida, S.; Bahling, J.; *et al.* Age-related macular degeneration: A high-resolution genome scan for susceptibility loci in a population enriched for late-stage disease. *Am. J. Hum. Genet.* **2004**, *74*, 482–494.
15. Iyengar, S.K.; Song, D.; Klein, B.E.; Klein, R.; Schick, J.H.; Humphrey, J.; Millard, C.; Liptak, R.; Russo, K.; Jun, G.; *et al.* Dissection of genomewide-scan data in extended families reveals a major locus and oligogenic susceptibility for age-related macular degeneration. *Am. J. Hum. Genet.* **2004**, *74*, 20–39.
16. Klein, M.L.; Schultz, D.W.; Edwards, A.; Matise, T.C.; Rust, K.; Berselli, C.B.; Trzupek, K.; Weleber, R.G.; Ott, J.; Wirtz, M.K.; *et al.* Age-related macular degeneration. Clinical features in a large family and linkage to chromosome 1q. *Arch. Ophthalmol.* **1998**, *116*, 1082–1088.
17. Majewski, J.; Schultz, D.W.; Weleber, R.G.; Schain, M.B.; Edwards, A.O.; Matise, T.C.; Acott, T.S.; Ott, J.; Klein, M.L. Age-related macular degeneration—A genome scan in extended families. *Am. J. Hum. Genet.* **2003**, *73*, 540–550.
18. Seddon, J.M.; Santangelo, S.L.; Book, K.; Chong, S.; Cote, J. A genomewide scan for age-related macular degeneration provides evidence for linkage to several chromosomal regions. *Am. J. Hum. Genet.* **2003**, *73*, 780–790.
19. Tuo, J.; Bojanowski, C.M.; Chan, C.C. Genetic factors of age-related macular degeneration. *Prog. Retin. Eye Res.* **2004**, *23*, 229–249.
20. Weeks, D.E.; Conley, Y.P.; Mah, T.S.; Paul, T.O.; Morse, L.; Ngo-Chang, J.; Dailey, J.P.; Ferrell, R.E.; Gorin, M.B. A full genome scan for age-related maculopathy. *Hum. Mol. Genet.* **2000**, *9*, 1329–1349.
21. Weeks, D.E.; Conley, Y.P.; Tsai, H.J.; Mah, T.S.; Rosenfeld, P.J.; Paul, T.O.; Eller, A.W.; Morse, L.S.; Dailey, J.P.; Ferrell, R.E.; *et al.* Age-related maculopathy: An expanded

genome-wide scan with evidence of susceptibility loci within the 1q31 and 17q25 regions. *Am. J. Ophthalmol.* **2001**, *132*, 682–692.

22. Weeks, D.E.; Conley, Y.P.; Tsai, H.J.; Mah, T.S.; Schmidt, S.; Postel, E.A.; Agarwal, A.; Haines, J.L.; Pericak-Vance, M.A.; Rosenfeld, P.J.; *et al.* Age-related maculopathy: A genomewide scan with continued evidence of susceptibility loci within the 1q31, 10q26, and 17q25 regions. *Am. J. Hum. Genet.* **2004**, *75*, 174–189.
23. Allikmets, R. A photoreceptor cell-specific ATP-binding transporter gene (ABCR) is mutated in recessive Stargardt macular dystrophy. *Nat. Genet.* **1997**, *17*, 122.
24. Allikmets, R. Further evidence for an association of ABCR alleles with age-related macular degeneration. The International ABCR Screening Consortium. *Am. J. Hum. Genet.* **2000**, *67*, 487–491.
25. Allikmets, R.; Shroyer, N.F.; Singh, N.; Seddon, J.M.; Lewis, R.A.; Bernstein, P.S.; Peiffer, A.; Zabriskie, N.A.; Li, Y.; Hutchinson, A.; *et al.* Mutation of the Stargardt disease gene (ABCR) in age-related macular degeneration. *Science* **1997**, *277*, 1805–1807.
26. Meyers, S.M.; Greene, T.; Gutman, F.A. A twin study of age-related macular degeneration. *Am. J. Ophthalmol.* **1995**, *120*, 757–766.
27. Shroyer, N.F.; Lewis, R.A.; Yatsenko, A.N.; Wensel, T.G.; Lupski, J.R. Cosegregation and functional analysis of mutant ABCR (ABCA4) alleles in families that manifest both Stargardt disease and age-related macular degeneration. *Hum. Mol. Genet.* **2001**, *10*, 2671–2678.
28. Fritsche, L.G.; Fleckenstein, M.; Fiebig, B.S.; Schmitz-Valckenberg, S.; Bindewald-Wittich, A.; Keilhauer, C.N.; Renner, A.B.; Mackensen, F.; Mossner, A.; Pauleikhoff, D.; *et al.* A subgroup of age-related macular degeneration is associated with mono-allelic sequence variants in the *ABCA4* gene. *Invest. Ophthalmol. Vis. Sci.* **2012**, *53*, 2112–2118.
29. Guymer, R.H.; Heon, E.; Lotery, A.J.; Munier, F.L.; Schorderet, D.F.; Baird, P.N.; McNeil, R.J.; Haines, H.; Sheffield, V.C.; Stone, E.M. Variation of codons 1961 and 2177 of the Stargardt disease gene is not associated with age-related macular degeneration. *Arch. Ophthalmol.* **2001**, *119*, 745–751.
30. Rivera, A.; White, K.; Stohr, H.; Steiner, K.; Hemmrich, N.; Grimm, T.; Jurklies, B.; Lorenz, B.; Scholl, H.P.; Apfelstedt-Sylla, E.; *et al.* A comprehensive survey of sequence variation in the ABCA4 (ABCR) gene in Stargardt disease and age-related macular degeneration. *Am. J. Hum. Genet.* **2000**, *67*, 800–813.
31. Webster, A.R.; Heon, E.; Lotery, A.J.; Vandenburgh, K.; Casavant, T.L.; Oh, K.T.; Beck, G.; Fishman, G.A.; Lam, B.L.; Levin, A.; *et al.* An analysis of allelic variation in the ABCA4 gene. *Invest. Ophthalmol. Vis. Sci.* **2001**, *42*, 1179–1189.
32. Fisher, S.A.; Abecasis, G.R.; Yashar, B.M.; Zareparsi, S.; Swaroop, A.; Iyengar, S.K.; Klein, B.E.; Klein, R.; Lee, K.E.; Majewski, J.; *et al.* Meta-analysis of genome scans of age-related macular degeneration. *Hum. Mol. Genet.* **2005**, *14*, 2257–2264.
33. dbSNP. Available online: http://www.ncbi.nlm.nih.gov/SNP/ (accessed on 7 March 2014).
34. Thorisson, G.A.; Stein, L.D. The SNP Consortium website: Past, present and future. *Nucleic Acids Res.* **2003**, *31*, 124–127.

35. Klein, R.J.; Zeiss, C.; Chew, E.Y.; Tsai J.Y.; Sackler R.S.; Haynes, C.; Henning, A.K.; SanGiovanni J.P.; Mane, S.M.; Mayne, S.T.; *et al.* Complement factor H polymorphism in age-related macular degeneration. *Science* **2005**, *308*, 385–389.
36. Haines, J.L.; Hauser, M.A.; Schmidt, S.; Scott, W.K.; Olson, L.M.; Gallins, P.; Spencer, K.L.; Kwan, S.Y.; Noureddine, M.; Gilbert, J.R.; *et al.* Complement factor H variant increases the risk of age-related macular degeneration. *Science* **2005**, *308*, 419–421.
37. Edwards, A.O.; Ritter, R., III; Abel, K.J.; Manning, A.; Panhuysen, C.; Farrer, L.A. Complement factor H polymorphism and age-related macular degeneration. *Science* **2005**, *308*, 421–424.
38. Hageman, G.S.; Anderson, D.H.; Johnson, L.V.; Hancox, L.S.; Taiber, A.J.; Hardisty, L.I.; Hageman, J.L.; Stockman, H.A.; Borchardt, J.D.; Gehrs, K.M.; *et al.* A common haplotype in the complement regulatory gene factor H (HF1/CFH) predisposes individuals to age-related macular degeneration. *Proc. Natl. Acad. Sci. USA* **2005**, *102*, 7227–7232.
39. Zerhouni, E. House subcommittee of labor-HHS-Education appropriations. Available online: http://legislative.csancer.gov/files/appropriations-2006-04-06.pdf (accessed on 1 March 2014).
40. Welter, D.; Macarthur, J.; Morales, J.; Burdett, T.; Hall, P.; Junkins, H.; Klemm, A.; Flicek, P.; Manolio, T.; Hindorff, L.; *et al.* The NHGRI GWAS Catalog, a curated resource of SNP-trait associations. *Nucleic Acids Res.* **2014**, *42*, D1001–D1006.
41. Goate, A.; Chartier-Harlin, M.C.; Mullan, M.; Brown, J.; Crawford, F.; Fidani, L.; Giuffra, L.; Haynes, A.; Irving, N.; James, L.; *et al.* Segregation of a missense mutation in the amyloid precursor protein gene with familial Alzheimer's disease. *Nature* **1991**, *349*, 704–706.
42. Jakobsdottir, J.; Conley, Y.P.; Weeks, D.E.; Mah, T.S.; Ferrell, R.E.; Gorin, M.B. Susceptibility genes for age-related maculopathy on chromosome 10q26 29. *Am. J. Hum. Genet.* **2005**, *77*, 389–407.
43. Schmidt, S.; Hauser, M.A.; Scott, W.K.; Postel, E.A.; Agarwal, A.; Gallins, P.; Wong, F.; Chen, Y.S.; Spencer, K.; Schnetz-Boutaud, N.; *et al.* Cigarette smoking strongly modifies the association of LOC387715 and age-related macular degeneration. *Am. J. Hum. Genet.* **2006**, *78*, 852–864.
44. Schwartz, S.G.; Agarwal, A.; Kovach, J.L.; Gallins, P.J.; Cade, W.; Postel, E.A.; Wang, G.; Ayala-Haedo, J.; Spencer, K.M.; Haines, J.L.; *et al.* The ARMS2 A69S variant and bilateral advanced age-related macular degeneration. *Retina* **2012**, *32*, 1486–1491.
45. Shuler, R.K., Jr.; Hauser, M.A.; Caldwell, J.; Gallins, P.; Schmidt, S.; Scott, W.K.; Agarwal, A.; Haines, J.L.; Pericak-Vance, M.A.; Postel, E.A. Neovascular age-related macular degeneration and its association with LOC387715 and complement factor H polymorphism. *Arch. Ophthalmol.* **2007**, *125*, 63–67.
46. Wang, G. Chromosome 10q26 locus and age-related macular degeneration: A progress update. *Exp. Eye Res.* **2014**, *119*, 1–7.
47. Wang, G.; Dubovy, S.R.; Kovach, J.L.; Schwartz, S.G.; Agarwal, A.; Scott, W.K.; Haines, J.L.; Pericak-Vance, M.A. Variants at chromosome 10q26 locus and the expression of HTRA1 in the retina. *Exp. Eye Res.* **2013**, *112*, 102–105.
48. Dewan, A.; Liu, M.; Hartman, S.; Zhang, S.S.; Liu, D.T.; Zhao, C.; Tam, P.O.; Chan, W.M.; Lam, D.S.; Snyder, M.; *et al.* HTRA1 promoter polymorphism in wet age-related macular degeneration. *Science* **2006**, *314*, 989–992.

49. Yang, Z.; Camp, N.J.; Sun, H.; Tong, Z.; Gibbs, D.; Cameron, D.J.; Chen, H.; Zhao, Y.; Pearson, E.; Li, X.; *et al.* A variant of the HTRA1 gene increases susceptibility to age-related macular degeneration. *Science* **2006**, *314*, 992–993.
50. Hughes, A.E.; Orr, N.; Esfandiary, H.; Diaz-Torres, M.; Goodship, T.; Chakravarthy, U. A common CFH haplotype, with deletion of CFHR1 and CFHR3, is associated with lower risk of age-related macular degeneration. *Nat. Genet.* **2006**, *38*, 1173–1177.
51. Fritsche, L.G.; Lauer, N.; Hartmann, A.; Stippa, S.; Keilhauer, C.N.; Oppermann, M.; Pandey, M.K.; Kohl, J.; Zipfel, P.F.; Weber, B.H.; *et al.* An imbalance of human complement regulatory proteins CFHR1, CFHR3 and factor H influences risk for age-related macular degeneration (AMD). *Hum. Mol. Genet.* **2010**, *19*, 4694–4704.
52. Sawitzke, J.; Im, K.M.; Kostiha, B.; Dean, M.; Gold, B. Association assessment of copy number polymorphism and risk of age-related macular degeneration. *Ophthalmology* **2011**, *118*, 2442–2446.
53. Spencer, K.L.; Hauser, M.A.; Olson, L.M.; Schmidt, S.; Scott, W.K.; Gallins, P.; Agarwal, A.; Postel, E.A.; Pericak-Vance, M.A.; Haines, J.L. Protective effect of complement factor B and complement component 2 variants in age-related macular degeneration. *Hum. Mol. Genet.* **2007**, *16*, 1986–1992.
54. Gold, B.; Merriam, J.E.; Zernant, J.; Hancox, L.S.; Taiber, A.J.; Gehrs, K.; Cramer, K.; Neel, J.; Bergeron, J.; Barile, G.R.; *et al.* Variation in factor B (BF) and complement component 2 (C2) genes is associated with age-related macular degeneration. *Nat. Genet.* **2006**, *38*, 458–462.
55. Maller, J.B.; Fagerness, J.A.; Reynolds, R.C.; Neale, B.M.; Daly, M.J.; Seddon, J.M. Variation in complement factor 3 is associated with risk of age-related macular degeneration. *Nat. Genet.* **2007**, *39*, 1200–1201.
56. Spencer, K.L.; Olson, L.M.; Anderson, B.M.; Schnetz-Boutaud, N.; Scott, W.K.; Gallins, P.; Agarwal, A.; Postel, E.A.; Pericak-Vance, M.A.; Haines, J.L. C3 R102G polymorphism increases risk of age-related macular degeneration. *Hum. Mol. Genet.* **2008**, *17*, 1821–1824.
57. Yates, J.R.; Sepp, T.; Matharu, B.K.; Khan, J.C.; Thurlby, D.A.; Shahid, H.; Clayton, D.G.; Hayward, C.; Morgan, J.; Wright, A.F.; *et al.* Complement C3 variant and the risk of age-related macular degeneration. *N. Engl. J. Med.* **2007**, *357*, 553–561.
58. Fagerness, J.A.; Maller, J.B.; Neale, B.M.; Reynolds, R.C.; Daly, M.J.; Seddon, J.M. Variation near complement factor I is associated with risk of advanced AMD. *Eur. J. Hum. Genet.* **2009**, *17*, 100–104.
59. Stanton, C.M.; Yates, J.R.; den Hollander, A.I.; Seddon, J.M.; Swaroop, A.; Stambolian, D.; Fauser, S.; Hoyng, C.; Yu, Y.; Atsuhiro, K.; *et al.* Complement factor D in age-related macular degeneration. *Invest. Ophthalmol. Vis. Sci.* **2011**, *52*, 8828–8834.
60. Chen, W.; Stambolian, D.; Edwards, A.O.; Branham, K.E.; Othman, M.; Jakobsdottir, J.; Tosakulwong, N.; Pericak-Vance, M.A.; Campochiaro, P.A.; Klein, M.L.; *et al.* Genetic variants near TIMP3 and high-density lipoprotein-associated loci influence susceptibility to age-related macular degeneration. *Proc. Natl. Acad. Sci. USA* **2010**, *107*, 7401–7406.
61. Fritsche, L.G.; Chen, W.; Schu, M.; Yaspan, B.L.; Yu, Y.; Thorleifsson, G.; Zack, D.J.; Arakawa, S.; Cipriani, V.; Ripke, S.; *et al.* Seven new loci associated with age-related macular degeneration. *Nat. Genet.* **2013**, *45*, 433–439.

62. Neale, B.M.; Fagerness, J.; Reynolds, R.; Sobrin, L.; Parker, M.; Raychaudhuri, S.; Tan, P.L.; Oh, E.C.; Merriam, J.E.; Souied, E.; *et al.* Genome-wide association study of advanced age-related macular degeneration identifies a role of the hepatic lipase gene (LIPC). *Proc. Natl. Acad. Sci. USA* **2010**, *107*, 7395–7400.
63. Helgason, H.; Sulem, P.; Duvvari, M.R.; Luo, H.; Thorleifsson, G.; Stefansson, H.; Jonsdottir, I.; Masson, G.; Gudbjartsson, D.F.; Walters, G.B.; *et al.* A rare nonsynonymous sequence variant in C3 is associated with high risk of age-related macular degeneration. *Nat. Genet.* **2013**, *45*, 1371–1374.
64. Ryu, E.; Fridley, B.L.; Tosakulwong, N.; Bailey, K.R.; Edwards, A.O. Genome-wide association analyses of genetic, phenotypic, and environmental risks in the age-related eye disease study. *Mol. Vis.* **2010**, *16*, 2811–2821.
65. Scheetz, T.E.; Fingert, J.H.; Wang, K.; Kuehn, M.H.; Knudtson, K.L.; Alward, W.L.; Boldt, H.C.; Russell, S.R.; Folk, J.C.; Casavant, T.L.; *et al.* A genome-wide association study for primary open angle glaucoma and macular degeneration reveals novel Loci. *PLoS One* **2013**, *8*, e58657.
66. Manolio, T.A.; Collins, F.S.; Cox, N.J.; Goldstein, D.B.; Hindorff, L.A.; Hunter, D.J.; McCarthy, M.I.; Ramos, E.M.; Cardon, L.R.; Chakravarti, A.; *et al.* Finding the missing heritability of complex diseases. *Nature* **2009**, *461*, 747–753.
67. Li, Y.; Willer, C.; Sanna, S.; Abecasis, G. Genotype imputation. *Annu. Rev. Genomics Hum. Genet.* **2009**, *10*, 387–406.
68. Marchini, J.; Howie, B. Genotype imputation for genome-wide association studies. *Nat. Rev. Genet.* **2010**, *11*, 499–511.
69. Seddon, J.M.; Yu, Y.; Miller, E.C.; Reynolds, R.; Tan, P.L.; Gowrisankar, S.; Goldstein, J.I.; Triebwasser, M.; Anderson, H.E.; Zerbib, J.; *et al.* Rare variants in CFI, C3 and C9 are associated with high risk of advanced age-related macular degeneration. *Nat. Genet.* **2013**, *45*, 1366–1370.
70. Yaspan, B.L.; Bush, W.S.; Torstenson, E.S.; Ma, D.; Pericak-Vance, M.A.; Ritchie, M.D.; Sutcliffe, J.S.; Haines, J.L. Genetic analysis of biological pathway data through genomic randomization. *Hum. Genet.* **2011**, *129*, 563–571.
71. Lee, P.H.; O'Dushlaine, C.; Thomas, B.; Purcell, S.M. INRICH: Interval-based enrichment analysis for genome-wide association studies. *Bioinformatics* **2012**, *28*, 1797–1799.
72. Yang, J.; Li, Y.; Chan, L.; Tsai, Y.T.; Wu, W.H.; Nguyen, H.V.; Hsu, C.W.; Li, X.; Brown, L.M.; Egli, D.; *et al.* Validation of genome-wide association study (GWAS)-identified disease risk alleles with patient-specific stem cell lines. *Hum. Mol. Genet.* **2014**, *23*, 3445–3455.
73. Van Lookeren, C.M.; Le Couter, J.; Yaspan, B.L.; Ye, W. Mechanisms of age-related macular degeneration and therapeutic opportunities. *J. Pathol.* **2014**, *232*, 151–164.
74. Herrup, K. Reimagining Alzheimer's disease—An age-based hypothesis. *J. Neurosci.* **2010**, *30*, 16755–16762.
75. Querfurth, H.W.; LaFerla, F.M. Alzheimer's disease. *N. Engl. J. Med.* **2010**, *362*, 329–344.
76. Breitner, J.C.; Folstein, M.F. Familial Alzheimer Dementia: A prevalent disorder with specific clinical features. *Psychol. Med.* **1984**, *14*, 63–80.
77. Breitner, J.C.; Folstein, M.F. Familial nature of Alzheimer's disease. *N. Engl. J. Med.* **1984**, *311*, 192.

78. Folstein, M. Alzheimer's disease: Challenge to psychiatry. *Hosp. Community Psychiatry* **1984**, *35*, 111.
79. Pericak-Vance, M.A.; Haines, J.L. Genetic susceptibility to Alzheimer disease. *Trends Genet.* **1995**, *11*, 504–508.
80. Powell, D.; Folstein, M.F. Pedigree study of familial Alzheimer disease. *J. Neurogenet.* **1984**, *1*, 189–197.
81. St George-Hyslop, P.H.; Tanzi, R.E.; Polinsky, R.J.; Haines, J.L.; Nee, L.; Watkins, P.C.; Myers, R.H.; Feldman, R.G.; Pollen, D.; Drachman, D.; *et al.* The genetic defect causing familial Alzheimer's disease maps on chromosome 21. *Science* **1987**, *235*, 885–890.
82. Levy-Lahad, E.; Lahad, A.; Wijsman, E.M.; Bird, T.D.; Schellenberg, G.D. Apolipoprotein E genotypes and age of onset in early-onset familial Alzheimer's disease. *Ann. Neurol.* **1995**, *38*, 678–680.
83. Levy-Lahad, E.; Wasco, W.; Poorkaj, P.; Romano, D.M.; Oshima, J.; Pettingell, W.H.; Yu, C.E.; Jondro, P.D.; Schmidt, S.D.; Wang, K.; *et al.* Candidate gene for the chromosome 1 familial Alzheimer's disease locus. *Science* **1995**, *269*, 973–977.
84. Levy-Lahad, E.; Wijsman, E.M.; Nemens, E.; Anderson, L.; Goddard, K.A.; Weber, J.L.; Bird, T.D.; Schellenberg, G.D. A familial Alzheimer's disease locus on chromosome 1. *Science* **1995**, *269*, 970–973.
85. Sherrington, R.; Rogaev, E.I.; Liang, Y.; Rogaeva, E.A.; Levesque, G.; Ikeda, M.; Chi, H.; Lin, C.; Li, G.; Holman, K.; *et al.* Cloning of a gene bearing missense mutations in early-onset familial Alzheimer's disease. *Nature* **1995**, *375*, 754–760.
86. St George-Hyslop, P.; Haines, J.; Rogaev, E.; Mortilla, M.; Vaula, G.; Pericak-Vance, M.; Foncin, J.F.; Montesi, M.; Bruni, A.; Sorbi, S.; *et al.* Genetic evidence for a novel familial Alzheimer's disease locus on chromosome 14. *Nat. Genet.* **1992**, *2*, 330–334.
87. Ridge, P.G.; Mukherjee, S.; Crane, P.K.; Kauwe, J.S. Alzheimer's disease: Analyzing the missing heritability. *PLoS One* **2013**, *8*, e79771.
88. Pericak-Vance, M.A.; Bebout, J.L.; Gaskell, P.C., Jr.; Yamaoka, L.H.; Hung, W.Y.; Alberts, M.J.; Walker, A.P.; Bartlett, R.J.; Haynes, C.A.; Welsh, K.A.; *et al.* Linkage studies in familial Alzheimer disease: Evidence for chromosome 19 linkage. *Am. J. Hum. Genet.* **1991**, *48*, 1034–1050.
89. Pericak-Vance, M.A.; Yamaoka, L.H.; Haynes, C.S.; Speer, M.C.; Haines, J.L.; Gaskell, P.C.; Hung, W.Y.; Clark, C.M.; Heyman, A.L.; Trofatter, J.A.; *et al.* Genetic linkage studies in Alzheimer's disease families. *Exp. Neurol.* **1988**, *102*, 271–279.
90. Corder, E.H.; Saunders, A.M.; Strittmatter, W.J.; Schmechel, D.E.; Gaskell, P.C.; Small, G.W.; Roses, A.D.; Haines, J.L.; Pericak-Vance, M.A. Gene dose of apolipoprotein E type 4 allele and the risk of Alzheimer's disease in late onset families. *Science* **1993**, *261*, 921–923.
91. Corder, E.H.; Saunders, A.M.; Risch, N.J.; Strittmatter, W.J.; Schmechel, D.E.; Gaskell, P.C., Jr.; Rimmler, J.B.; Locke, P.A.; Conneally, P.M.; Schmader, K.E.; *et al.* Protective effect of apolipoprotein E type 2 allele for late onset Alzheimer disease. *Nat. Genet.* **1994**, *7*, 180–184.
92. Beecham, G.W.; Martin, E.R.; Li, Y.J.; Slifer, M.A.; Gilbert, J.R.; Haines, J.L.; Pericak-Vance, M.A. Genome-wide association study implicates a chromosome 12 risk locus for late-onset Alzheimer disease. *Am. J. Hum. Genet.* **2009**, *84*, 35–43.

93. Coon, K.D.; Myers, A.J.; Craig, D.W.; Webster, J.A.; Pearson, J.V.; Lince, D.H.; Zismann, V.L.; Beach, T.G.; Leung, D.; Bryden, L.; *et al.* A high-density whole-genome association study reveals that APOE is the major susceptibility gene for sporadic late-onset Alzheimer's disease. *J. Clin. Psychiatry*. **2007**, *68*, 613–618.

94. Li, H.; Wetten, S.; Li, L.; St Jean, P.L.; Upmanyu, R.; Surh, L.; Hosford, D.; Barnes, M.R.; Briley, J.D.; Borrie, M.; *et al.* Candidate single-nucleotide polymorphisms from a genomewide association study of Alzheimer disease. *Arch. Neurol.* **2008**, *65*, 45–53.

95. Lambert, J.C.; Ibrahim-Verbaas, C.A.; Harold, D.; Naj, A.C.; Sims, R.; Bellenguez, C.; Jun, G.; Destefano, A.L.; Bis, J.C.; Beecham, G.W.; *et al.* Meta-analysis of 74,046 individuals identifies 11 new susceptibility loci for Alzheimer's disease. *Nat. Genet.* **2013**, *45*, 1452–1458.

96. Naj, A.C.; Jun, G.; Beecham, G.W.; Wang, L.S.; Vardarajan, B.N.; Buros, J.; Gallins, P.J.; Buxbaum, J.D.; Jarvik, G.P.; Crane, P.K.; *et al.* Common variants at MS4A4/MS4A6E, CD2AP, CD33 and EPHA1 are associated with late-onset Alzheimer's disease. *Nat. Genet.* **2011**, *43*, 436–441.

97. Seshadri, S.; Fitzpatrick, A.L.; Ikram, M.A.; Destefano, A.L.; Gudnason, V.; Boada, M.; Bis, J.C.; Smith, A.V.; Carassquillo, M.M.; Lambert, J.C.; *et al.* Genome-wide analysis of genetic loci associated with Alzheimer disease. *JAMA* **2010**, *303*, 1832–1840.

98. Gourraud, P.A.; Sdika, M.; Khankhanian, P.; Henry, R.G.; Beheshtian, A.; Matthews, P.M.; Hauser, S.L.; Oksenberg, J.R.; Pelletier, D.; Baranzini, S.E. A genome-wide association study of brain lesion distribution in multiple sclerosis. *Brain* **2013**, *136*, 1012–1024.

99. Hauser, S.L.; Chan, J.R.; Oksenberg, J.R. Multiple sclerosis: Prospects and promise. *Ann. Neurol.* **2013**, *74*, 317–327.

100. Nylander, A.; Hafler, D.A. Multiple sclerosis. *J. Clin. Invest.* **2012**, *122*, 1180–1188.

101. Oksenberg, J.R. Decoding multiple sclerosis: An update on genomics and future directions. *Expert. Rev. Neurother.* **2013**, *13*, 11–19.

102. Compston, A.; Coles, A. Multiple sclerosis. *Lancet* **2002**, *359*, 1221–1231.

103. Sadovnick, A.D.; Ebers, G.C. Epidemiology of multiple sclerosis: A critical overview. *Can. J. Neurol. Sci.* **1993**, *20*, 17–29.

104. Sawcer, S.; Compston, A. Multiple sclerosis: Light at the end of the tunnel. *Eur. J. Hum. Genet.* **2006**, *14*, 257–258.

105. Compston, A.; Sawcer, S. Genetic analysis of multiple sclerosis. *Curr. Neurol. Neurosci. Rep.* **2002**, *2*, 259–266.

106. Sadovnick, A.D. Familial recurrence risks and inheritance of multiple sclerosis. *Curr. Opin. Neurol. Neurosurg.* **1993**, *6*, 189–194.

107. Sadovnick, A.D.; Armstrong, H.; Rice, G.P.; Bulman, D.; Hashimoto, L.; Paty, D.W.; Hashimoto, S.A.; Warren, S.; Hader, W.; Murray, T.J.; *et al.* A population-based study of multiple sclerosis in twins: Update. *Ann. Neurol.* **1993**, *33*, 281–285.

108. Sadovnick, A.D.; Yee, I.M.; Guimond, C.; Reis, J.; Dyment, D.A.; Ebers, G.C. Age of onset in concordant twins and other relative pairs with multiple sclerosis. *Am. J. Epidemiol.* **2009**, *170*, 289–296.

109. Sawcer, S.; Ban, M.; Maranian, M.; Yeo, T.W.; Compston, A.; Kirby, A.; Daly, M.J.;

de Jager, P.L.; Walsh, E.; Lander, E.S.; *et al.* A high-density screen for linkage in multiple sclerosis. *Am. J. Hum. Genet.* **2005**, *77*, 454–467.

110. Haines, J.L.; Ter-Minassian, M.; Bazyk, A.; Gusella, J.F.; Kim, D.J.; Terwedow, H.; Pericak-Vance, M.A.; Rimmler, J.B.; Haynes, C.S.; Roses, A.D.; *et al.* A complete genomic screen for multiple sclerosis underscores a role for the major histocompatability complex. The Multiple Sclerosis Genetics Group. *Nat. Genet.* **1996**, *13*, 469–471.
111. Ebers, G.C.; Kukay, K.; Bulman, D.E.; Sadovnick, A.D.; Rice, G.; Anderson, C.; Armstrong, H.; Cousin, K.; Bell, R.B.; Hader, W.; *et al.* A full genome search in multiple sclerosis. *Nat. Genet.* **1996**, *13*, 472–476.
112. Haines, J.L.; Bradford, Y.; Garcia, M.E.; Reed, A.D.; Neumeister, E.; Pericak-Vance, M.A.; Rimmler, J.B.; Menold, M.M.; Martin, E.R.; Oksenberg, J.R.; *et al.* Multiple susceptibility loci for multiple sclerosis. *Hum. Mol. Genet.* **2002**, *11*, 2251–2256.
113. Haines, J.L.; Terwedow, H.A.; Burgess, K.; Pericak-Vance, M.A.; Rimmler, J.B.; Martin, E.R.; Oksenberg, J.R.; Lincoln, R.; Zhang, D.Y.; Banatao, D.R.; *et al.* Linkage of the MHC to familial multiple sclerosis suggests genetic heterogeneity. The Multiple Sclerosis Genetics Group. *Hum. Mol. Genet.* **1998**, *7*, 1229–1234.
114. Kenealy, S.J.; Herrel, L.A.; Bradford, Y.; Schnetz-Boutaud, N.; Oksenberg, J.R.; Hauser, S.L.; Barcellos, L.F.; Schmidt, S.; Gregory, S.G.; Pericak-Vance, M.A.; *et al.* Examination of seven candidate regions for multiple sclerosis: Strong evidence of linkage to chromosome 1q44. *Genes Immun.* **2006**, *7*, 73–76.
115. McCauley, J.L.; Zuvich, R.L.; Bradford, Y.; Kenealy, S.J.; Schnetz-Boutaud, N.; Gregory, S.G.; Hauser, S.L.; Oksenberg, J.R.; Mortlock, D.P.; Pericak-Vance, M.A.; *et al.* Follow-up examination of linkage and association to chromosome 1q43 in multiple sclerosis. *Genes Immun.* **2009**, *10*, 624–630.
116. Pericak-Vance, M.A.; Rimmler, J.B.; Martin, E.R.; Haines, J.L.; Garcia, M.E.; Oksenberg, J.R.; Barcellos, L.F.; Lincoln, R.; Goodkin, D.E.; Hauser, S.L. Linkage and association analysis of chromosome 19q13 in multiple sclerosis. *Neurogenetics* **2001**, *3*, 195–201.
117. Gregory, S.G.; Schmidt, S.; Seth, P.; Oksenberg, J.R.; Hart, J.; Prokop, A.; Caillier, S.J.; Ban, M.; Goris, A.; Barcellos, L.F.; *et al.* Interleukin 7 receptor alpha chain (IL7R) shows allelic and functional association with multiple sclerosis. *Nat. Genet.* **2007**, *39*, 1083–1091.
118. Hafler, D.A.; Compston, A.; Sawcer, S.; Lander, E.S.; Daly, M.J.; de Jager, P.L.; de Bakker, P.I.; Gabriel, S.B.; Mirel, D.B.; Ivinson, A.J.; *et al.* Risk alleles for multiple sclerosis identified by a genomewide study. *N. Engl. J. Med.* **2007**, *357*, 851–862.
119. Beecham, A.H.; Patsopoulos, N.A.; Xifara, D.K.; Davis, M.F.; Kemppinen, A.; Cotsapas, C.; Shah, T.S.; Spencer, C.; Booth, D.; Goris, A.; *et al.* Analysis of immune-related loci identifies 48 new susceptibility variants for multiple sclerosis. *Nat. Genet.* **2013**, *45*, 1353–1360.
120. Damotte, V.; Guillot-Noel, L.; Patsopoulos, N.A.; Madireddy, L.; El, B.M.; Ban, M.; Baranzini, S.; Barcellos, L.; Beecham, G.; Beecham, A.; *et al.* A gene pathway analysis highlights the role of cellular adhesion molecules in multiple sclerosis susceptibility. *Genes Immun.* **2014**, *15*, 126–132.
121. Yan, J.; Greer, J.M. NF-kappa B, a potential therapeutic target for the treatment of multiple sclerosis. *CNS Neurol. Disord. Drug Targets* **2008**, *7*, 536–557.

122. Metacore. Available online: http://thomsonreuters.com/metacore/ (accessed on 7 March 2014).
123. Cytoscape. Available online: http://www.cytoscape.org/ (accessed on 7 March 2014).
124. Ratnapriya, R.; Chew, E.Y. Age-related macular degeneration-clinical review and genetics update. *Clin. Genet.* **2013**, *84*, 160–166.
125. ESP. Available online: https://esp.gs.washington.edu/drupal/ (accessed on 7 March 2014).
126. 1000Genomes. Available online: http://www.1000genomes.org/ (accessed on 7 March 2014).
127. ENCODE. Available online: https://genome.ucsc.edu/ENCODE/ (accessed on 7 March 2014).
128. International HapMap Consortium. A haplotype map of the human genome. *Nature* **2005**, *437*, 1299–1320.
129. HapMap. Available online: http://hapmap.ncbi.nlm.nih.gov/ (accessed on 7 March 2014).

4

The Challenges of Genome Analysis in the Health Care Setting

Anneke Lucassen [1,*] and Richard S. Houlston [2]

[1] Clinical Ethics and Law Unit, Wessex Clinical Genetics Service, The Princess Anne Hospital, Southampton, SO16 5YA, UK

[2] Division of Genetics and Epidemiology, The Institute of Cancer Research, Sutton, Surrey SM2 5NG, UK; E-Mail: richard.houlston@icr.ac.uk

* Author to whom correspondence should be addressed; E-Mail: A.M.Lucassen@soton.ac.uk;

Abstract: Genome sequencing is now a sufficiently mature and affordable technology for clinical use. Its application promises not only to transform clinicians' diagnostic and predictive ability, but also to improve preventative therapies, surveillance regimes, and tailor patient treatment to an individual's genetic make-up. However, as with any technological advance, there are associated fresh challenges. While some of the ethical, legal and social aspects resulting from the generation of data from genome sequencing are generic, several nuances are unique. Since the UK government recently announced plans to sequence the genomes of 100,000 Health Service patients, and similar initiatives are being considered elsewhere, a discussion of these nuances is timely and needs to go hand in hand with formulation of guidelines and public engagement activities around implementation of sequencing in clinical practice.

Keywords: genome sequencing; clinical practice; incidental findings; consent; disclosure

---

1. Introduction

The speed by which a person's genome can be analysed has increased phenomenally over recent years, while the attendant costs have plummeted. As a result, genetic testing is shifting from a targeted approach analysing specific genes based on particular symptoms or family histories to sequencing of an entire exome or genome (whole exome sequencing [WES], whole genome sequencing [WGS]).

Targeted approaches characteristically have a high yield for penetrant monogenic conditions; whole genome approaches have the potential to unravel a much larger proportion of genetic disease burden. Whole genome analyses, therefore, are likely not only to transform a clinician's diagnostic and predictive ability, but also to improve preventative therapies, surveillance regimes, and tailor patient treatment to an individual's genetic make-up.

The improved diagnostic yields of genome sequencing are to be welcomed; however as with any technological advance there are associated fresh challenges. While the ethical, legal and social aspects resulting from the generation of data from genome sequencing are not unique, several nuances merit serious consideration. Since the UK government recently announced plans to sequence the genomes of 100,000 Health Service patients [1], and similar initiatives are being considered elsewhere, a discussion of these nuances is timely and needs to go hand in hand with formulation of guidelines and public engagement activities around the implementation of genome sequencing. Box 1 lists some of the ethical, legal, and social and practical issues that we consider merit consideration.

Box 1. Some of the overlapping ethical, legal, social and practical issues that need to be addressed as genome analysis enters clinical practice.

Complexity

- Genome analysis can provide many different predictions about diagnoses, or susceptibilities to conditions. However, it will do so with varying degrees of certainty or confidence intervals around the predictions. Such predictions are likely to change substantially over time as evidence about epistatic factors accumulates.
- Providing consent to genomic testing is therefore complex. Should consent be sought to any answer that genome analysis might provide? Or should there be cut-off for levels of certainty? Or should a genome analysis be used solely to answering a current clinical question? Should some results be staged? (e.g., risk of adult onset conditions diagnosed in children?)

Familial Aspects

- Although genomic information is on the one hand very personal, on the other, it may be relevant to relatives who have not sought medical advice but may be identified as being at risk from the results in another person. How can health services best record, store and communicate such familial information?

Re-Contacting/Follow Up Policies

- Who should be re-contacted and when, in the light of evolving knowledge? Who might be liable if a patient remains unaware of new evidence and therefore interpretation of previous test results?

Data Management

- What should be stored: the DNA sample, the DNA sequence, the interpretation of the sequence? Or combinations of these? What is to be stored in medical record systems, and how can these be compliant with relevant data protection—and other—legislation? How can/should these be linked with biobank or research databases, and how can the security issues around identifiable data best be managed?

Research/Clinical Divide

- The traditional route of research to clinic evolution is not necessarily applicable in rapidly evolving technologies.

Public Perceptions of Genetics

- Currently, this is often thought as a clear cut, or deterministic result than there is evidence for.
- Analytical validity not the same as clinical validity or utility; $1000 genome analysis is a reality soon, yet the cost of interpretation is much greater.

2. The Promise of Whole Genome Analysis

Genetic testing has traditionally been restricted to analysing small numbers of genes usually picked on the basis of a high prior probability of being mutated. However, this approach has several limitations. Firstly, many inherited diseases are genetically heterogeneous and sequential mutational analysis of individual genes is slow and expensive. Secondly, while subsets of some common diseases can be caused by mutations in a single gene, traditional methods of selecting whom to test on basis of disease characteristics or family history are crude and have a high false negative rate. Finally, analysis may not ultimately be diagnostic if the disease is a consequence of a hitherto unknown disease-causing gene. Collectively these issues make whole genome approaches at competitive prices an attractive proposition.

Most next-generation sequencing (NGS) technologies are based on the fragmentation of genomic DNA with the oversampling of reads providing the necessary linking information for whole-genome assembly algorithms. For analysis of a gene to be of diagnostic quality using NGS there needs to be sufficient read depth for any mutation to be called with a high degree of confidence. While WGS or WES typically provide good overall coverage for most regions of the genome, for other regions it may be poor; sequencing some regions of the genome is problematic because of repetitive sequence and other features leading to systematic error [2]. Such limitations have, in part, been the motivation for developing targeted sequencing approaches focusing on panels of genes relevant to specific diseases states; for example, cancer gene and nervous system disease panels. Such technical shortcomings are likely to be addressed in the near future so that a "one-stop-shop" test will replace the sequential approaches to genetic diagnoses which were time and labour intensive.

3. Analytical Validity *versus* Interpretation of WGS Approaches

Whilst the analytical validity of WGS approaches is high, and improving at a rapid pace, the clinical validity of the output from WGS is much more complex than commonly perceived and the utility has often been evaluated only in very small groups. There is much genome variation that is either: uninterpretable; probably benign; or only pathogenic in certain circumstances, for example, in the presence of as yet unknown epistatic factors. This gap between technological advances and the interpretation of any NGS output, is neatly encapsulated by the phrase "$1000 genome; $1 million interpretation [3], yet, little recognised in the popular discourse around whole genome technologies.

In the clinical setting, certainly in the short term, diagnostic accuracy will therefore continue to depend on additional factors such as clinical history and, therefore, pre-test probability. Attempts to overcome these issues include use of gene panels or analyses of selected portions of the genetic code; an apparently anachronistic step in the evolution of whole genome approaches. However, if WGS approaches are to be used to answer clinical questions, some sort of filtering of sequence output will need to take place. Although targeted approaches are commonplace in health care, this has usually involved a targeting of the investigations. In WGS the targeting will have to be at the analysis stage—the results require targeted analysis—and this raises novel issues about what constitutes a result, what is disclosed to the patient, and what is recorded in a patient's medical records.

4. The Data Interpretation Problem

Much of the misperception about the diagnostic value of genome sequencing results from an oversimplification in which it is assumed there is "a gene" for the condition, when in fact any increase in risk conferred by a mutation may be subtle, or only manifest in the context of specific genetic background or environmental exposure. For many common diseases there are multiple risk factors and while the identification of susceptibility genes has often provided novel insights in disease biology, their clinical utility in an individual may be very low because their predictive power in isolation is very poor.

There is, however, also a risk of over-interpretation even for mutations with seemingly large effects. For affected patients where there is a strong prior probability of the gene mutation being causal because of a positive family history and or specific clinical phenotype, interpretation can be straightforward. However, if mutations are not fully penetrant, there will be carriers in the population who are healthy. Much of our knowledge about the penetrance of mutations to date is based on family data and, hence, suffers from ascertainment bias [4]. Without unbiased knowledge of the effect of mutations, interpretation at the population level will be inherently problematic. Whilst policies to restrict genetic testing to high risk populations were initially driven by budget restraints, and the more widespread availability of testing thought to be an advantage of declining costs, another consequence is that the interpretation of the clinical significance of a mutation is much more difficult if found without the ascertainment bias noted above. That is to say, predicting the effects of a novel *BRCA2* mutation in the context of a strong family history of the mutation segregating with disease in the family, is far easier than when it is discovered in a population screen (see Box 2 for an illustrative example).

Box 2. Difficulty of clinical interpretation of genomic findings in absence of clear clinical phenotype or family history.

A two-year old boy was investigated for "absence spells". He had no loss of consciousness, was investigated in detail for epilepsy and no abnormalities were found. Paediatric cardiologists also found no abnormalities, his baseline ECG was defined as within normal limits and he had no family history (to 3rd degree relatives) of any cardiac problems. The cardiologist had been to a presentation about mainstreaming genetics and realised that long QT (LQT) interval gene carriers can be difficult to detect in childhood. He therefore requested genetic testing of LQT genes "to exclude LQT syndrome". A LQT1—associated mutation was identified, described on the laboratory report as "highly likely to be pathogenic". A reveal device was inserted but no abnormalities in his QT interval were recorded during subsequence "absence spells". Nevertheless, it was thought appropriate to treat him with beta blockers. Cascade testing of his family revealed his three-year-old sister, father, paternal aunt (and her two children, aged four and eight) and paternal grandfather all carried the same mutation. Cardiac investigations of their phenotype, at rest, with exercise, and pharmacological challenge were normal or equivocal. All carriers in the family were prescribed beta blockade and two members of the family were referred for possible implantable cardiac defibrillator insertion.

In Box 2 the assumption that this LQT1 mutation depicts a high future risk of clinical symptoms from LQT syndrome is based on the laboratory description of its likely pathogenicity and the previous finding in families with symptomatic LQT. The intensive therapy is in part because the first presentation of LQT can be sudden cardiac death. However, this family was not ascertained on the basis of any relevant clinical symptoms and clear clinical predictions for the seven asymptomatic carriers are extremely difficult. However, if the mutation was found in a family with a segregating LQT phenotype, preventative therapy would be justifiable on clinical grounds. These cases serve to illustrate that the predictive powers of genetics require more than information about genotype, for the effects of any genotype are dependent on a range of other factors. Importantly, the penetrance of different mutations in the same gene can vary substantially and assigning a likelihood of a mutation being disease-causing will increasingly be based on the synthesis of multiple forms of evidence.

5. Determining Clinical Utility of Sequence Variants

The translation of genome sequence into medically actionable information is a key challenge. Without support from segregation in families, assigning pathogenicity can be problematic; notably large duplications, most synonymous and some missense mutations, intronic variants, and most variants in promoter and enhancers are particularly difficult to interpret. Predicting the functional consequences of variants which disrupt protein-coding sequence can also be challenging. A variant might affect a transcription factor binding site, a microRNA target site, affect RNA-splicing or stability or truncate a protein. Finally the issue of linkage disequilibrium (where benign variants lie close to a disease predisposing variant) can complicate interpretation of recurrent risk variants.

Irrespective of whether animal models can adequately mimic human disease such model systems are inherently unsuited to determining the consequences of specific mutations as a routine activity. While yeast and cell line systems can be used to assess the functionality of DNA repair gene mutations the general applicability of such model systems is limited. In view of these factors increasing reliance will be placed on the implementation of *in silico* tools to infer the functional consequences of mutations. Although such algorithms can help to predict the likely pathogenicity of variants, often different tools conclude in opposite directions and without an established relationship between gene dysfunction and disease phenotype, robust risk prediction is problematic.

6. The Need to Systematically Catalogue Sequence Variation with Phenotype

Several initiatives are cataloguing and assigning pathogenicity to variants/mutations in various specific genes. Examples of such databases include InSiGHT (International Society for Gastrointestinal Hereditary Tumours Incorporated) [5], LoVd (Leiden open variant database) [6] Decipher [7] and DMuDB [8] (the diagnostic mutation database), and the Locus Reference Genomic Collaboration [9]. These resources provide health care professionals with valuable information for decision making processes. While published reports are valuable sources for such databases their stewardship depends heavily on the submission of individual variants and associated clinic-pathological data by sequencing laboratories using some form of incentivization. Currently these databases are limited to curation of restricted number of genes. Even here translating genomic sequence into medically actionable information can be highly time consuming.

To meet the future needs, comprehensive resources with a far more overarching remit will need to be developed and maintained. This needs to be coupled with adoption of automated machine learning, support vector machines and other technologies to create systematic and efficient mechanisms to assess the impact of variants found by genomic sequencing. All of this will require substantial investment before it becomes a reality and has not been factored into the $1000 genome analysis headlines.

7. Diagnostics *versus* Population Screening

Given the significant limitations to our current understanding of the impact of genetic variation, we believe that clinical genome sequencing should for now be focused on particular clinical presentations compatible with a genetic aetiology, rather than engaging in opportunistic population screening. For example, the identification of an *APC* mutation in a person with colonic polyposis is diagnostic and highly predictive for family members. In contrast the identification of variants, such as *LQT1* described in Box 1, in a population screen do not have sufficient certainty to infer as much, resulting in difficult clinical management issues. Such contextual differences may be difficult to grasp if genetics is portrayed as being clear cut, and clinical interventions may therefore be offered without sufficient evidence for their benefit.

Intelligent interrogation of genomic outputs in the clinic should initially therefore be restricted to specific genes or diseases for which there is a high prior likelihood of diagnosis. Any opportunistic screening should in the first instance be limited to known epistatic factors for particular conditions, e.g., low risk genes for breast cancer in the investigation of a family history of breast cancer, and formal evaluation of the benefits should not be leap frogged just because of the rapidly decreasing costs of the technologies involved.

8. The Need for Large Scale Genotype-Phenotype Linkages

Before more widespread population genome screening is to be contemplated, large-scale systematic and longitudinal investigation of variants in categorised populations would need to take place and their penetrance robustly determined. Depending on variant prevalence the ongoing international biobank sequencing projects are likely to provide a rich source of such data. Additionally, variants identified through clinical testing or research projects, could together with associated phenotypic information, be submitted to publicly accessible databases cataloguing genomic variation. Many of the current databases are however relatively ad hoc affairs and disease-specific. If the full potential of genomics is to be realised there is a need for the development of big data centres which have an overarching remit. However, the development and establishment of such initiatives brings with it the significant issue of data-storage and allied security requirements. These linkages will have to be undertaken within legislative frameworks relating to data protection within host countries and adapted to any changes to such legislation. For example, proposed changes by the European Commission to the data protection directive may have far reaching consequences for the gathering of such linkages [10].

9. The Need for Public-Professional Engagement

In parallel with the acquisition and curation of genetic data there needs to be an ongoing dialogue with health care professionals and the public around understandings and interpretations of genomic data so that expectations of new WGS approaches are realistic and grounded in evidence. In the wake of public anxiety around large scale databases, e.g., care.data in UK [11], this dialogue urgently needs to incorporate the importance of data sharing to realise the clinical utility of whole genome approaches. It also needs to incorporate the issues around shifting the point of targeting, as outlined in Section 3. For example, international recommendations suggest that children should not be offered genetic testing for adult-onset conditions (unless a result would alter their medical management). However, once such a result is available many would opine it should be disclosed, even if they would not have tested for it in the first place [12,13].

10. Incidental Findings

Any broad, highly sensitive investigation has the ability to occasionally detect abnormalities that are incidental to the reason for the test. Whole genome approaches are much more likely to detect asymptomatic or silent abnormalities that have nothing to do with the current clinical reason for a test. Such findings have been variably termed "secondary", "non-pertinent", "unexpected" or "incidental" belying the fact that the appropriate adjective may vary according to the situation [14]. A genome test can, however, only have an incidental finding (IF) if it is used to answer, for example, a particular clinical question. If the question is "what are the abnormalities in this genome?" then there can be no IFs.

There has been much recent debate about the management of IFs in clinical applications of WG technologies [15–20]. The American College of Medical Genetics and Genomics (ACMG) produced guidelines that recommended the active search for particular IFs if using WGS/WES approaches [21]. The heated debate that ensued was largely focused on patient/parental choice regarding such IF searches with their purported "right not to know" being exercised by such guidance. A subsequent amendment now argues for decision about IF search to be made at the time of testing, but still recommends search for additional mutations not indicated by the clinical symptoms. The European Society of Human Genetics (ESHG) responded that WG approaches should be targeted to the clinical question, but there is still widespread debate about the management of IFs in practice and whether real up-front patient choice is feasible or preferable.

11. Familial Consequences of IFs

A family history of a particular disease usually means that unaffected relatives have some idea they too might be at risk. In contrast, if something is found incidentally there is unlikely to be awareness of the suspected condition. Furthermore, a new variant may only be found to be clinically significant once it has been studied alongside phenotypes in a family and the absence of a family history is likely to make the need for such cascade screening more difficult to comprehend. Furthermore, professionals may be uncertain what, if any, duties they have to alert relatives about risks that may only be clarified after cascade screening.

12. Return of Results from Genomic Testing

As the pace and scale of genetic testing increases, it is inevitable there will be less time to prepare individuals for potential test results. Since the implications of some variants, particularly IFs, may fall outside the expertise of the professional who requested the test, referral to another health care professional may be necessary. This process is likely to add to anxiety of families burdened with unexpected genetic information and means that consent and disclosure practices become dissociated. Training in genomic medicine should be expanded to all medical specialties so that the complexities of genomic information can be adequately communicated but we do not underestimate the size of this task in a rapidly changing environment. We suggest that clinical genetic professionals, although relatively few in number, will need to take on greater liaison activities to facilitate this training.

Opinions about disclosure of IFs vary, ranging from full disclosure to disclosing only those with established clinical significance, and/or which have an intervention can impact on disease. In reality clinically significant, because further investigations of the patient, and their relatives, may be required it can be extremely difficult to withhold details of IFs, even if a conclusion is that they are not to arrive at this conclusion. Even if the pathogenicity of an IF is established, disease onset may not be for many years. Hence robust mechanisms are required to identify, re-contact, and review family members when health care interventions become appropriate. Current health-care systems are, however ill-equipped to deal with the recording of familial information, future risks to health, or the monitoring of multiple family members. We consider that genome results need to be considered as a resource that can be accessed over time [22], rather than as one result that needs to be disclosed as one at the point of testing.

13. Consent for Genome Testing

Providing individuals with sufficient information in order to make decisions about investigations or interventions is a key element of good clinical practice. Achieving a balance between providing sufficient information but avoiding overload can be a challenge, especially for tests where multiple different outcomes are possible. Individuals need to understand what genome tests can reveal, but also that some degree of uncertainty is likely. The possible need to investigate relatives to assign pathogenicity of variants found is a difficult issue to incorporate into any consent process. Obtaining adequate consent to disclose an IF for which there is no prior suspicion on the basis of family history or symptoms is likewise problematic, especially if such an IF is unlikely to have clinical consequences for some time. All of this is set against a background of media coverage that generally portrays genetics as clear-cut and highly determinative.

14. Is Personalised Medicine a Helpful Term to Promote Genomics?

Although genomic analyses will help to stratify individuals into subpopulations with common characteristics so that particular variants might have greater predictive value, this is not the same as individualisation. A concern about describing genomics as leading to personalised medicine is that it may encourage views of genetic determinism or reductionism. There has been much professional and public discussion regarding which parts of a whole genome sequence should be communicated, with emotive discussions about rights to personal information. On the one hand there is a public perception that some form of medical paternalism might be exhibited where useful information would be withheld, on the other there is acceptance that most of the three billion letter output of a genome sequence has no personal clinical relevance [23]. Some advocate that anyone sequenced should have the right to be appraised of "all of" the test results, even if the clinical relevance is indeterminate.

Whilst full disclosure is thought to respect a person's autonomy it may do the opposite if it delivers outputs that are uninterpretable.

15. Conclusions

Whilst the technology of genome sequencing is now a sufficiently mature and affordable technology for it to be implemented clinically, significant challenges around interpretation and implementation remain. We believe that clinical genome analyses should be directed to delivering diagnoses for patients and that integration or linkage with biobanks and other research ventures will be crucial for better clinical translation in the future. We do not underestimate the practical challenges such a statement results in but hope that by delineating some of the complexities and aligning them next to common perceptions of genetics will lead to intelligent international debate about consent and disclosure practices, long-term follow-up arrangements, appropriate communication with relatives and linkages between clinical practice and research.

Author Contributions

This paper was jointly conceived and written by the authors.

References

1. Government. HMs. Strategy for UK Life Sciences: One Year on. Available online: http://www.bisgovuk/assests/biscore/inovation/docs/s/12-1346-stragy-for-uk-life-sciences-one-year-onpdf/ (accessed on 4 July 2014).
2. Meacham, F.; Boffelli, D.; Dhahbi, J.; Martin, D.I.; Singer, M.; Pachter, L. Identification and correction of systematic error in high-throughput sequence data. *BMC Bioinform.* **2011**, *12*, 451.
3. Davies, K. The $1,000,000 genome interpretation. *Bio-IT World* **2010**, *8*, 50–54.
4. Zollner, S.; Pritchard, J.K. Overcoming the winner's curse: Estimating penetrance parameters from case-control data. *Am. J. Hum. Genet.* **2007**, *80*, 605–615.
5. Ou, J.; Niessen, R.C.; Vonk, J.; Westers, H.; Hofstra, R.M.; Sijmons, R.H. A database to support the interpretation of human mismatch repair gene variants. *Hum. Mutat.* **2008**, *29*, 1337–1341.
6. Fokkema, I.F.; Taschner, P.E.; Schaafsma, G.C.; Celli, J.; Laros, J.F.; den Dunnen, J.T. LOVD v.2.0: The next generation in gene variant databases. *Hum. Mutat.* **2011**, *32*, 557–563.
7. Decipher Database. Available online: https://decipher.sanger.ac.uk/ (accessed on 4 July 2014).
8. Diagnostic Mutation Database. Available online: https://secure.dmudb.net/ngrl-rep/Home.do/ (accessed 4 July 2014).
9. Dalgleish, R.; Flicek, P.; Cunningham, F.; Astashyn, A.; Tully, R.E.; Proctor, G.; Chen, Y.; McLaren, W.M.; Larsson, P.; Vaughan, B.W.; *et al.* Locus Reference Genomic sequences: An improved basis for describing human DNA variants. *Genome Med.* **2010**, *2*, 24.
10. Act EDP. EU Data Protection Act Proposed Changes & Reforms. Updates to the Current EU Data Protection Acts Look Set to Radically Change the Way We Deal with Data. Available online: http://www.eudataprotectionlaw.com/ (accessed on 4 July 2014).

11. Handling of NHS patient data. Available online: http://www.parliament.uk/business/committees/committees-a-z/commons-select/health-committee/inquiries/parliament-2010/cdd-2014/ (accessed on 4 July 2014).
12. Clayton, E.W.; McCullough, L.B.; Biesecker, L.G.; Joffe, S.; Ross, L.F.; Wolf, S.M.; Clinical Sequencing Exploratory Research (CSER) Consortium Pediatrics Working Group. Addressing the ethical challenges in genetic testing and sequencing of children. *Am. J. Bioeth.* **2014**, *14*, 3–9.
13. Lucassen, A.; Widdershoven, G.; Metselaar, S.; Fenwick, A.; Parker, M. Genetic testing of children: The need for a family perspective. *Am. J. Bioeth.* **2014**, *14*, 26–28.
14. Crawford, G.; Fenwick, A.; Lucassen, A. A more fitting term in the incidental findings debate: One term does not fit all situations. *Eur. J. Hum. Genet.* **2013**, doi:10.1038/ejhg.2013.266.
15. Wolf, S.M. The past, present, and future of the debate over return of research results and incidental findings. *Genet. Med.* **2012**, *14*, 355–357.
16. Wolf, S.M.; Lawrenz, F.P.; Nelson, C.A.; Kahn, J.P.; Cho, M.K.; Clayton, E.W.; Fletcher, J.G.; Georgieff, M.K.; Hammerschmidt, D.; Hudson, K.; *et al.* Managing incidental findings in human subjects research: Analysis and recommendations. *J. Law Med. Ethics* **2008**, *36*, 219–248.
17. Bredenoord, A.L.; Kroes, H.Y.; Cuppen, E.; Parker, M.; van Delden, J.J. Disclosure of individual genetic data to research participants: The debate reconsidered. *Trends Genet.* **2011**, *27*, 41–47.
18. Gliwa, C.; Berkman, B.E. Do researchers have an obligation to actively look for genetic incidental findings? *Am. J. Bioeth.* **2013**, *13*, 32–42.
19. Holtzman, N.A. ACMG recommendations on incidental findings are flawed scientifically and ethically. *Genet. Med.* **2013**, *15*, 750–751.
20. Crawford, G.; Foulds, N.; Fenwick, A.; Hallowell, N.; Lucassen, A. Genetic medicine and incidental findings: It is more complicated than deciding whether to disclose or not. *Genet Med.* **2013**, *15*, 896–899.
21. Green, R.C.; Berg, J.S.; Grody, W.W.; Kalia, S.S.; Korf, B.R.; Martin, C.L.; McGuire, A.L.; Nussbaum, R.L.; O'Daniel, J.M.; Ormond, K.E.; *et al.* ACMG recommendations for reporting of incidental findings in clinical exome and genome sequencing. *Genet. Med.* **2013**, *15*, 565–574.
22. Biesecker, L.G. Opportunities and challenges for the integration of massively parallel genomic sequencing into clinical practice: Lessons from the ClinSeq project. *Genet. Med.* **2012**, *14*, 393–398.
23. Genomics England. Genomics England Town Hall Event. Available online: http://www.genomicsengland.co.uk/town-hall-engagement-event/ (accessed on 4 July 2014).

5

The Epigenome View: An Effort towards Non-Invasive Prenatal Diagnosis

Elisavet A. Papageorgiou [1,2], George Koumbaris [1,2], Elena Kypri [1], Michael Hadjidaniel [2] and Philippos C. Patsalis [2,*]

1 NIPD Genetics Ltd., Neas Engomis 31, Engomi, Nicosia 2409, Cyprus; E-Mails: e.papageorgiou@nipd.com (E.A.P.); g.koumbaris@nipd.com (G.K.); e.kypri@nipd.com (E.K.)

2 The Cyprus Institute of Neurology and Genetics, 6 International Airport Avenue, Ayios Dometios, Nicosia 2370, Cyprus; E-Mail: michaelh@cing.ac.cy

* Author to whom correspondence should be addressed; E-Mail: patsalis@cing.ac.cy;

Abstract: Epigenetic modifications have proven to play a significant role in cancer development, as well as fetal development. Taking advantage of the knowledge acquired during the last decade, great interest has been shown worldwide in deciphering the fetal epigenome towards the development of methylation-based non-invasive prenatal tests (NIPT). In this review, we highlight the different approaches implemented, such as sodium bisulfite conversion, restriction enzyme digestion and methylated DNA immunoprecipitation, for the identification of differentially methylated regions (DMRs) between free fetal DNA found in maternal blood and DNA from maternal blood cells. Furthermore, we evaluate the use of selected DMRs identified towards the development of NIPT for fetal chromosomal aneuploidies. In addition, we perform a comparison analysis, evaluate the performance of each assay and provide a comprehensive discussion on the potential use of different methylation-based technologies in retrieving the fetal methylome, with the aim of further expanding the development of NIPT assays.

Keywords: non-invasive prenatal diagnosis; epigenetic modifications; free fetal DNA; maternal circulation; chromosomal aneuploidies

---

1. Introduction

The discovery of free fetal DNA in maternal circulation [1] was a landmark towards the development of non-invasive prenatal diagnostic assays, and remarkable advances have taken place since then. The revolution was initiated in 1997 with the determination of the fetal fraction, which was estimated to be 3% during the early stages of the pregnancy [2]. In the following years, more advanced technologies were used (e.g., digital PCR) to re-evaluate the fetal DNA fraction, which is now estimated to be 10%–20% [3].

Deciphering the critical characteristics of the fetal genome has been the main goal for the development of non-invasive prenatal tests (NIPT). Studies have shown that the origin of maternal free DNA present in maternal peripheral blood is the hematopoietic system of the mother [4]. On the other hand, free fetal DNA (ffDNA) is derived from embryonic cell degradation in maternal peripheral blood [5,6] or from apoptotic placental cells [7–9]. More recent studies have confirmed the above, using bisulfite sequencing technologies and provided convincing evidence for the origin of both fetal and maternal free DNA in maternal plasma [10]. It has also been demonstrated that free fetal DNA from maternal plasma is cleared immediately (within a few hours) after pregnancy [11]. These findings were confirmed by more recent studies [12–15] and is a finding of great importance, since the presence of fetal DNA from previous pregnancies would interfere with the correct interpretation of subsequent pregnancies. A number of independent studies have also demonstrated that the amount of fetal DNA released in maternal circulation increases with pregnancy progression [2,16].

Other studies characterizing ffDNA have found that the size of fetal DNA fragments were estimated to be <0.3 kb, whereas that of maternal DNA was >1 kb [17,18]. Follow-up studies have demonstrated that the release of fetal DNA is due to the apoptosis of no more than three nucleosomal complexes, and it has been shown that the average fetal fragment size is 286 ± 28 bp with a maximum ffDNA fragment size ranging from 219 to 313 bp [19]. However, better determination and characterization of free fetal DNA fragment sizes will allow further evaluation of the diagnostic limitations that are introduced because of fragment size.

The first attempts towards NIPT were based on the use of fetal-specific markers, which were easily distinguishable in maternal circulation, as they were fetal-specific. Such markers were Y-chromosome-specific loci for fetal sex determination, such as DYS14 [1,20], as well as fetal Rhesus D found in maternal circulation in pregnancies in which the mother was Rhesus D negative [21,22]. These methods were readily and rapidly introduced in the clinical setting of diagnostic laboratories worldwide [23], and within a few years, the field of NIPT evolved even further with the use of Y-chromosome-specific markers or paternally inherited polymorphic loci for the NIPT of X-linked inherited diseases, as well as through the identification of fetal-specific chromosomal translocations [24] and trinucleotide repeats in muscular dystrophy (DMPK) [25].

The above successful developments relied on the presence or absence of a fetal-specific marker. However, further developments and advances were needed for the identification of fetal specific-markers that are independent of gender and polymorphic sites and would allow direct discrimination of the free fetal DNA from the free maternal DNA [23,26]. The challenge of the field was the development of NIPT for the detection of chromosomal aneuploidies in the fetus. The need for the identification of fetal-specific markers that would enable the discrimination of a diploid pregnancy from an aneuploid pregnancy was urgent, because aneuploidies are among the most frequent fetal abnormalities, the most common of which are trisomy 21, trisomy 18, trisomy 13 and aneuploidies associated with chromosomes X and Y [23,27]. Major efforts took place from a number of independent research groups towards the NIPT of the most common chromosomal aneuploidies [23,26,28]. One such area that was extensively investigated was epigenetic modifications during development and how such changes could be taken into consideration for the identification of methylation fetal-specific markers that could potentially be used for the development of NIPT of fetal chromosomal abnormalities. In this review, we describe, compare and evaluate the different epigenetic-based approaches that have been implemented in the field of NIPT of fetal aneuploidies.

2. DNA Methylation in Fetal Development

DNA methylation is an enzymatic chemical modification of the genome, which includes the addition of a methyl group to the carbon-5 position of the cytokines of CpG dinucleotides [29].

The methylation pattern of the cell is reset during embryogenesis, and it is established early during development [30,31]. After its establishment, the methylation pattern is inherited from one cell generation to the next [29]. The methylation occurs in CpG dinucleotides non-uniformly distributed in the genome. In contrast, areas rich in CpG dinucleotides (CpG Islands) are usually found in promoter regions of genes, and the majority of them are presented as non-methylated [29]. It is estimated that the human genome consists of approximately 30,000 CpG islands, of which, a proportion of 50%–60% lies within promoters [32]. Although the majority of these sequences are non-methylated, the CpG islands of imprinted genes and the inactive X chromosome are predominantly methylated [33].

DNA methylation is a dynamic process and may change during the post-developmental stage [34]. It is believed that 60% of tissue-specific differentially methylated regions (TDMRs) are methylated in embryonic cells, while during the differentiation of embryonic tissues to adult tissues, they undergo de-methylation [35–39]. More recent studies confirm the above, indicating that some of the methylated TDMRs undergo de-methylation in embryonic cells during the transformation into adult tissues, while a large proportion remains methylated in newborn tissues [40]. Therefore, the de-methylation of TDMRs occurs at a later developmental stage. In addition, the results indicated that specific regions of the genome show a different methylation pattern in different tissues and at different stages of development. The above findings provided convincing evidence that fetal DNA will present different methylation patterns from the methylation pattern of the maternal DNA.

Several independent research groups argued that methylation patterns are different between different tissues [41–44]. In 2008, a team of researchers led by Beck implemented a newly developed methodology known as MeDIP (methylated DNA immunoprecipitation), which was used in combination with whole genome microarray technologies to investigate the methylation status of all known promoter regions and CpG islands in different tissues [44]. Based on the above study, the phenomenon of CpG islands' methylation in normal cells and their contribution to normal cellular functions is more frequent than ever anticipated.

An epigenetic modification is a dynamic process and has been proven to play a very important role in the development of cancer cells [45,46]. More interestingly, the identification of tumor-specific DNA methylation patterns in the plasma of patients has led to great efforts towards the non-invasive diagnosis of cancer [47,48]. These developments in the field of cancer investigation have provided additional convincing support that epigenetic differences may be present between the fetal DNA and the maternal DNA in maternal circulation during pregnancy.

3. DNA Methylation Biomarkers Discovery

The aim of DNA methylation-based approaches was first to identify fetal-specific methylation markers that would allow the discrimination of fetal DNA from the maternal DNA in maternal circulation and that have the potential to be developed into non-invasive prenatal diagnostic markers. The approaches that have been used for investigating the DNA methylation patterns in fetal DNA and maternal DNA are of three main categories: sodium bisulfite-based approaches, restriction enzyme-based approaches and methylated DNA immunoprecipitation-based approaches.

3.1. Sodium Bisulfite-Based Approaches

Sodium bisulfite conversion leads to the transformation of an epigenetic modification into a genetic sequence change for further investigation. More specifically, the treatment of DNA with sodium bisulfite results in the conversion of unmethylated cytosines to uracils, leaving methylated cytosines unchanged [49]. The genetic composition of the converted sequences of interest could be investigated using methylation-specific PCR (MSP) in which the amplification process is separate for the methylated (non-converted) fragments and the non-methylated (converted) fragments [50]. Alternatively, the methylation status of bisulfite converted sequences could be assessed through the implementation of sequencing technologies [49,51]. In 2002, Poon and his colleagues demonstrated for the first time the potential for the presence of epigenetic differences between the fetus and the mother by performing sodium bisulfite conversion of placental DNA and female peripheral blood DNA followed by MSP [50,52]. The first differentially methylated region was identified in 2005 by the use of sodium bisulfite conversion in combination with MSP and sequencing. The differentially methylated gene, known as *SERPINB5*, was found to be hypomethylated in fetal DNA and hypermethylated in maternal DNA [12]. The identification of hypomethylated fetal-specific *SERPINB5* sequences was also achieved in maternal plasma during pregnancy. This genomic region was used to demonstrate that fetal DNA is not detectable in maternal plasma 24 h after delivery [28].

Since then, great efforts have taken place from independent groups towards the identification of fetal-specific methylation markers. The initial attempts were based on the investigation of promoter regions and CpG islands. In 2008, a bisulfite based systematic search for placental DNA methylation markers on chromosome 21 was described. In this study, the methylation-sensitive single nucleotide extension (Ms-SNuPE) method was used to assess the methylation differences of CpG sites [53,54]. The above study performed an evaluation of the methylation status of 114 CpG islands (based on bioinformatics criteria) in five first trimester placental tissues and two samples of non-pregnant female blood. Among them, 22 CpG islands were identified as having the potential to be developed into biomarkers for the NIPT of trisomy [54].

In 2010, a second study was performed with the aim of identifying a panel of fetal-specific hypermethylated markers on chromosome 21, and it used the methylation pattern of a previously characterized gene, *RASSF1A*. The *RASSF1A* gene is located on chromosome 3 and has been found to be completely methylated in fetal DNA and completely unmethylated in maternal DNA. This characteristic allowed the use of the *RASSF1A* gene as a fetal universal marker [28,55]. The study was performed using the combined bisulfite restriction analysis (COBRA) [56] to investigate 35 gene promoter regions on chromosome 21. The analysis demonstrated that the *HLCS* gene located on chromosome 21 is fully methylated in placenta and unmethylated in maternal blood cells [15].

A recent report published in 2013 illustrates the potential of retrieving the methylation profiles of placental tissues and maternal blood cells using sodium bisulfite in combination with next generation sequencing technologies [10]. The investigators were able to retrieve the fetal methylome through the identification of single nucleotide polymorphism (SNP) genotype differences between the mother and the fetus in maternal plasma and to identify differentially methylated regions (DMRs). They identified 44,455 loci as being fetal-specific hypomethylated and 3081 regions as being fetal-specific hypermethylated. The above findings are in agreement with previous studies in which it was clearly evident that the fetal genome is mostly hypomethylated in contrast to the adult peripheral blood, which is greatly hypermethylated, indicating a regulatory role of the methylation patterns and gene expression profiles [44,57,58]. Interestingly, it has also been reported that hypomethylated sequences tend to be of a smaller fragment size. These findings could indicate a contribution of the fetal methylation status to the small fetal DNA fragments size in maternal plasma [10].

3.2. Restriction Enzyme-Based Approaches

Methylation patterns of CG dinucleotides can also be assessed using restriction enzymes, which have recognition sites containing CG sequences. Methylation-sensitive restriction enzymes can digest their recognition site only when unmethylated, whereas methylation insensitive restriction enzymes digest their recognition sites only when the cytokines of the CGs within their recognition site are methylated. In 2007, the team headed by Old reported for the first time the investigation and identification of a panel of differentially methylated regions on chromosome 21 using methylation-sensitive enzymes [59]. More specifically, the team used the HpaII enzyme, and the underlying idea was based on the fact that the enzyme would digest only the unmethylated type of its recognition site (CCGG). Therefore, this would allow them to identify regions containing the above recognition sites, which are differentially methylated between placenta and maternal blood cells. The study was focused on the investigation of promoters from highly expressed genes, randomly selected promoters, as well as randomly selected non-promoter regions. Among the 200 pre-selected regions, three promoter regions of the *AIRE*, *SIM2* and *ERG* genes were found to be methylated in the placenta and unmethylated in the maternal blood cells. The methylation status of those regions was confirmed by sodium bisulfite followed by MSP [59].

In 2011 a study performed by Peters and his team demonstrated that the use of methylation-based restriction enzymes, such as HpaII and MSpI, in combination with high-resolution arrays can distinguish differentially methylated regions between the placenta and maternal blood cells [58]. They presented a large panel of DMRs consisting of 6311 DMRs across chromosomes 13, 18 and 21 [58,60] and demonstrated that the fetal DNA is mostly hypomethylated, whereas the maternal blood cells are mostly hypermethylated, findings which are in agreement with previous reports [44,57]. Moreover, they illustrated that the majority of the hypomethylated regions of both fetal and maternal origin are located within CpG islands, promoters and exons, indicating a potential correlation with expression profiles [58].

3.3. Methylated DNA Immunoprecipitation-Based Approaches

One of the most modern methods of studying the levels of DNA methylation is the MeDIP (methylated DNA immunoprecipitation) approach. The method was first described in 2005 by Weber *et al.* with the aim of investigating the methylation pattern of cancer cells in a genome-wide fashion using microarray platforms [45]. In 2007, Beck and his team introduced linker-mediated PCR amplification (LM-PCR) in combination with the MeDIP methodology. They obtained large amounts of immunoprecipitated DNA and generated the first whole genome mammalian methylome using a large panel of different tissues [44,61]. The principles of the MeDIP methodology includes fragmentation of the DNA (through sonication or enzymatic digestion) into short DNA fragments of 300–1000 bp. The sample is denatured and incubated with a monoclonal antibody, which recognizes and attaches to the 5-methylcytosines of CpG dinucleotides. Immunoprecipitation of methylated sequences is accomplished with the addition of magnetic beads. Through the implementation of the MeDIP methodology, you can achieve direct enrichment of methylated fragments. Enrichment of methylated target sequences is easily retrieved through the use of a large number of different technologies, such as PCR, qPCR (quantitative Polymerase Chain Reaction), microarray and sequencing. Since its development, MeDIP has been extensively used for the investigation of the methylation status/patterns of cancer tissues with great success either in combination with microarray technologies (MeDIP-chip) [42,44,45] or, more recently, in conjunction with next generation sequencing (MeDIP-seq) [62–65].

The MeDIP methodology was first introduced to the field of NIPT by our team in 2009 with the aim of investigating and identifying DMRs between placenta and female peripheral blood towards the development of NIPT for the identification of common aneuploidies [57]. Our team used MeDIP in combination with chromosome-specific high-resolution oligo arrays for the investigation of the methylation pattern of chromosomes 13, 18, 21, X and Y. Although previous studies solely investigated promoter regions and CpG islands for DMR identification, we were the first to screen entire chromosomes of interest irrespective of the genomic position or CG content. At the time, we reported the largest panel of DMRs with the potential to be developed into NIPT biomarkers for the most common fetal aneuploidies. More specifically, we identified around 2000 DMRs on each of the chromosomes investigated, and interestingly, we noticed that the vast majority of the DMRs were located within non-genic regions and in relatively poor CG regions. More specifically, we illustrated that 56%–83% of the DMRs were located within non-genic regions, whereas only 1%–11% were located within CpG islands. Our findings were concordant with previous studies performed by other groups investigating a panel of different tissues [44] and were also in agreement with more recent reports using bisulfite sequencing technologies [3,58]. We were also able to report the presence of inter-individual variability and the changes in the methylation patterns during the progression of the pregnancy, findings which have recently been confirmed by independent groups [10].

Following our study, the group headed by Chim used MeDIP in combination with a microarray platform targeting promoter regions and CpG islands. The group identified a panel of eight DMRs with the potential of being developed into biomarkers for diagnostic purposes [66], most of which are among the DMRs identified previously by our group [57]. Any discrepancies reported regarding the identification of DMRs, such as the failure to have the exact same methylation status of all DMRs reported by independent studies, are not uncommon, since different platforms and different methylation-based technologies were used.

4. Implementation of Methyl-Biomarkers in NIPT

The discovery of DMRs has mainly been focused on chromosomes 13, 18, 21, X and Y with the aim of identifying as a priority methylation-based biomarkers (methyl-biomarkers) suitable for the development of NIPT for the most common chromosomal fetal aneuploidies. The first attempt was reported back in 2006 for the NIPT of trisomy 18 (Edward's syndrome) [67]. In this study, the authors implemented a combination of sodium bisulfite conversion with MSP using maternal plasma samples from normal and trisomy 18 pregnancies. To achieve discrimination, they used the information of an SNP located within the *SERPINB5* gene. The cases were considered informative if the SNP was homozygous in the mother and heterozygous in the fetus, and only those cases could be used for NIPT of trisomy 18 (T18). To achieve this, the team introduced the so-called epigenetic allelic ratio (EAR) in which the chromosome 18 copy number was assessed based on the allele ratio calculation of an informative SNP. The challenge in this study was to have informative SNPs, and because there was only a single SNP in the target sequence, it was extremely difficult to be informative in all cases tested (Table 1). The results showed that among the 173 euploid placentas and 14 trisomy18 placentas genotyped for the polymorphism, only 31 and seven placentas, respectively, were informative. The rarity of having an informative SNP in this study does not allow this approach to be implemented population-wide [23,26].

To overcome the above limitations, in 2010, the same group developed an SNP-free methylation-based assay for NIPT of trisomy 21 (Down syndrome). Methylation-sensitive restriction digestion was used followed by digital PCR to investigate DMRs identified on chromosome 21 [15]. The copy number of chromosome 21 was determined through the epigenetic-genetic (EGG) chromosome dosage approach using the fetal-specific hypermethylated promoter region of the *HLCS* gene located on chromosome 21 and the *ZFY* locus on chromosome Y. The assay tested 24 maternal plasma samples from euploid pregnancies and five maternal plasma samples from trisomy 21 pregnancies. All but one euploid pregnancy were correctly classified (Table 1) [15].

The EGG chromosome dosage approach was also implemented for the NIPT of trisomy 18 in which the fetal-specific methylated *VAPA-APCDD1* loci on chromosome 18 and the *ZFY* on chromosome Y were quantified with digital PCR after HinP1I- and HpaII sample digestion [66]. The study was performed on nine maternal plasma samples from male trisomy 18 pregnancies and 27 maternal plasma samples from male euploid pregnancies. Among them, eight out of nine and one out of 27 trisomy 18 and euploid pregnancies, respectively, were correctly identified, which corresponds to 88.9% sensitivity and 96.3% specificity (Table 1) [66].

Table 1. Comparison of different methylation-based approaches towards the non-invasive prenatal tests (NIPT) of aneuploidies. EAR, epigenetic allelic ratio (EAR); EGG, epigenetic-genetic; SNP, single nucleotide polymorphism.

Assay	Technology	Sample size	Sensitivity/Specificity (%)	Advantages	Disadvantages	Reproduced by others
EAR on chromosome 18 [67]	Sodium bisulfite, digital PCR	2 normal 2 trisomy 18	Not defined/not applicable population-wide	Applicable irrespective of gender	Requires informative SNP, depends on the bisulfite conversion performance	No
EGG on chromosome 21 using *ZFY* [15]	* COBRA, digital PCR	24 normal 5 trisomy 21	95.8% specificity 100% sensitivity	SNP-free assay	Applicable only to male pregnancies, depends on the digestion and bisulfite conversion efficiency	No
EGG on chromosome 18 using *ZFY* [66]	* COBRA, digital PCR	27 normal 9 Trisomy 18	96.3% specificity 88.9% sensitivity	SNP-free assay	Applicable only to male pregnancies, depends on the digestion and bisulfite conversion efficiency	No
EGG on chromosome 21 using *TMED8* [68]	** MRED digestion, digital PCR	33 normal 14 trisomy 21	Variable depending on the fetal allele	Applicable irrespective of gender	Requires informative SNP, applicable only to male pregnancies, depends on the digestion efficiency	No
Fetal-specific DNA methylation ratio on chromosome 21 (1st study) [69]	*** MeDIP, real-time qPCR	40 normal 40 trisomy 21	100% specificity 100% sensitivity	Applicable irrespective of gender and SNPs	Depends on MeDIP performance	Yes [70,71]
Fetal-specific DNA methylation ratio on chromosome 21 (2nd study) [72]	*** MeDIP, real-time qPCR	125 normal 50 trisomy 21	99.2% specificity 100% sensitivity	Applicable irrespective of gender and SNPs	Depends on MeDIP performance	No
Bisulfite sequencing [10]	Sodium bisulfite, next generation sequencing	7 normal 5 trisomy 21	100% specificity 100% sensitivity	Applicable irrespective of gender and SNPs	Depends on bisulfite conversion efficiency	No

* Combined bisulfite restriction analysis; ** methylation restriction enzymatic digestion; *** methylated DNA immunoprecipitation.

Although the results from the studies using the EGG chromosome dosage approach were promising, the technology was restricted to male pregnancies, because the EGG calculation involved the use of the *ZFY* gene (Table 1). To overcome the above difficulties, a modification was introduced in the EGG calculation to be able to include the testing of female pregnancies, as well. The study was performed using 14 maternal plasma from trisomy 21 pregnancies and were compared to 33 cases with a euploid fetus [68]. For calculation purposes, the *ZFY* gene was replaced with an autosomal genetic reference marker. Interpretation of the results was achieved using a paternally-inherited SNP allele on the *TMED8* gene located on chromosome 14, which served as a baseline for the EGG chromosome dosage calculation. The sensitivity of the assay varied depending on which of the two alleles of an SNP was fetal-specific, making the evaluation of the assay performance even more challenging. Overall, although the limitation of testing only male pregnancies was overcome, the assessment of the copy number of chromosome 21 remained a challenge, as the presence of at least one informative SNP was necessary (Table 1).

A different approach was proposed by our group in 2011 and was based on using the MeDIP methodology in combination with real-time quantitative PCR (real time-qPCR) for the quantification of selected DMRs located on chromosome 21 [69]. We selected 12 previously identified DMRs located on chromosome 21 [57], which were hypermethylated in fetal DNA and hypomethylated in female peripheral blood cells. We used in our study a total of 40 maternal blood samples from euploid pregnancies and 40 maternal blood samples from trisomy 21 cases. We developed a diagnostic formula by calculating the DNA methylation ratio of the selected DMRs using 20 normal pregnancies and 20 trisomy 21 pregnancies. Eight specific DMRs were the most statistically significant markers in discriminating normal from trisomy 21 pregnancies. The MeDIP-qPCR methodology was used to then test 40 additional pregnancies, of which 20 were obtained from trisomy 21 pregnancies and showed 100% specificity and 100% sensitivity [69]. We also demonstrated that diagnostic accuracy can only be achieved through the combination of multiple DMRs from chromosome 21, which was an important finding for further NIPT developments [23].

Our team continued to improve the MeDIP-qPCR assay with a larger validation study of 175 pregnancies that included 50 trisomy 21 pregnancies [72]. In this larger-scale validation, we re-evaluated our diagnostic assay, taking into consideration the genomic composition of our DMRs and by selectively excluding those DMRs located in copy number variable (CNV) regions. Based on the above, we re-designed our diagnostic formula and then evaluated its performance using 100 new cases, which included 25 trisomy 21 pregnancies. The results demonstrated 100% sensitivity and 99.2% specificity (Table 1) [72]. Our group also investigated whether the variability of the fetal fraction present in maternal plasma has a negative effect in our assay's diagnostic efficiency. Although previous reports demonstrated an effect of different fetal amounts present in maternal plasma [73–75], our study has shown no significant association between cffDNA fraction, absolute fetal amount or the concentration present in maternal plasma with the test result classification using our diagnostic formula [20,72]. We speculate that this is due to the fact that maternal blood contains <1% of fetal DNA [20,72] in contrast to maternal plasma, which contains ~10%–15% fetal DNA [10,76].

More importantly, the results of our studies have been reproduced by two independent groups, which have reported their results using the MeDIP-qPCR methodology and the published diagnostic formula [70,71]. In addition, independent groups have also commented positively on the potential prospects or application of the MeDIP-qPCR assay towards the NIPT of chromosomal aneuploidies. The low cost of the technology and the ease of implementing it, in combination with the use of equipment common to every laboratory, allows its implementation in any diagnostic laboratory setting [77]. A major strength of the MeDIP-qPCR assay is that it is a gender- and polymorphism-independent assay that could be implemented population-wide. Nevertheless, a different independent group has failed to reproduce the MeDIP-qPCR results by performing a small scale validation study [78]. Lack of reproducibility of the results would not be a surprise to our team, since, as stated in our reply to the above manuscript, very stringent quality control criteria must be applied to critical reagents and conditions throughout the method [79].

A very interesting recent development of investigating DNA methylation for use in NIPT has been the implementation of sodium bisulfite DNA treatment in combination with next generation sequencing technologies (NGS) [10]. The study is presented as a proof of principle and demonstrates one use of the assay with the detection of trisomy 21. NGS technologies have already been introduced in the field of NIPT by different independent groups with the primary aim of detecting the most common chromosomal aneuploidies [73–76,80–82]. Biotechnology companies have already introduced in the market their NGS-based NIPT of the most common chromosomal fetal aneuploidies [83–85]. However, sequencing of maternal plasma can turn out to be very challenging, due to the restrictions of the very low amount of fetal DNA available. Furthermore, such technology is not yet available in all clinical laboratories. Sequencing technologies are still considered to be of a high cost, requiring significant infrastructure, are labor intensive and require highly trained personnel, and the bioinformatics analysis can be very challenging, especially when the target sequence is of a very low amount, such as fetal DNA present in maternal plasma.

5. Evaluating the Efficiency of Methylation Assays

Developments towards methyl-biomarker discovery and their applications in the NIPT of fetal chromosomal abnormalities were achieved through a number of independent groups, as described above, using different methylation-based approaches. Different analytical tools and a variety of quantitative approaches (e.g., MSP, digital PCR, real-time qPCR, microarray platforms and next generation sequencing) were used, of which the statistical power in discriminating normal from abnormal pregnancies has been extensively assessed [23,26,86]. Nevertheless, the statistical discriminating power of each of the end point analytical tools relies on the efficiency of the methylation-based technology used to enrich the fetal DNA in maternal circulation (Table 1). Therefore, the evaluation and assessment of the efficiency of the methylation-based enrichment technology used is of significant importance.

One of the most commonly used approaches is the treatment of DNA with sodium bisulfite. Sodium bisulfite conversion is considered the gold standard in the evaluation of the methylation status of different tissues and has been extensively used, especially in the field of cancer [87,88]. However, it is well known that this chemical treatment of the DNA is associated with a high degree of DNA degradation, reaching >90% of the template DNA [89]. This major drawback of the technology is undesirable for its implementation in plasma samples of pregnant women. During pregnancy, the amount of fetal DNA in maternal plasma is very low [10,76], and further degradation will result in even fewer fetal DNA molecules available for quantification; therefore, the accuracy and sensitivity of the test will be reduced. To compensate for the degradation effect, much larger amounts of maternal plasma are required, which makes the testing of maternal plasma even more complicated. Furthermore, bisulfite conversion can be challenging, since 100% conversion of the unmethylated cytosines to uracils is rarely achieved, and purification is required to remove the sodium bisulfite [90]. Such an effect will bias the correct interpretation of the results [23]. On the other hand, bisulfite conversion strategies are not sensitive to low purity and low integrity samples, an advantage especially for samples with very low starting DNA amounts. Nevertheless, bisulfite conversion in combination with sequencing technologies can provide a comprehensive analysis of the methylation status at the base pair composition, which can make it a very powerful tool (Table 2).

Table 2. Comparison of different methylation assays.

Methylation assay	Advantages	Disadvantages	Analytical tool used for NIPT
Sodium bisulfite	Not sensitive to sample impurities, methylation analysis at the base pair level	DNA degradation (>90%), 100% conversion is rarely achieved	* MSP, microarrays, Digital PCR, ** COBRA, *** NGS
Restriction enzyme digestion	Easy to perform and low cost	Sensitive to sample impurities, requires high amount of starting DNA, applicable to a limited number of DNA sequences	** COBRA, digital PCR
**** MeDIP	Ideal for investigating low CG content regions, low cost assay, not sensitive to sample impurities, can be applied with low starting DNA amounts	Depends on antibody efficiency and ideal combination of affinity reagents	Real time-qPCR, microarrays

* Methylation-specific PCR; ** combined bisulfite restriction analysis; *** next generation sequencing; **** methylated DNA immunoprecipitation.

A different approach implemented by a number of independent groups towards methyl-biomarker discovery and methylation-based NIPT developments has been the use of methylation restriction enzymes, as described above. Through methylation restriction enzymatic digestions (MRED), the unmethylated maternal origin sequences, present in maternal plasma, are digested to achieve indirect enrichment for the corresponding sequences of fetal origin, which are methylated. The efficiency of the MRED assays depends on the purity of the sample, and for this reason, they require high purity and high integrity samples [90]. Additionally, MRED assays require fairly high quantities of starting material, which is a restriction to its implementation in plasma samples, because not only the target fetal DNA sequences are of a low amount, but also the total plasma DNA is very low (around 10 ng/4 mL plasma) [20]. An additional drawback of the assay is that it can only evaluate the methylation status of a specific and very limited number of genomic sequences. Only those sequences that include a recognition site of a methylation-dependent restriction enzyme could be evaluated. Such inherent restrictions do not allow efficient and detailed genome-wide methylation assessment [23,26]. An example is the recognition sites of the HpaII restriction enzyme, which are presented in only 3.9% of CGs across non-repetitive sequences of the human genome [91]. Moreover, the efficiency of digestion should always be carefully evaluated for an unbiased interpretation of the results. Nevertheless, it is a very easy to perform assay and low cost.

The MeDIP assay, an affinity-enrichment method, was also utilized towards DMR identification and characterization to discriminate fetal DNA from maternal DNA in maternal circulation during pregnancy. Based on studies performed by several independent groups, it is clearly evident that the vast majority of DMRs identified between different tissues are located within non-genic and CG poor regions [44,58]. Based on recent reports, the MeDIP methodology is ideal for the investigation of low CpG density regions [92]. Indeed, the DMRs identified and selected for NIPT of trisomy 21 using MeDIP-qPCR are located in low CpG sites and are mostly found within non-genic regions [57,69,72]. Therefore, we strongly feel that MeDIP is the choice of selection for the investigation of DMRs towards NIPT. MeDIP is an efficient method for genome-wide methylation assessment [42,44,45], as it can evaluate the methylation levels irrespective of genomic composition and overcomes limitations of the previously described methodologies. The MeDIP assay can tolerate sample impurities, and thus, no prior sample purification is required. Furthermore, it has recently been proven to be applicable for low starting DNA templates, generating sufficiently enriched outputs [64,65], a development that simplifies and makes possible its implementation with plasma samples. Moreover, it is a technically robust methodology, easy to use and affordable. Nevertheless, the efficiency and performance of MeDIP greatly depends on determining the ideal combination of affinity reagents. This is very important, especially in regions with varying methylcytosine density, such as the DMRs identified for the NIPT of common chromosomal aneuploidies [57,69,72]. The advantages and disadvantages of all the different methylation-based assays implemented towards the NIPT of fetal chromosomal abnormalities are summarized in Table 2.

6. Conclusions and Future Directions

Deciphering the epigenome and understanding the underlying mechanisms that lead to epigenetic modifications has been one of the most interesting fields under investigation for the last decade. Since 2002, a large panel of DMRs has been identified by independent groups, with the potential of being developed into diagnostic markers having as a primary goal the development of NIPT for common fetal chromosomal abnormalities.

We speculate that epigenetic approaches towards NIPT will soon dominate the field of NIPT, because they are easy to perform, are fast and inexpensive compared to existing NIPT approaches, which are based on next generation sequencing technologies [73–75,81,82]. We speculate that one of the first epigenetic-based approaches that will be launched for the NIPT of common chromosomal fetal abnormalities will be a MeDIP-based approach. NIPD Genetics Ltd., a company in which three of the authors are employed, is dedicated to developing a MeDIP-qPCR-based diagnostic assay. The company will be soon ready to launch the first epigenetic-based NIPT for trisomy 21 following completion of a large-scale validation study [23,72,93].

Methylation-based approaches could also be used for retrieving the methylation status of abnormal tissues, such as placental tissues from aneuploid pregnancies. A very recent study has shown that trisomy 21 placentas are characterized by a global hypermethylation in contrast to normal placentas, which are mainly hypomethylated [94]. Identifying such disease-associated characteristics can benefit and contribute to more robust and sensitive NIPT. Furthermore, methylation differences during fetal development have also been shown to be associated with transcription. It has been demonstrated that the early gestational placental methylome is significantly associated with gene expression [58]. Such structural and regulatory characteristics of the placental epigenome are of great importance and could be used to determine the role of aberrant or altered methylation in placental dysfunction.

In addition to the methods described in this review, the implementation of alternative methylation-based approaches, such as MBD (methylated binding domain) [92] and McrBC (a GTP-requiring, modification-dependent endonuclease of *Escherichia coli* K-12) fragmentation, as well as HELP (HpaII tiny fragment enrichment by ligation-mediated PCR) [95,96], in combination with the development of bioinformatics-based algorithms, will contribute to a better understanding of the fetal methylome. We envision that epigenetic-based enrichment methods will have a major contribution to fetal methylome analysis through direct testing of maternal plasma. Looking ahead, we predict that epigenetic-based approaches in combination with genetic-based approaches and advanced technological approaches, such as digital PCR and next generation sequencing, will contribute to the development of NIPT of more subtle fetal abnormalities, such as point mutations, microdeletion/microduplication syndromes, *etc.*

Acknowledgments

The work performed by the author's laboratories is supported by the Cyprus Institute of Neurology and Genetics, NIPD Genetics Ltd., EU 7th Framework Programme, as part of the ANGELAB (A New GEnetic LABoratory) project (#317635), and the European Research Council (ERC), as part of the European Research Council program, ERC-2012-AdG 322953-NIPD.

Author Contributions

Designed the structure and content of the manuscript: EAP, PCP. Contributed materials for writing the manuscript: GK, EK, MH. Wrote the manuscript: EAP, PCP.

References

1. Lo, Y.M.; Corbetta, N.; Chamberlain, P.F.; Rai, V.; Sargent, I.L.; Redman, C.W.; Wainscoat, J.S. Presence of fetal DNA in maternal plasma and serum. *Lancet* **1997**, *350*, 485–487.
2. Lo, Y.M.; Tein, M.S.; Lau, T.K.; Haines, C.J.; Leung, T.N.; Poon, P.M.; Wainscoat, J.S.; Johnson, P.J.; Chang, A.M.; Hjelm, N.M. Quantitative analysis of fetal DNA in maternal plasma and serum: implications for noninvasive prenatal diagnosis. *Am. J. Hum. Genet.* **1998**, *62*, 768–775.
3. Lun, F.M.; Chiu, R.W.; Chan, A.K.C.; Yeung Leung, T.; Kin Lau, T.; Lo, D.Y.M. Microfluidics digital PCR reveals a higher than expected fraction of fetal DNA in maternal plasma. *Clin. Chem.* **2008**, *54*, 1664–1672.

4. Partsalis, T.; Chan, L.Y.; Hurworth, M.; Willers, C.; Pavlos, N.; Kumta, N.; Wood, D.; Xu, J.; Kumta, S.; Lo, Y.M.; *et al.* Evidence of circulating donor genetic material in bone allotransplantation. *Int. J. Mol. Med.* **2006**, *17*, 1151–1155.
5. Bianchi, D.W.; Shuber, A.P.; DeMaria, M.A.; Fougner, A.C.; Klinger, K.W. Fetal cells in maternal blood: determination of purity and yield by quantitative polymerase chain reaction. *Am. J. Obstet. Gynecol.* **1994**, *171*, 922–926.
6. Lo, Y.M.; Lau, T.K.; Chan, L.Y.; Leung, T.N.; Chang, A.M. Quantitative analysis of the bidirectional fetomaternal transfer of nucleated cells and plasma DNA. *Clin. Chem.* **2000**, *46*, 1301–1309.
7. Alberry, M.; Maddocks, D.; Jones, M.; Abdel Hadi, M.; Abdel-Fattah, S.; Avent, N.; Soothill, P.W. Free fetal DNA in maternal plasma in anembryonic pregnancies: Confirmation that the origin is the trophoblast. *Prenat. Diagn.* **2007**, *27*, 415–418.
8. Tjoa, M.L.; Cindrova-Davies, T.; Spasic-Boskovic, O.; Bianchi, D.W.; Burton, G.J. Trophoblastic oxidative stress and the release of cell-free feto-placental DNA. *Am. J. Pathol.* **2006**, *169*, 400–404.
9. Smid, M.; Galbiati, S.; Lojacono, A.; Valsecchi, L.; Platto, C.; Cavoretto, P.; Calza, S.; Ferrari, A.; Ferrari, M.; Cremonesi, L. Correlation of fetal DNA levels in maternal plasma with Doppler status in pathological pregnancies. *Prenat. Diagn.* **2006**, *26*, 785–790.
10. Lun, F.M.; Chiu, R.W.; Sun, K.; Leung, T.Y.; Jiang, P.; Chan, K.C.; Sun, H.; Lo, Y.M. Noninvasive prenatal methylomic analysis by genomewide bisulfite sequencing of maternal plasma DNA. *Clin. Chem.* **2013**, *59*, 1583–1594.
11. Lo, Y.M.; Zhang, J.; Leung, T.N.; Lau, T.K.; Chang, A.M.; Hjelm, N.M. Rapid clearance of fetal DNA from maternal plasma. *Am. J. Hum. Genet.* **1999**, *64*, 218–224.
12. Chim, S.S.; Tong, Y.K.; Chiu, R.W.; Lau, T.K.; Leung, T.N.; Chan, L.Y.; Oudejans, C.B.; Ding, C.; Lo, Y.M. Detection of the placental epigenetic signature of the maspin gene in maternal plasma. *Proc. Natl. Acad. Sci. USA* **2005**, *102*, 14753–14758.
13. Tsumita, T.; Iwanaga, M. Fate of injected deoxyribonucleic acid in mice. *Nature* **1963**, *198*, 1088–1089.
14. Emlen, W.; Mannik, M. Kinetics and mechanisms for removal of circulating single-stranded DNA in mice. *J. Exp. Med.* **1978**, *147*, 684–699.
15. Tong, Y.K.; Jin, S.; Chiu, R.W.; Ding, C.; Chan, K.C.; Leung, T.Y.; Yu, L.; Lau, T.K.; Lo, Y.M. Noninvasive prenatal detection of trisomy 21 by an epigenetic-genetic chromosome-dosage approach. *Clin. Chem.* **2010**, *56*, 90–98.
16. Smith, S.C.; Baker, P.N.; Symonds, E.M. Placental apoptosis in normal human pregnancy. *Am. J. Obstet. Gynecol.* **1997**, *177*, 57–65.
17. Chan, K.C.; Zhang, J.; Hui, A.B.; Wong, N.; Lau, T.K.; Leung, T.N.; Lo, K.W.; Huang, D.W.; Lo, Y.M. Size distributions of maternal and fetal DNA in maternal plasma. *Clin. Chem.* **2004**, *50*, 88–92.
18. Li, Y.; Zimmermann, B.; Rusterholz, C.; Kang, A.; Holzgreve, W.; Hahn, S. Size separation of circulatory DNA in maternal plasma permits ready detection of fetal DNA polymorphisms. *Clin. Chem.* **2004**, *50*, 1002–1011.

19. Kimura, M.; Hara, M.; Itakura, A.; Sato, C.; Ikebuchi, K.; Ishihara, O. Fragment size analysis of free fetal DNA in maternal plasma using Y-STR loci and SRY gene amplification. *Nagoya J. Med. Sci.* **2011**, *73*, 129–135.
20. Kyriakou, S.; Kypri, E.; Spyrou, C.; Tsaliki, E.; Velissariou, V.; Papageorgiou, E.A.; Patsalis, P.C. Variability of ffDNA in maternal plasma does not prevent correct classification of trisomy 21 using MeDIP-qPCR methodology. *Prenat. Diagn.* **2013**, *33*, 650–655.
21. Lo, Y.M.; Hjelm, N.M.; Fidler, C.; Sargent, I.L.; Murphy, M.F.; Chamberlain, P.F.; Poon, P.M.; Redman, C.W.; Wainscoat, J.S. Prenatal diagnosis of fetal RhD status by molecular analysis of maternal plasma. *N. Engl. J. Med.* **1998**, *339*, 1734–1738.
22. Daniels, G.; Finning, K.; Martin, P.; Summers, J. Fetal blood group genotyping: Present and future. *Ann. N. Y. Acad. Sci.* **2006**, *1075*, 88–95.
23. Papageorgiou, E.A.; Patsalis, P.C. Non-invasive prenatal diagnosis of aneuploidies: New technologies and clinical applications. *Genome Med.* **2012**, *4*, 46.
24. Chen, C.P.; Chern, S.R.; Wang, W. Fetal DNA analyzed in plasma from a mother's three consecutive pregnancies to detect paternally inherited aneuploidy. *Clin. Chem.* **2001**, *47*, 937–939.
25. Amicucci, P.; Gennarelli, M.; Novelli, G.; Dallapiccola, B. Prenatal diagnosis of myotonic dystrophy using fetal DNA obtained from maternal plasma. *Clin. Chem.* **2000**, *46*, 301–302.
26. Patsalis, P.C.; Tsaliki, E.; Koumbaris, G.; Karagrigoriou, A.; Velissariou, V.; Papageorgiou, E.A. A new non-invasive prenatal diagnosis of Down syndrome through epigenetic markers and real-time qPCR. *Exp. Opin. Biol. Ther.* **2012**, *12*, S155– S 161.
27. Driscoll, D.A.; Gross, S. Clinical practice. Prenatal screening for aneuploidy. *N. Engl. J. Med.* **2009**, *360*, 2556–2562.
28. Chiu, R.W.; Lo, Y.M. Non-invasive prenatal diagnosis by fetal nucleic acid analysis in maternal plasma: the coming of age. *Semin. Fetal Neonatal Med.* **2011**, *16*, 88–93.
29. Raedle, J.; Trojan, J.; Brieger, A.; Weber, N.; Schafer, D.; Plotz, G.; Staib-Sebler, E.; Kriener, S.; Lorenz, M.; Zeuzem, S. Bethesda guidelines: Relation to microsatellite instability and MLH1 promoter methylation in patients with colorectal cancer. *Ann. Intern. Med.* **2001**, *135*, 566–576.
30. Bird, A.P. The relationship of DNA methylation to cancer. *Cancer Surv.* **1996**, *28*, 87–101.
31. Monk, M. Changes in DNA methylation during mouse embryonic development in relation to X-chromosome activity and imprinting. *Philos. Trans. R. Soc. Lond.* **1990**, *326*, 299–312.
32. Szyf, M. DNA methylation and demethylation as targets for anticancer therapy. *Biochemistry* **2005**, *70*, 533–549.
33. Costello, J.F.; Plass, C. Methylation matters. *J. Med. Genet.* **2001**, *38*, 285–303.
34. Reik, W.; Dean, W.; Walter, J. Epigenetic reprogramming in mammalian development. *Science* **2001**, *293*, 1089–1093.
35. Kawai, J.; Hirotsune, S.; Hirose, K.; Fushiki, S.; Watanabe, S.; Hayashizaki, Y. Methylation profiles of genomic DNA of mouse developmental brain detected by restriction landmark genomic scanning (RLGS) method. *Nucleic Acids Res.* **1993**, *21*, 5604–5608.
36. Watanabe, S.; Kawai, J.; Hirotsune, S.; Suzuki, H.; Hirose, K.; Taga, C.; Ozawa, N.; Fushiki, S.; Hayashizaki, Y. Accessibility to tissue-specific genes from methylation profiles of mouse brain genomic DNA. *Electrophoresis* **1995**, *16*, 218–226.

37. Shiota, K. DNA methylation profiles of CpG islands for cellular differentiation and development in mammals. *Cytogenet. Genome Res.* **2004**, *105*, 325–334.
38. Song, F.; Smith, J.F.; Kimura, M.T.; Morrow, A.D.; Matsuyama, T.; Nagase, H.; Held, W.A. Association of tissue-specific differentially methylated regions (TDMs) with differential gene expression. *Proc. Natl. Acad. Sci. USA* **2005**, *102*, 3336–3341.
39. Ching, T.T.; Maunakea, A.K.; Jun, P.; Hong, C.; Zardo, G.; Pinkel, D.; Albertson, D.G.; Fridlyand, J.; Mao, J.H.; Shchors, K.; *et al.* Epigenome analyses using BAC microarrays identify evolutionary conservation of tissue-specific methylation of SHANK3. *Nat. Genet.* **2005**, *37*, 645–651.
40. Song, F.; Mahmood, S.; Ghosh, S.; Liang, P.; Smiraglia, D.J.; Nagase, H.; Held, W.A. Tissue specific differentially methylated regions (TDMR): Changes in DNA methylation during development. *Genomics* **2009**, *93*, 130–139.
41. Eckhardt, F.; Lewin, J.; Cortese, R.; Rakyan, V.K.; Attwood, J.; Burger, M.; Burton, J.; Cox, T.V.; Davies, R.; Down, T.A.; *et al.* DNA methylation profiling of human chromosomes 6, 20 and 22. *Nat. Genet.* **2006**, *38*, 1378–1385.
42. Weber, M.; Hellmann, I.; Stadler, M.B.; Ramos, L.; Paabo, S.; Rebhan, M.; Schubeler, D. Distribution, silencing potential and evolutionary impact of promoter DNA methylation in the human genome. *Nat. Genet.* **2007**, *39*, 457–466.
43. Illingworth, R.; Kerr, A.; Desousa, D.; Jorgensen, H.; Ellis, P.; Stalker, J.; Jackson, D.; Clee, C.; Plumb, R.; Rogers, J.; *et al.* A novel CpG island set identifies tissue-specific methylation at developmental gene loci. *PLoS Biol.* **2008**, *6*, e22.
44. Rakyan, V.K.; Down, T.A.; Thorne, N.P.; Flicek, P.; Kulesha, E.; Graf, S.; Tomazou, E.M.; Backdahl, L.; Johnson, N.; Herberth, M.; *et al.* An integrated resource for genome-wide identification and analysis of human tissue-specific differentially methylated regions (tDMRs). *Genome Res.* **2008**, *18*, 1518–1529.
45. Weber, M.; Davies, J.J.; Wittig, D.; Oakeley, E.J.; Haase, M.; Lam, W.L.; Schubeler, D. Chromosome-wide and promoter-specific analyses identify sites of differential DNA methylation in normal and transformed human cells. *Nat. Genet.* **2005**, *37*, 853–862.
46. Jones, P.A.; Baylin, S.B. The epigenomics of cancer. *Cell* **2007**, *128*, 683–692.
47. Esteller, M.; Sanchez-Cespedes, M.; Rosell, R.; Sidransky, D.; Baylin, S.B.; Herman, J.G. Detection of aberrant promoter hypermethylation of tumor suppressor genes in serum DNA from non-small cell lung cancer patients. *Cancer Res.* **1999**, *59*, 67–70.
48. Lo, Y.M.; Wong, I.H.; Zhang, J.; Tein, M.S.; Ng, M.H.; Hjelm, N.M. Quantitative analysis of aberrant p16 methylation using real-time quantitative methylation-specific polymerase chain reaction. *Cancer Res.* **1999**, *59*, 3899–3903.
49. Frommer, M.; McDonald, L.E.; Millar, D.S.; Collis, C.M.; Watt, F.; Grigg, G.W.; Molloy, P.L.; Paul, C.L. A genomic sequencing protocol that yields a positive display of 5-methylcytosine residues in individual DNA strands. *Proc. Natl. Acad. Sci. USA* **1992**, *89*, 1827–1831.
50. Herman, J.G.; Graff, J.R.; Myohanen, S.; Nelkin, B.D.; Baylin, S.B. Methylation-specific PCR: A novel PCR assay for methylation status of CpG islands. *Proc. Natl. Acad. Sci. USA* **1996**, *93*, 9821–9826.

51. Clark, S.J.; Harrison, J.; Paul, C.L.; Frommer, M. High sensitivity mapping of methylated cytosines. *Nucleic Acids Res.* **1994**, *22*, 2990–2997.
52. Poon, L.L.; Leung, T.N.; Lau, T.K.; Chow, K.C.; Lo, Y.M. Differential DNA methylation between fetus and mother as a strategy for detecting fetal DNA in maternal plasma. *Clin. Chem.* **2002**, *48*, 35–41.
53. Gonzalgo, M.L.; Jones, P.A. Rapid quantitation of methylation differences at specific sites using methylation-sensitive single nucleotide primer extension (Ms-SNuPE). *Nucleic Acids Res.* **1997**, *25*, 2529–2531.
54. Chim, S.S.; Jin, S.; Lee, T.Y.; Lun, F.M.; Lee, W.S.; Chan, L.Y.; Jin, Y.; Yang, N.; Tong, Y.K.; Leung, T.Y.; *et al.* Systematic search for placental DNA-methylation markers on chromosome 21: Toward a maternal plasma-based epigenetic test for fetal trisomy 21. *Clin. Chem.* **2008**, *54*, 500–511.
55. Chan, K.C.; Ding, C.; Gerovassili, A.; Yeung, S.W.; Chiu, R.W.; Leung, T.N.; Lau, T.K.; Chim, S.S.; Chung, G.T.; Nicolaides, K.H.; *et al.* Hypermethylated RASSF1A in maternal plasma: A universal fetal DNA marker that improves the reliability of noninvasive prenatal diagnosis. *Clin. Chem.* **2006**, *52*, 2211–2218.
56. Xiong, Z.; Laird, P.W. COBRA: A sensitive and quantitative DNA methylation assay. *Nucleic Acids Res.* **1997**, *25*, 2532–2534.
57. Papageorgiou, E.A.; Fiegler, H.; Rakyan, V.; Beck, S.; Hulten, M.; Lamnissou, K.; Carter, N.P.; Patsalis, P.C. Sites of differential DNA methylation between placenta and peripheral blood: Molecular markers for noninvasive prenatal diagnosis of aneuploidies. *Am. J. Pathol.* **2009**, *174*, 1609–1618.
58. Chu, T.; Handley, D.; Bunce, K.; Surti, U.; Hogge, W.A.; Peters, D.G. Structural and regulatory characterization of the placental epigenome at its maternal interface. *PLoS One* **2011**, *6*, e14723.
59. Old, R.W.; Crea, F.; Puszyk, W.; Hulten, M.A. Candidate epigenetic biomarkers for non-invasive prenatal diagnosis of Down syndrome. *Reprod. Biomed. Online* **2007**, *15*, 227–235.
60. Chu, T.; Burke, B.; Bunce, K.; Surti, U.; Allen Hogge, W.; Peters, D.G. A microarray-based approach for the identification of epigenetic biomarkers for the noninvasive diagnosis of fetal disease. *Prenat. Diagn.* **2009**, *29*, 1020–1030.
61. Down, T.A.; Rakyan, V.K.; Turner, D.J.; Flicek, P.; Li, H.; Kulesha, E.; Graf, S.; Johnson, N.; Herrero, J.; Tomazou, E.M.; *et al.* A Bayesian deconvolution strategy for immunoprecipitation-based DNA methylome analysis. *Nat. Biotechnol.* **2008**, *26*, 779–785.
62. Ruike, Y.; Imanaka, Y.; Sato, F.; Shimizu, K.; Tsujimoto, G. Genome-wide analysis of aberrant methylation in human breast cancer cells using methyl-DNA immunoprecipitation combined with high-throughput sequencing. *BMC Genomics* **2010**, *11*, 137.
63. Feber, A.; Wilson, G.A.; Zhang, L.; Presneau, N.; Idowu, B.; Down, T.A.; Rakyan, V.K.; Noon, L.A.; Lloyd, A.C.; Stupka, E.; *et al.* Comparative methylome analysis of benign and malignant peripheral nerve sheath tumors. *Genome Res.* **2011**, *21*, 515–524.
64. Taiwo, O.; Wilson, G.A.; Morris, T.; Seisenberger, S.; Reik, W.; Pearce, D.; Beck, S.; Butcher, L.M. Methylome analysis using MeDIP-seq with low DNA concentrations. *Nat. Protoc.* **2012**, *7*, 617–636.

65. Borgel, J.; Guibert, S.; Weber, M. Methylated DNA immunoprecipitation (MeDIP) from low amounts of cells. *Methods Mol. Biol.* **2012**, *925*, 149–158.
66. Tsui, D.W.Y.; Lam, Y.M.D.; Lee, W.S.; Leung, T.Y.; Lau, T.K.; Lau, E.T.; Tang, M.H.Y.; Akolekar, R.; Nicolaides, K.H.; Chiu, R.W.K.; *et al.* Systematic Identification of Placental Epigenetic Signatures for the Noninvasive Prenatal Detection of Edwards Syndrome. *PLoS One* **2010**, *5*, e15069.
67. Tong, Y.K.; Ding, C.; Chiu, R.W.; Gerovassili, A.; Chim, S.S.; Leung, T.Y.; Leung, T.N.; Lau, T.K.; Nicolaides, K.H.; Lo, Y.M. Noninvasive prenatal detection of fetal trisomy 18 by epigenetic allelic ratio analysis in maternal plasma: Theoretical and empirical considerations. *Clin. Chem.* **2006**, *52*, 2194–2202.
68. Tong, Y.K.; Chiu, R.W.; Akolekar, R.; Leung, T.Y.; Lau, T.K.; Nicolaides, K.H.; Lo, Y.M. Epigenetic-genetic chromosome dosage approach for fetal trisomy 21 detection using an autosomal genetic reference marker. *PLoS One* **2010**, *5*, e15244.
69. Papageorgiou, E.A.; Karagrigoriou, A.; Tsaliki, E.; Velissariou, V.; Carter, N.P.; Patsalis, P.C. Fetal specific DNA methylation ratio permits non-invasive prenatal diagnosis of trisomy 21. *Nat. Med.* **2011**, *17*, 510–513.
70. Gorduza, E.V.; Popescu, R.; Caba, L.; Ivanov, I.; Martiniuc, V.; Nedelea, F.; Militaru, M.; Socolov, D.G. Prenatal diagnosis of 21 trisomy by quantification of methylated fetal DNA in maternal blood: Study on 10 pregnancies. *Rev. Rom. Med. Lab.* **2013**, *21*, 275–284.
71. Qin, H.; Bonifacio, M.; McArthur, S.; McLennan, A.; Boogert, T.; Bowman, M. Comment on "MeDIP real-time qPCR of maternal peripheral blood reliably identifies trisomy 21". *Prenat. Diagn.* **2013**, *33*, 403.
72. Tsaliki, E.; Papageorgiou, E.A.; Spyrou, C.; Koumbaris, G.; Kypri, E.; Kyriakou, S.; Sotiriou, C.; Touvana, E.; Keravnou, A.; Karagrigoriou, A.; *et al.* MeDIP real-time qPCR of maternal peripheral blood reliably identifies trisomy 21. *Prenat. Diagn.* **2012**, *32*, 996–1001.
73. Chiu, R.W.; Akolekar, R.; Zheng, Y.W.; Leung, T.Y.; Sun, H.; Chan, K.C.; Lun, F.M.; Go, A.T.; Lau, E.T.; To, W.W.; *et al.* Non-invasive prenatal assessment of trisomy 21 by multiplexed maternal plasma DNA sequencing: large scale validity study. *BMJ* **2011**, *342*, c7401.
74. Ehrich, M.; Deciu, C.; Zwiefelhofer, T.; Tynan, J.A.; Cagasan, L.; Tim, R.; Lu, V.; McCullough, R.; McCarthy, E.; Nygren, A.O.; *et al.* Noninvasive detection of fetal trisomy 21 by sequencing of DNA in maternal blood: a study in a clinical setting. *Am. J. Obstet. Gynecol.* **2011**, *204*, 205.e1–205.e11.
75. Palomaki, G.E.; Kloza, E.M.; Lambert-Messerlian, G.M.; Haddow, J.E.; Neveux, L.M.; Ehrich, M.; van den Boom, D.; Bombard, A.T.; Deciu, C.; Grody, W.W.; *et al.* DNA sequencing of maternal plasma to detect Down syndrome: An international clinical validation study. *Genet. Med.* **2011**, *13*, 913–920.
76. Chiu, R.W.; Chan, K.C.; Gao, Y.; Lau, V.Y.; Zheng, W.; Leung, T.Y.; Foo, C.H.; Xie, B.; Tsui, N.B.; Lun, F.M.; *et al.* Noninvasive prenatal diagnosis of fetal chromosomal aneuploidy by massively parallel genomic sequencing of DNA in maternal plasma. *Proc. Natl. Acad. Sci. USA* **2008**, *105*, 20458–20463.
77. Ladha, S. A new era of non-invasive prenatal genetic diagnosis: Exploiting fetal epigenetic differences. *Clin. Genet.* **2012**, *81*, 362–363.

78. Tong, Y.K.; Chiu, R.W.; Chan, K.C.; Leung, T.Y.; Lo, Y.M. Technical concerns about immunoprecipitation of methylated fetal DNA for noninvasive trisomy 21 diagnosis. *Nat. Med.* **2012**, *18*, 1327–1328; author reply 1328–1329.
79. Patsalis, P.C. Reply to: Technical concerns about immunoprecipitation of methylated fetal DNA for noninvasive trisomy 21 diagnosis. *Nat. Med.* **2012**, *18*, 1328–1329.
80. Fan, H.C.; Quake, S.R. Sensitivity of noninvasive prenatal detection of fetal aneuploidy from maternal plasma using shotgun sequencing is limited only by counting statistics. *PLoS One* **2010**, *5*, e10439.
81. Palomaki, G.E.; Deciu, C.; Kloza, E.M.; Lambert-Messerlian, G.M.; Haddow, J.E.; Neveux, L.M.; Ehrich, M.; van den Boom, D.; Bombard, A.T.; Grody, W.W.; *et al.* DNA sequencing of maternal plasma reliably identifies trisomy 18 and trisomy 13 as well as Down syndrome: an international collaborative study. *Genet. Med.* **2012**, *14*, 296–305.
82. Chen, E.Z.; Chiu, R.W.K.; Sun, H.; Akolekar, R.; Chan, K.C.A.; Leung, T.Y.; Jiang, P.; Zheng, Y.W.L.; Lun, F.M.F.; Chan, L.Y.S.; *et al.* Noninvasive Prenatal Diagnosis of Fetal Trisomy 18 and Trisomy 13 by Maternal Plasma DNA Sequencing. *PLoS One* **2011**, *6*, e21791.
83. Aria Diagnostics, Inc. Available online: http://www.ariadx.com/ (accessed on 5 December 2013).
84. SEQUENOM, Inc. Available online: http://www.sequenom.com/ (accessed on 5 December 2013).
85. Verinata Health, Inc. Available online: http://www.verinata.com/ (accessed on 5 December 2013).
86. Tsui, D.W.; Chiu, R.W.; Lo, Y.D. Epigenetic approaches for the detection of fetal DNA in maternal plasma. *Chimerism* **2010**, *1*, 30–35.
87. Korshunova, Y.; Maloney, R.K.; Lakey, N.; Citek, R.W.; Bacher, B.; Budiman, A.; Ordway, J.M.; McCombie, W.R.; Leon, J.; Jeddeloh, J.A.; *et al.* Massively parallel bisulphite pyrosequencing reveals the molecular complexity of breast cancer-associated cytosine-methylation patterns obtained from tissue and serum DNA. *Genome Res.* **2008**, *18*, 19–29.
88. Comprehensive genomic characterization defines human glioblastoma genes and core pathways. *Nature* **2008**, *455*, 1061–1068.
89. Grunau, C.; Clark, S.J.; Rosenthal, A. Bisulfite genomic sequencing: Systematic investigation of critical experimental parameters. *Nucleic Acids Res.* **2001**, *29*, E65.
90. Laird, P.W. Principles and challenges of genomewide DNA methylation analysis. *Nat. Rev. Genet.* **2010**, *11*, 191–203.
91. Fazzari, M.J.; Greally, J.M. Epigenomics: beyond CpG islands. *Nat. Rev. Genet.* **2004**, *5*, 446–455.
92. Nair, S.S.; Coolen, M.W.; Stirzaker, C.; Song, J.Z.; Statham, A.L.; Strbenac, D.; Robinson, M.D.; Clark, S.J. Comparison of methyl-DNA immunoprecipitation (MeDIP) and methyl-CpG binding domain (MBD) protein capture for genome-wide DNA methylation analysis reveal CpG sequence coverage bias. *Epigenetics* **2011**, *6*, 34–44.
93. NIPD Genetics Ltd. Available online: http://www.nipd.com/ (accessed on 5 December 2013).
94. Jin, S.; Lee, Y.K.; Lim, Y.C.; Zheng, Z.; Lin, X.M.; Ng, D.P.; Holbrook, J.D.; Law, H.Y.; Kwek, K.Y.; Yeo, G.S.; *et al.* Global DNA hypermethylation in down syndrome placenta. *PLoS Genet.* **2013**, *9*, e1003515.
95. Stewart, F.J.; Panne, D.; Bickle, T.A.; Raleigh, E.A. Methyl-specific DNA binding by McrBC, a modification-dependent restriction enzyme. *J. Mol. Biol.* **2000**, *298*, 611–622.

96. Khulan, B.; Thompson, R.F.; Ye, K.; Fazzari, M.J.; Suzuki, M.; Stasiek, E.; Figueroa, M.E.; Glass, J.L.; Chen, Q.; Montagna, C.; *et al*. Comparative isoschizomer profiling of cytosine methylation: The HELP assay. *Genome Res.* **2006**, *16*, 1046–1055.

6

Functional Gene-Set Analysis does not Support a Major Role for Synaptic Function in Attention Deficit/Hyperactivity Disorder (ADHD)

Anke R. Hammerschlag [1], Tinca J. C. Polderman [1], Christiaan de Leeuw [1,2], Henning Tiemeier [3], Tonya White [3], August B. Smit [4], Matthijs Verhage [5,6] and Danielle Posthuma [1,3,6,*]

[1] Department of Complex Trait Genetics, Center for Neurogenomics and Cognitive Research, Neuroscience Campus Amsterdam, VU University Amsterdam, De Boelelaan 1085, 1081 HV Amsterdam, The Netherlands; E-Mails: a.r.hammerschlag@vu.nl (A.R.H.); tinca.polderman@vu.nl (T.J.C.P.); c.deleeuw@science.ru.nl (C.L.)

[2] Institute for Computing and Information Sciences, Radboud University Nijmegen, P.O. Box 9010, 6500 GL Nijmegen, The Netherlands

[3] Department of Child and Adolescent Psychiatry, Erasmus University Medical Center and Sophia Children's Hospital, P.O. Box 2060, 3000 CB Rotterdam, The Netherlands; E-Mails: h.tiemeier@erasmusmc.nl (H.T.); t.white@erasmusmc.nl (T.W.)

[4] Department of Molecular and Cellular Neurobiology, Center for Neurogenomics and Cognitive Research, Neuroscience Campus Amsterdam, VU University Amsterdam, De Boelelaan 1085, 1081 HV Amsterdam, The Netherlands; E-Mail: guus.smit@vu.nl

[5] Department of Functional Genomics, Center for Neurogenomics and Cognitive Research, Neuroscience Campus Amsterdam, VU University Amsterdam, De Boelelaan 1085, 1081 HV Amsterdam, The Netherlands; E-Mail: matthijs.verhage@cncr.vu.nl

[6] Department of Clinical Genetics, VU University Medical Center, P.O. Box 7057, 1007 MB Amsterdam, The Netherlands

* Author to whom correspondence should be addressed; E-Mail: danielle.posthuma@vu.nl;

Abstract: Attention Deficit/Hyperactivity Disorder (ADHD) is one of the most common childhood-onset neuropsychiatric disorders. Despite high heritability estimates, genome-wide association studies (GWAS) have failed to find significant genetic associations, likely due to the polygenic character of ADHD. Nevertheless, genetic studies suggested the involvement of several processes important for synaptic function. Therefore, we applied a functional gene-set analysis to formally test whether synaptic functions are associated with ADHD. Gene-set analysis tests the joint effect of multiple genetic variants in groups of functionally related genes. This method provides increased statistical power compared to conventional GWAS. We used data from the Psychiatric Genomics Consortium including 896 ADHD cases and 2455 controls, and 2064 parent-affected offspring trios, providing sufficient statistical power to detect gene sets representing a genotype relative risk of at least 1.17. Although all synaptic genes together showed a significant association with ADHD, this association was not stronger than that of randomly generated gene sets matched for same number of genes. Further analyses showed no association of specific synaptic function categories with ADHD after correction for multiple testing. Given current sample size and gene sets based on current knowledge of genes related to synaptic function, our results do not support a major role for common genetic variants in synaptic genes in the etiology of ADHD.

Keywords: complex trait; polygenic; gene network; biological pathway; synapse; GWAS; PGC

---

1. Introduction

Attention Deficit/Hyperactivity Disorder (ADHD) is one of the most common childhood-onset neuropsychiatric disorders. The worldwide prevalence is estimated at ~5% [1], and remained relatively stable across the last three decades [2]. ADHD is characterized by a persistent pattern of inattention and/or impulsiveness and hyperactivity. Despite high heritability estimates for ADHD, averaging 70% [3], the identification of genes has been difficult. Most likely this is mainly due to the polygenic character of ADHD, similar to that of other complex traits, meaning that many genetic variants with small effects contribute to ADHD risk [4].

Genome-wide association studies (GWAS) of ADHD have yielded no significant single nucleotide polymorphism (SNP) associations thus far [5]. However, it has been reported that the top hits of GWAS point to the involvement of synaptic processes such as neurotransmission, cell-cell communication systems, potassium channel subunits and regulators, and more basic processes like neuronal migration, neurite outgrowth, spine formation, neuronal plasticity, cell division, and adhesion [6–8]. Furthermore, many genes previously implicated in ADHD [9] are expressed in the synapse (*i.e.*, *DBH*, *SLC6A2*, *ADRA2A*, *HTR1B*, *HTR2A*, *TPH1/2*, *MAOA*, *CHRNA4*, *SNAP25*, and *BDNF*), suggesting the involvement of synaptic function in the etiology of ADHD.

In addition to common genetic variants, rare variants may contribute to ADHD risk. Increased structural variation burden has been reported, particularly in subjects with intellectual disability [10–13]. Interestingly, biological pathways enriched for GWAS SNP associations with low *p*-values overlap with pathways enriched for rare structural variants, including pathways important for synaptic function [12]. Of special interest are SNPs and duplications spanning the *CHRNA7* gene, which is primarily involved in modulation of rapid synaptic transmission and which has been associated with other neuropsychiatric phenotypes in addition to ADHD [12,13]. Furthermore, strong associations have been reported for structural variation affecting metabotropic glutamate receptor genes and genes that interact with them. Several of these genes are important modulators of synaptic transmission and neurogenesis [11].

Given the polygenic nature of ADHD, it is likely that non-random combinations of genetic variants are involved in the etiology of ADHD. Genes do not work in isolation; rather, they form complex molecular networks and cellular pathways. Therefore, it is plausible that the numerous genetic variants of small effect aggregate in genes that share a similar cellular function. Evaluating the joint effect of multiple SNPs in functionally related genes increases the statistical power to detect associations with ADHD compared to single SNP methods, as it reduces multiple testing. Moreover, single SNP associations do not necessarily lead to knowledge about underlying biological mechanisms, while a set of genes with the same function could result in more insight in the molecular or cellular mechanisms of ADHD [14].

Prior studies that tested the joint effect of genetic variants generally grouped genes based on biological pathways. However, grouping genes based on cellular function ("horizontal grouping") instead of biological pathways ("vertical grouping") may be especially powerful in synaptic protein networks [15,16]. Many different pathways regulate synaptic function, but act not independent, as many proteins act across pathways. For example, different neuromodulator pathways (e.g., dopamine or serotonin) include receptors that are activated by the specific neuromodulators, but are functionally and often structurally similar to each other. It may well be that genetic variants influencing complex traits like ADHD concentrate at similar cellular function, by which they influence different pathways leading to similar consequences in synaptic function.

The majority of gene-set analyses that have been conducted have used publicly available gene sets. However, currently available public gene sets are generally incomplete and neither error-free nor unbiased, especially with regard to genes active in the brain [17,18]. Fortunately, expert-curated sets of genes are increasingly becoming available, such as the mir-137 gene set [19], specific synaptic gene sets [15], and gene sets for glial function [20].

As the results of previous GWAS and genes affected by structural variation suggested involvement of synaptic function, we hypothesized that synaptic processes play a role in the etiology of ADHD. Collective testing of genetic variants in genes grouped according to similar synaptic functions may be the most optimal way to test this. Therefore, we applied a functional gene-set analysis for ADHD using 18 previously published, expert-curated pre- and postsynaptic gene sets [15]. To our knowledge, this is the first study to conduct hypothesis-driven gene-set analysis for ADHD by grouping synaptic genes according to cellular function. We used ADHD data from the Psychiatric Genomics Consortium (PGC) [5].

2. Methods

2.1. Sample

We used GWAS summary statistics from the currently largest publicly available ADHD data set, as provided by the PGC [5]. Details on the data set have been described previously [5]. In short, the data set consisted of four projects: the Children's Hospital of Philadelphia (CHOP), phase I of the International Multicenter ADHD Genetics Project (IMAGE), phase II of IMAGE (IMAGE II), and the Pfizer-funded study from the University of California, Los Angeles, Washington University, and Massachusetts General

Hospital (PUWMa). The total sample consisted of 896 unrelated cases and 2455 controls, and 2064 trio samples (alleles transmitted to offspring were considered as "trio cases", and non-transmitted alleles as "pseudo-controls"). All samples were of European ancestry and met diagnostic criteria for ADHD as defined by the DSM-IV. All samples underwent the same quality control and analysis steps. The strongest single SNP association with ADHD in this data set was $p = 1.10 \times 10^{-6}$ [5].

2.2. Defining Functional Gene Sets

Generation of the synaptic gene sets has been described previously [15]. Briefly, synaptic gene grouping was based on cellular function as determined by previous synaptic protein identification experiments and data mining for synaptic genes and gene function. This resulted in the inclusion of 1028 genes, expressed in either the pre- or postsynapse or in both, divided over 17 synaptic gene sets with a specific synaptic function, and one synaptic gene set with unassigned cellular function. The gene sets with gene IDs are available at the Complex Trait Genetics webpage [21].

2.3. Power Analysis

The Genetic Power Calculator (GPC) [22,23] was used to define the minimal genotype relative risk that could reliably be detected for a gene set given the current sample size. Because the PGC data set consists of both case-control samples and trio samples, power was calculated using the weighted mean of the noncentrality parameters of the samples. To use the GPC for gene-set analysis, we assumed that the risk allele frequency represents the average allele frequency of all contributing risk variants in a gene set, and that the relative risk is representing the global effect of the gene set. We further used a disease prevalence of 5% (as estimated by Polanczyk *et al.* [1]), and a multiplicative model (power calculation based on the allelic test). Tests were corrected for the number of gene sets ($\alpha = 0.05/18 = 2.8 \times 10^{-3}$).

2.4. Gene-Set Analysis

Gene-set analysis was conducted using JAG [24]. To test the hypothesis that synaptic function was associated with ADHD, we conducted self-contained tests for each gene set and one overall test including all synaptic gene sets. For each gene set, the test statistic was defined as the sum over the $-\log_{10}$ of SNP p-values annotated to genes in that gene set. These SNP p-values were taken from the PGC association results. To allow for unbiased interpretation of the test statistic, 10,000 permutations were conducted in which any relation between a genetic variant and affection status was disconnected. As such, linkage disequilibrium (LD), and number of SNPs and genes within each gene set stayed intact. For each permutation of the data set, the test statistics of the gene sets were computed. The self-contained p-value was calculated as the proportion of test statistics in the permuted data sets that was higher than the original test statistic. Bonferroni correction was applied to account for multiple testing with a corrected significance threshold of $\alpha = 0.05/18 = 2.8 \times 10^{-3}$.

For the permutations of the data set, we used the genotype data of the European ancestry samples from the 1000 Genomes project [25] with a simulated binary phenotype (as we had no access to raw data of the PGC). Using this as reference data, we could appropriately account for LD effects on correlations in SNP p-values in the PGC association data. For the test statistics of the original gene sets, only SNPs that were also available in the 1000 Genomes genotype data were used.

Competitive tests were performed for gene sets found to be significant in the self-contained test. While self-contained tests evaluate whether a gene set is associated with ADHD under the null hypothesis of no association, a competitive test shows whether the observed (self-contained) association is stronger than expected by chance for gene sets with the same number of genes. To this end, 150 random gene sets were generated, matching for the same number of genes. JAG calculated a self-contained p-value for each of these random gene sets. The competitive p-value was then computed as the proportion of random gene sets with self-contained p-values lower than the self-contained p-value for the gene set itself. Only gene sets with a competitive p-value < 0.05 were considered to be significant.

3. Results

3.1. Power Analysis

Power analyses showed that for gene sets containing on average SNPs with a risk allele frequency (RAF) of at least 0.1, our sample had sufficient power (≥0.80) to detect gene sets with a genotype relative risk (GRR) of 1.23 (Figure 1). For gene sets containing a mean RAF of at least 0.2, we had sufficient power to detect gene sets with a GRR of 1.17.

Figure 1. Statistical power to detect gene sets in the Psychiatric Genomics Consortium (PGC) Attention Deficit/Hyperactivity Disorder (ADHD) sample. Power is displayed for different genotype relative risks (GRR), and risk allele frequencies (RAF) of 0.1 and 0.2. The weighted mean of the noncentrality parameters of the case-control sample (896 cases and 2455 controls) and trio sample (2064 trios) was used to calculate power. Power analyses assume a disease prevalence of 5% and a multiplicative model. We assumed that gene sets behave as individual single nucleotide polymorphisms (SNPs). Tests are corrected for number of gene sets ($\alpha = 2.8 \times 10^{-3}$). Dotted horizontal line represents power of 0.80.

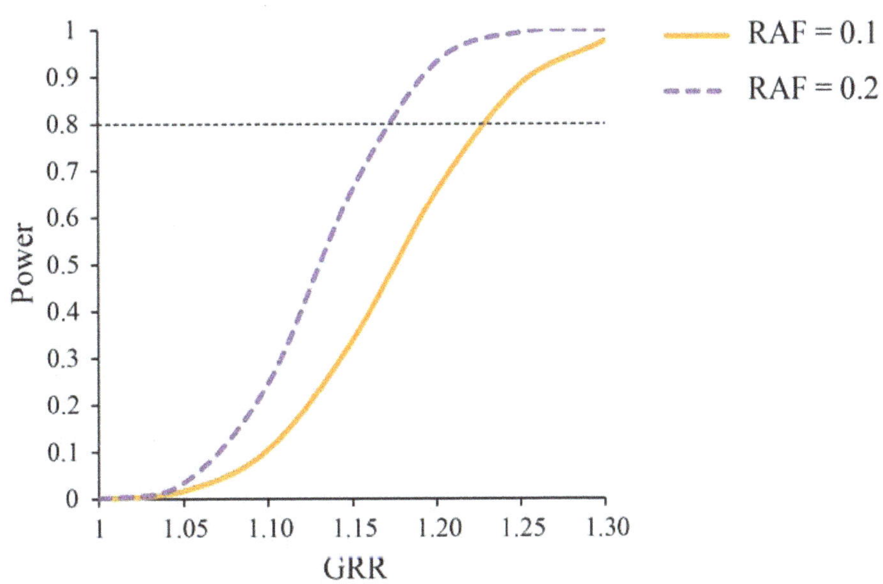

3.2. Gene-Set Analysis

A total number of 1,206,461 SNPs were available for gene-set analysis. Of these, 61,413 SNPs mapped to 956 genes (out of 1028) within our gene sets. All 956 synaptic genes together were significantly associated with ADHD in the self-contained test (Table 1). However, the competitive test showed that the synaptic genes were not more strongly associated with ADHD than randomly generated gene sets matched for same number of genes, suggesting that the self-contained *p*-value was significant merely due to a large number of SNPs being evaluated, which did not particularly aggregate in genes involved in synaptic function.

Table 1. Association findings between synaptic gene sets and ADHD.

Gene Set	Number of Genes in Original Set	Number of Genes Present in GWAS Data	Number of SNPs Present in GWAS Data	Self-Contained *p*-Value ($\alpha = 2.8 \times 10^{-3}$)	Competitive *p*-Value ($\alpha = 0.05$)
All synaptic genes	1028	956	61413	0.0393 *	0.1733
Ion balance/transport	43	40	1454	0.0118	NA
Cell metabolism	57	51	1059	0.0429	NA
Endocytosis	26	26	1075	0.0554	NA
Cell adhesion and trans-synaptic signaling	81	76	13550	0.0709	NA
Exocytosis	87	83	4855	0.0962	NA
Protein cluster	47	42	4182	0.1491	NA
Peptide/neurotrophin signals	28	25	1742	0.1659	NA
Structural plasticity	98	90	4655	0.1764	NA
Tyrosine kinase signaling	7	7	1281	0.2030	NA
Neurotransmitter metabolism	29	27	1059	0.2959	NA
RNA and protein synthesis, folding and breakdown	71	64	1152	0.4994	NA
Ligand-gated ion channel signaling	36	32	2935	0.6500	NA
G-protein-coupled receptor signaling	41	40	3129	0.6578	NA
Unassigned	61	53	2258	0.6644	NA
Intracellular signal transduction	150	145	9563	0.7001	NA
G-protein relay	27	25	946	0.7047	NA
Intracellular trafficking	80	75	2024	0.7334	NA
Excitability	59	56	4508	0.7914	NA

* $\alpha = 0.05$.

Self-contained tests for the specific synaptic gene sets showed associations at nominal significance levels for the involvement of *ion balance/transport* and *cell metabolism* in ADHD (Table 1). However, these associations did not survive Bonferroni correction. All other self-contained *p*-values were >0.05. We thus conclude that no significant associations were found between any of the specific synaptic gene sets and ADHD. Consequently, no subsequent competitive tests were performed for the synaptic gene sets of specific functions.

4. Discussion

Results from previous GWAS have led to the conclusion that ADHD is a heritable, yet polygenic disorder influenced by many genetic variants of small effect. Top hits from previous studies have suggested a role for synaptic processes in the etiology of ADHD. In the current study, we tested the hypothesis that genetic variants that influence the risk for ADHD cluster in synaptic gene sets. We used expert-curated gene sets of pre- and postsynaptic genes. Using the largest public ADHD GWAS sample currently available, our study had sufficient statistical power to detect gene sets representing a GRR of at least 1.17 (or 1.23 for less common alleles) for the liability to develop ADHD. The self-contained test of all synaptic genes together showed a significant association with ADHD. However, for complex traits that are polygenic, any large group of genes is likely to be associated due to background polygenic effects. The competitive test showed that the association was not stronger compared to that of randomly generated gene sets with the same number of genes. This suggests that the association was not a result of the selection of synaptic genes, but merely because of the large number of genes. Hence, our results support the idea that ADHD is a polygenic disorder, and suggest that overall synaptic function does not play a major role in the etiology of ADHD, given current synaptic genes.

In addition, no specific synaptic function categories were associated with ADHD after correction for multiple testing. These results suggest that if common genetic variants in the current synaptic gene sets with a specific function play a role in the etiology of ADHD, their effect is modest at most, even when considering the joint effect of multiple genetic variants.

Although previous analyses suggested involvement of several synaptic processes in ADHD [6,7,11–13], it should be kept in mind that the majority of previous results reported non-significant, suggestive results, and hence no strong conclusions could be drawn regarding the impact of those processes on ADHD. For example, a recent study used a different type of categorization of gene sets: they constructed gene sets based on pathways and candidate genes, and did report significant associations of dopamine/norepinephrine and serotonin pathways, and genes involved in neuritic outgrowth, with the hyperactive/impulsive component of ADHD [26]. However, in this study competitive tests to investigate if reported associations were stronger than can be expected by the polygenic nature of ADHD were not performed. Consequently, it remains unclear whether the reported associations are due to the background polygenic effects like our apparent association of synaptic genes with ADHD.

Synaptic function has been implicated and confirmed for other psychiatric disorders, especially schizophrenia [19,24] and bipolar disorder [27,28]. For example, gene sets of *cell adhesion and trans-synaptic signaling* and *excitability* showed replicated associations with schizophrenia [19,24]. Recent cross-disorder analyses by the PGC reported overlap in genetic liability between psychiatric disorders (schizophrenia, bipolar disorder, major depressive disorder, autism spectrum disorder, and ADHD) [29,30]. However, of all five psychiatric disorders, ADHD showed the weakest genetic overlap with other psychiatric disorders, having only a moderate genetic correlation with major depressive disorder, and showing no overlap with schizophrenia, bipolar disorder, and autism spectrum disorder. Our current findings fit into this overall picture of a separate genetic etiology of ADHD, by showing no evidence for an association with common variants in the current curated list of synaptic genes.

The list of genes involved in synaptic function is however a dynamic list: it depends on available experimental data and expert curation. When more experimental data is generated more genes may be included, which may have been missed in the current analyses. However, if genetic variants with an effect on ADHD risk aggregate in genes that are active in the synapse, it is expected that many genes within this gene set play a role in ADHD. Thus, an indication of association should be present if any of our current gene sets has a strong effect on ADHD risk, even when the current gene sets are not complete. Our results do not show any clear trends of association between the gene sets and ADHD.

An alternative explanation for the lack of association in our study could be the heterogeneous nature of ADHD. It is known that ADHD is characterized by a heterogeneous manifestation of symptoms, possibly reflecting genetic heterogeneity [31]. Genetic heterogeneity makes it more challenging to detect genetic variation that plays a role in the etiology of ADHD, as the heterogeneity results in an apparent reduction of the effect sizes of true genetic variants. The current lack of association of synaptic functions with ADHD diagnosis together with previous reports that implicate a role of synaptic function based on smaller scaled samples, may reflect the involvement of synaptic function in only very specific sub-populations of ADHD symptoms. Future studies focusing on ADHD symptom profiles are needed to detect such specific associations between synaptic function and ADHD subtypes.

5. Conclusions

We find no evidence for involvement of specific synaptic functions in the etiology of ADHD, given current sample size and gene sets based on current knowledge of genes related to synaptic function. Our results suggest that if common genetic variants in the current synaptic gene sets play a role in the etiology of ADHD, their effect is modest at most, even when considering the joint effect of multiple genetic variants.

Acknowledgments

This work is supported by the Netherlands Organization for Scientific Research (NWO Brain & Cognition 433-09-228, and NWO Complexity Project 645-000-003) and the European Union Seventh Framework Program ("SynSys" project) (HEALTHF2-2009-242167). The data set used for the analysis described in this paper was obtained from the Psychiatric Genomics Consortium (PGC) webpage [32]. Statistical analyses were carried out on the Genetic Cluster Computer [33] hosted by SURFsara and financially supported by the Netherlands Organization for Scientific Research (NWO 480-05-003 PI: Posthuma) along with a supplement from the Dutch Brain Foundation and the VU University Amsterdam. Scripts to modify JAG to be able to work with SNP *p*-values instead of raw data as input can be obtained through the first (Anke R. Hammerschlag) or third author (Christiaan de Leeuw) of the current paper.

Author Contributions

Danielle Posthuma conceived the project. Anke R. Hammerschlag performed the analyses and drafted the first version of the manuscript. Anke R. Hammerschlag, Tinca J. C. Polderman and Danielle Posthuma wrote the final version. Matthijs Verhage and August B. Smit constructed the synaptic gene sets. Christiaan de Leeuw wrote scripts to modify JAG to be able to work with SNP *p*-values instead of raw data as input. All authors provided a critical review of the manuscript and approved the final version.

References

1. Polanczyk, G.; de Lima, M.S.; Horta, B.L.; Biederman, J.; Rohde, L.A. The worldwide prevalence of ADHD: A systematic review and metaregression analysis. *Am. J. Psychiatry* **2007**, *164*, 942–948.
2. Polanczyk, G.V.; Willcutt, E.G.; Salum, G.A.; Kieling, C.; Rohde, L.A. ADHD prevalence estimates across three decades: An updated systematic review and meta-regression analysis. *Int. J. Epidemiol.* **2014**, *43*, 434–442.
3. Posthuma, D.; Polderman, T.J.C. What have we learned from recent twin studies about the etiology of neurodevelopmental disorders? *Curr. Opin. Neurol.* **2013**, *26*, 111–121.
4. Visscher, P.M.; Brown, M.A.; McCarthy, M.I.; Yang, J. Five years of GWAS discovery. *Am. J. Hum. Genet.* **2012**, *90*, 7–24.
5. Neale, B.M.; Medland, S.E.; Ripke, S.; Asherson, P.; Franke, B.; Lesch, K.P.; Faraone, S.V.; Nguyen, T.T.; Schäfer, H.; Holmans, P.; *et al.* Meta-analysis of genome-wide association studies of attention-deficit/hyperactivity disorder. *J. Am. Acad. Child Adolesc. Psychiatry* **2010**, *49*, 884–897.
6. Lesch, K.P.; Timmesfeld, N.; Renner, T.J.; Halperin, R.; Röser, C.; Nguyen, T.T.; Craig, D.W.; Romanos, J.; Heine, M.; Meyer, J.; *et al.* Molecular genetics of adult ADHD: Converging evidence from genome-wide association and extended pedigree linkage studies. *J. Neural Transm.* **2008**, *115*, 1573–1585.
7. Franke, B.; Neale, B.M.; Faraone, S.V. Genome-wide association studies in ADHD. *Hum. Genet.* **2009**, *126*, 13–50.
8. Poelmans, G.; Pauls, D.L.; Buitelaar, J.K.; Franke, B. Integrated genome-wide association study findings: Identification of a neurodevelopmental network for attention deficit hyperactivity disorder. *Am. J. Psychiatry* **2011**, *168*, 365–377.
9. Gizer, I.R.; Ficks, C.; Waldman, I.D. Candidate gene studies of ADHD: A meta-analytic review. *Hum. Genet.* **2009**, *126*, 51–90.
10. Williams, N.M.; Zaharieva, I.; Martin, A.; Langley, K.; Mantripragada, K.; Fossdal, R.; Stefansson, H.; Stefansson, K.; Magnusson, P.; Gudmundsson, O.O.; *et al.* Rare chromosomal deletions and duplications in attention-deficit hyperactivity disorder: A genome-wide analysis. *Lancet* **2010**, *376*, 1401–1408.
11. Elia, J.; Glessner, J.T.; Wang, K.; Takahashi, N.; Shtir, C.J.; Hadley, D.; Sleiman, P.M.; Zhang, H.; Kim, C.E.; Robison, R.; *et al.* Genome-wide copy number variation study associates metabotropic glutamate receptor gene networks with attention deficit hyperactivity disorder. *Nat. Genet.* **2012**, *44*, 78–84.
12. Stergiakouli, E.; Hamshere, M.; Holmans, P.; Langley, K.; Zaharieva, I.; Hawi, Z.; Kent, L.; Gill, M.; Williams, N.; Owen, M.J.; *et al.* Investigating the contribution of common genetic variants to the risk and pathogenesis of ADHD. *Am. J. Psychiatry* **2012**, *169*, 186–194.

13. Williams, N.M.; Franke, B.; Mick, E.; Anney, R.J.L.; Freitag, C.M.; Gill, M.; Thapar, A.; O'Donovan, M.C.; Owen, M.J.; Holmans, P.; *et al.* Genome-wide analysis of copy number variants in attention deficit hyperactivity disorder: The role of rare variants and duplications at 15q13.3. *Am. J. Psychiatry* **2012**, *169*, 195–204.
14. Schadt, E.E. Molecular networks as sensors and drivers of common human diseases. *Nature* **2009**, *461*, 218–223.
15. Ruano, D.; Abecasis, G.R.; Glaser, B.; Lips, E.S.; Cornelisse, L.N.; de Jong, A.P.H.; Evans, D.M.; Davey Smith, G.; Timpson, N.J.; Smit, A.B.; *et al.* Functional gene group analysis reveals a role of synaptic heterotrimeric G proteins in cognitive ability. *Am. J. Hum. Genet.* **2010**, *86*, 113–125.
16. De Jong, A.P.H.; Verhage, M. Presynaptic signal transduction pathways that modulate synaptic transmission. *Curr. Opin. Neurobiol.* **2009**, *19*, 245–253.
17. Feldman, I.; Rzhetsky, A.; Vitkup, D. Network properties of genes harboring inherited disease mutations. *Proc. Natl. Acad. Sci. USA* **2008**, *105*, 4323–4328.
18. Rossin, E.J.; Lage, K.; Raychaudhuri, S.; Xavier, R.J.; Tatar, D.; Benita, Y.; International Inflammatory Bowel Disease Genetics Constortium; Cotsapas, C.; Daly, M.J. Proteins encoded in genomic regions associated with immune-mediated disease physically interact and suggest underlying biology. *PLoS Genet.* **2011**, *7*, e1001273.
19. Ripke, S.; O'Dushlaine, C.; Chambert, K.; Moran, J.L.; Kähler, A.K.; Akterin, S.; Bergen, S.E.; Collins, A.L.; Crowley, J.J.; Fromer, M.; *et al.* Genome-wide association analysis identifies 13 new risk loci for schizophrenia. *Nat. Genet.* **2013**, *45*, 1150–1159.
20. Goudriaan, A.; de Leeuw, C.; Ripke, S.; Hultman, C.M.; Sklar, P.; Sullivan, P.F.; Smit, A.B.; Posthuma, D.; Verheijen, M.H.G. Specific glial functions contribute to schizophrenia susceptibility. *Schizophr. Bull.* **2014**, *40*, 925–935.
21. Complex Trait Genetics. Available online: http://ctglab.nl/software/ (accessed on 4 April 2014).
22. Purcell, S.; Cherny, S.S.; Sham, P.C. Genetic power calculator: Design of linkage and association genetic mapping studies of complex traits. *Bioinformatics* **2003**, *19*, 149–150.
23. Purcell, S.; Sham, P.C. Genetic Power Calculator. Available online: http://pngu.mgh.harvard.edu/~purcell/gpc/ (accessed on 3 March 2014).
24. Lips, E.S.; Cornelisse, L.N.; Toonen, R.F.; Min, J.L.; Hultman, C.M.; International Scizophrenia Consortium; Holmans, P.A.; O'Donovan, M.C.; Purcell, S.M.; Smit, A.B.; *et al.* Functional gene group analysis identifies synaptic gene groups as risk factor for schizophrenia. *Mol. Psychiatry* **2012**, *17*, 996–1006.
25. 1000 Genomes Project Consortium; Abecasis, G.R.; Auton, A.; Brooks, L.D.; DePristo, M.A.; Durbin, R.M.; Handsaker, R.E.; Kang, H.M.; Marth, G.T.; McVean, G.A. An integrated map of genetic variation from 1092 human genomes. *Nature* **2012**, *491*, 56–65.
26. Bralten, J.; Franke, B.; Waldman, I.; Rommelse, N.; Hartman, C.; Asherson, P.; Banaschewski, T.; Ebstein, R.P.; Gill, M.; Miranda, A.; *et al.* Candidate genetic pathways for attention-deficit/hyperactivity disorder (ADHD) show association to hyperactive/impulsive symptoms in children with ADHD. *J. Am. Acad. Child Adolesc. Psychiatry* **2013**, *52*, 1204.e1–1212.e1.
27. Ferreira, M.A.R.; O'Donovan, M.C.; Meng, Y.A.; Jones, I.R.; Ruderfer, D.M.; Jones, L.; Fan, J.;

Kirov, G.; Perlis, R.H.; Green, E.K.; *et al.* Collaborative genome-wide association analysis supports a role for ANK3 and CACNA1C in bipolar disorder. *Nat. Genet.* **2008**, *40*, 1056–1058.

28. Psychiatric GWAS Consortium Bipolar Disorder Working Group. Large-scale genome-wide association analysis of bipolar disorder identifies a new susceptibility locus near ODZ4. *Nat. Genet.* **2011**, *43*, 977–983.
29. Cross-Disorder Group of the Psychiatric Genomics Consortium. Identification of risk loci with shared effects on five major psychiatric disorders: A genome-wide analysis. *Lancet* **2013**, *381*, 1371–1379.
30. Cross-Disorder Group of the Psychiatric Genomics Consortium; Lee, S.H.; Ripke, S.; Neale, B.M.; Faraone, S.V.; Purcell, S.M.; Perlis, R.H.; Mowry, B.J.; Thapar, A.; Goddard, M.E.; *et al.* Genetic relationship between five psychiatric disorders estimated from genome-wide SNPs. *Nat. Genet.* **2013**, *45*, 984–994.
31. Faraone, S.V. Genetics of childhood disorders: XX. ADHD, Part 4: is ADHD genetically heterogeneous? *J. Am. Acad. Child Adolesc. Psychiatry* **2000**, *39*, 1455–1457.
32. Psychiatric Genomics Consortium. Available online: https://pgc.unc.edu/Sharing.php#SharingOpp/ (accessed on 23 October 2013).
33. Genetic Cluster Computer. Available online: http://www.geneticcluster.org/ (accessed on 4 April 2014).

7

Mouse ENU Mutagenesis to Understand Immunity to Infection: Methods, Selected Examples and Perspectives

Grégory Caignard [1,2], Megan M. Eva [1,2,3,†], Rebekah van Bruggen [2,4,†], Robert Eveleigh [5,†], Guillaume Bourque [5], Danielle Malo [1,2,3], Philippe Gros [2,4] and Silvia M. Vidal [1,2,6,*]

1 Department of Human Genetics, McGill University, Montréal, QC H3G 0B1, Canada; E-Mail: gcaignard@gmail.com

2 Complex Traits Group, McGill University, Montréal, QC H3G 0B1, Canada

3 Department of Medicine, McGill University, Montréal, QC, H3G 0B1, Canada; E-Mails: megan.eva@mail.mcgill.ca (M.M.E.); danielle.malo@mcgill.ca (D.M.)

4 Department of Biochemistry, McGill University, Montréal, QC, H3G 0B1, Canada; E-Mails: rebekah.vanbruggen@mail.mcgill.ca (R.B.); philippe.gros@mcgill.ca (P.G.)

5 McGill University and Genome Quebec Innovation Center, Montréal, QC, H3A 1A4, Canada; E-Mails: eveleigh.rjm@gmail.com, (R.E.); guil.bourque@mcgill.ca (G.B.)

6 McGill Life Sciences Complex, Bellini Building, 3649 Sir William Osler Promenade, Room 367, Montreal, QC, H3G 0B1, Canada

† These authors contributed equally to this work.

* Author to whom correspondence should be addressed; E-Mail: silvia.vidal@mcgill.ca;

Abstract: Infectious diseases are responsible for over 25% of deaths globally, but many more individuals are exposed to deadly pathogens. The outcome of infection results from a set of diverse factors including pathogen virulence factors, the environment, and the genetic make-up of the host. The completion of the human reference genome sequence in 2004 along with technological advances have tremendously accelerated and renovated the tools to study the genetic etiology of infectious diseases in humans and its best characterized mammalian model, the mouse. Advancements in mouse genomic resources have accelerated genome-wide functional approaches, such as gene-driven and phenotype-driven mutagenesis, bringing to the fore the use of mouse models that reproduce accurately many aspects of the pathogenesis of human infectious diseases. Treatment with the mutagen *N*-ethyl-*N*-nitrosourea (ENU) has become the most popular phenotype-driven approach. Our team and others have employed mouse ENU mutagenesis to identify host genes that directly impact susceptibility to pathogens of global significance. In this review, we first describe the strategies and tools used in mouse genetics to understand immunity to infection with special emphasis on chemical mutagenesis of the mouse germ-line together with current strategies to efficiently identify functional mutations using next generation sequencing. Then, we highlight illustrative examples of genes, proteins, and cellular signatures that have been revealed by ENU screens and have been shown to be involved in susceptibility or resistance to infectious diseases caused by parasites, bacteria, and viruses.

Keywords: infectious diseases; ENU; immunity; mouse genetic models

---

1. Introduction

The Neolithic Era, which began around 10,000 years B.C., constituted a turning point in human civilization. Its importance stems not only from the establishment of the first human settlements, but also from the development of farming activities involving the domestication of wild plants and animals. These changes in societal organization brought humans into close contact with animals and soil, exposing them to potential new pathogens, and with each other, allowing the spread of any new infection. It therefore comes as no surprise that the Neolithic Era saw the emergence of several human infectious diseases [1]. Indeed, given this close proximity, trans-species infections became more likely and ultimately resulted in the appearance of diseases such as measles and smallpox [2]. As a result, from the Neolithic Era until the Industrial Revolution, human life expectancy did not exceed 25 years of age [3]. Fortunately, life expectancy has been steadily increasing over the last 150 years for two main reasons. First, public hygiene measures implemented in the mid-19th century reduced the transmission of infection. Additionally, the advent of vaccination and antimicrobial drugs in the late 19th and early 20th century meant that many deadly infections were now curable or preventable. On a larger scale, diseases such as polio and measles were drastically reduced, while the dreaded smallpox was completely eradicated.

Nevertheless, infectious diseases remain directly responsible for close to 25% of all deaths globally and constitute a perpetual burden for humankind [4]. Numerous circumstances favor the emergence or reemergence of pathogens, or their spread to new ecological niches; these include pathogen virulence factors, as well as changing environmental conditions and host factors (e.g., aging populations, a heavier chronic disease burden, and therapeutic suppression of host defenses) (Figure 1). Of course, the eradication of most infectious diseases is highly unlikely. Instead, we are often involved in an unremitting struggle to control infection, for which a constant influx of novel countermeasure strategies is needed.

Figure 1. Factors involved in susceptibility to infectious diseases.

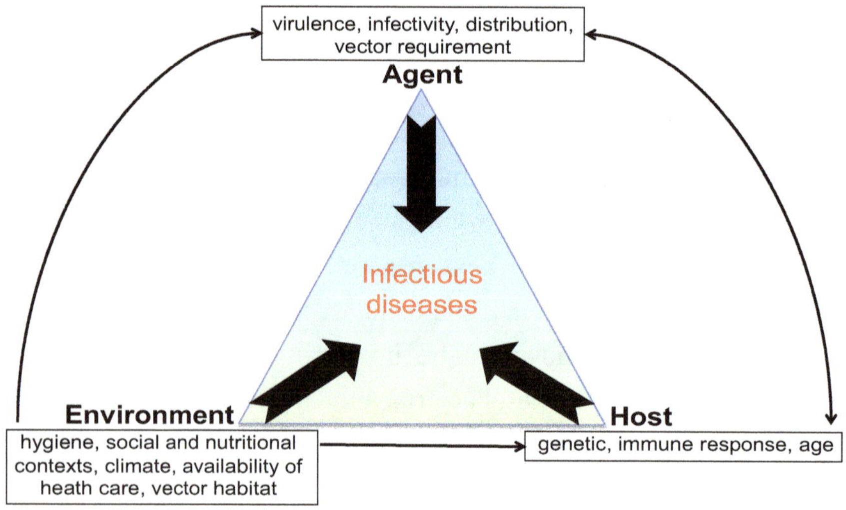

The development of these novel countermeasure strategies largely relies on a better understanding of the molecular mechanisms of disease pathogenesis. This requires not only basic research on the pathogen side but also on its interaction with the host. A possible approach is to exploit the observed variability in the outcome of infection, since at any given time, even during epidemics, clinical disease only develops in a subset of exposed persons. A large body of evidence indicates that the human genome is a major determinant of the variability in the onset, progression, and severity of infectious diseases [5–8]. In light of this evidence, research efforts aiming to better understand the pathogenesis of infectious diseases have shifted their focus from the pathogen to the host. Investigators are thus now attempting to identify host genes that are essential for successful pathogen infection, instead of focusing solely on pathogen genes. Candidate gene analysis studies have revealed a handful of single gene variants associated with increased susceptibility or resistance to specific infectious diseases (reviewed in [5]). Some remarkable examples identified in human populations include the malaria-protective effect of heterozygosity in the case of otherwise disease-causing hemoglobinopathies, such as sickle cell anemia and thalassemia [9], the protective effects of *CCR5* mutations against HIV [10], and resistance to norovirus infection conferred by loss-of-function alleles of the *FUT2* gene [11]. Further, the study of children with rare monogenic defects has revealed a considerable number of rare human genetic variations in innate immune pathways that underlie susceptibility to certain infectious diseases. For example, *IRAK* and *MYD88* deficiencies predispose to life-threatening infection by some bacterial species [12]. Another example is Mendelian Susceptibility to Mycobacterial Disease (MSMD), a primary immunodeficiency characterized by genetic defects in the IFNγ pathway, leading to susceptibility to *Mycobacterium bovis* (BCG) or other environmental mycobacteria species innocuous to the general population and to non-typhoidal, extra-intestinal salmonellosis (for review, see [5]). Thus, the fact that individuals exposed to life-threatening pathogens display differential susceptibility to infection and varying disease outcome not only reflects the genetic variability within the human population, but also the functional genetic diversity of the immune response itself.

The growing awareness of the importance of host genetic makeup in infectious disease outcome has motivated large-scale investigations of the human genome, made possible by recent technological advances. Namely, sequencing of the human genome [13], the International HapMap project [14], and microarray-based high-throughput genotyping technology have paved the way to Genome Wide Association Studies (GWAS) of major infectious diseases. In these GWAS, millions of single nucleotide polymorphisms (SNPs) can be tested for association with major infectious diseases, and this can be done simultaneously in thousands of individuals (for review, see [5]). Results emanating from these large datasets are certainly improving our understanding of infectious disease pathogenesis. However, full interpretation of the genes and pathways identified by GWAS studies is complicated by several factors including the modest effect size of most signals and the fact that even together these signals can explain only a fraction of the genetic predisposition to disease. Furthermore, the SNPs showing the strongest association are usually found near gene-coding regions rather than within obvious structural or regulatory regions making it difficult to pinpoint the gene directly involved in the disease phenotype. Such results are not entirely surprising given the inherent genetic heterogeneity of the human population, the variable exposure to the microbe during natural infection, the inherent variation in the microbe itself, and the difficulty associated with assembling the large cohorts required for GWAS. Yet, another key roadblock of GWAS studies is the lack of functional annotation for the majority of genes and encoded proteins, which is often limited to general ontology terms but lacks experimental validation for a possible role in an infectious disease phenotype.

2. Mice to the Rescue

An alternative and successful approach to identifying and characterizing the genetic component of the host response to infection in human studies has been the use of the mouse model. Owing to their striking physiological and genetic similarity with humans, mice have become a prime model for the study of human diseases. Numerous inbred strains exist that display natural resistance or susceptibility to a similar range of fungal, viral, parasitic, and bacterial pathogens, as well as the disease phenotypes associated with these infections [15–18]. These inbred strains represent homogeneous populations that serve to test different routes of inoculation, and various pathogen doses, all in a controlled environment, thus lessening many of the confounding effects encountered in human genetic studies. Due to its prominent role in biomedical research, the mouse was selected as the first non-human mammal to have its genome sequenced [19], revealing an astonishing genetic homology between the two species. The mouse and human genomes are approximately the same size, contain the same number of genes and show extensive conservation in gene order. Namely, 80% of human genes had 1:1 orthologous relationships with mouse genes, likely maintaining ancestral function in both species [20]. Mutations that cause diseases in humans often cause similar diseases in mice, including defects in the genes of the immune system [21]. Yet another advantage of the mouse is the string of unique technological advantages to manipulate the mouse genome.

Using the mouse model, two major genetic approaches have been employed to dissect the genetic architecture of the host defense against pathogens. The first is the so-called reverse genetic or gene-driven approach. In this approach, the sequence or expression of a gene of interest is altered, the effects of which are then investigated. Genetic modification of the mouse genome can be undertaken in various ways: (1) transgenesis or the introduction of gene DNA sequences into oocytes; (2) targeted mutation using embryonic stem cells (ES) which are modified to create knock-out alleles, whereby the function of the gene is abolished and equivalent to a null allele, or knock-in alleles resulting from the introduction of putative mutations in a given gene. In addition, recently developed genomic resources have further facilitated the use of genetically modified mice by the scientific community. These include large libraries of knock-out and conditional knock-out mice produced by international consortia aiming to target every gene in the mouse genome [22] and their accompanying large-scale phenotyping initiatives [23]; (3) targeted mutation in zygotes using the Clustered regularly interspaced short palindromic repeat (CRISPR)/CRISPR associated (Cas9) system [24]. With this approach it is possible to efficiently produce mice with mutations in both copies of multiple genes in a matter of weeks [25]. The phenotypes of these genetically modified mice can then be thoroughly scrutinized to determine the function of a gene in the context of the whole organism. These tools are dramatically improving our understanding of the genetic etiology of infectious diseases in both mice and humans. However, in many instances, these reverse genetics experiments can prove to be inconclusive. This is the case, for example, when the inactivation of a gene results in embryonic lethality or, conversely, when the resulting phenotypes are only slightly different from the wild-type or even undistinguishable because of gene redundancy. The reverse genetics approach also requires a preliminary hypothesis for gene function. Yet, as of 2014, less than 50% of about 34,000 known mouse genes (coding or not) have some form of functional annotation based on experimental evidence [26–28], which shows how our understanding of gene function still lags behind our knowledge of gene sequence.

The second approach is known as forward genetics, sometimes called phenotype-driven. The forward genetics approach begins with an inherited phenotype, with the aim of identifying the genomic regions and variant(s) underlying it. This involves the production of segregating crosses of inbred mouse strains or panels of specialized strains that display varying responses to infection, followed by linkage or association analyses. This approach is unbiased and requires no prior knowledge of gene function, allowing the discovery of unsuspected mechanisms. Numerous laboratory mouse resources are readily accessible for use in these studies: homozygous inbred strains, panels of selectively bred strains, consomic strains [29], recombinant congenic strains [30–32] or recombinant inbred strains from the collaborative cross [33]. A growing number of wild-derived inbred strains [34] or outbred crosses [35] can also be obtained, increasing the pool of genomic variation available for these studies. Whole genome sequencing has been performed on 18 of the most commonly used inbred mouse strains; the results are now public [36,37], facilitating the identification of candidate genes underlying a given phenotypic variation. Moreover, forward genetics studies in mice have already been shown to work; some elegant examples have allowed the identification of a number of genes and proteins that are essential for the early detection of and response to many invading pathogens (for review see [38]). In some cases, the human orthologues of these mouse genes (e.g., *NRAMP1*, *TLR4*, *IRF8*) have also been associated with predisposition to infection in humans, providing evidence of evolutionary conservation of immune defense mechanisms. However, there are limitations to this forward genetics strategy. Namely, a given genetic effect may be complex, making it difficult for investigators to determine the contribution of individual genes, as this requires subsequent breeding of congenic mice over several generations followed by positional cloning. Identifying the precise nature of a genetic lesion in a given candidate gene can also be complicated for other reasons, such as the presence of multigenic families or unrelated genes within the candidate interval bearing various coding polymorphisms, or predictive regulatory mutations or splicing variants rendering it difficult to identify the causative variant. Many of these drawbacks, however, can be overcome by the use of mutagens that introduce random mutations in the germ line. As presented later, in these models the causative mutation can be more easily identified by comparison with the parental non-mutagenized strain. This functional genomic strategy has successfully advanced our understanding of the intricate cellular and molecular cascades involved in immunodeficiency, autoimmunity, or behavioral disorders, which have already been well documented by others (see [39–42]). In the remainder of this review, we present the advantages and how-to of experiments using chemical mutagenesis of the mouse germ-line to dissect the genetic architecture of immunity to infection in mice. We also detail the procedures required to identify causal mutations underlying altered phenotypes using next generation sequencing. Finally, we highlight some of the most important findings from *in vivo* screens in the area of infectious disease research and discuss perspectives for mouse ENU approaches.

3. Chemical Mutagenesis and Generation of Mice Carrying Homozygous ENU-Induced Mutations

To better understand the link between genotypes and phenotypes, and ultimately gene function, mouse geneticists have elaborated upon several methods capable of introducing random mutations in the mouse germ-line, with the aim of expanding the phenotypic diversity in inbred mice and thus providing a wider range of research objects. These methods include the use of whole mouse radiation [43], infection of pre-implantation embryos with retroviruses [44], and injection with chemicals, such as procarbazine, methyl ethane sulfonate (MES), and *N*-ethyl-*N*-nitrosourea (ENU) [45]. ENU mutagenesis, however, has become the most popular technique to induce germ-line mutations due to its advantageous attributes: potency, preferential activity in spermatogonial stem cells, and a propensity to introduce point mutations.

As early as 1979, W. L. Russell demonstrated that a single dose of ENU was significantly more active than X-ray or procarbazine treatment, the most commonly used mouse mutagens at the time [46]. Later, studies showed that the mutation frequency could be increased if the ENU dose was fractionated and injected on a weekly schedule instead of being administered in one large dose, as this allowed a higher total dose to be tolerated [47]. In these conditions, the activity of ENU was 12 times that of X-rays and 36 times that of procarbazine, as well as being over 200 times the rate of spontaneous mutation [48]. The rate of ENU mutation appears variable for each gene, ranging from 1.5 to 10^{-3} per locus, which is equivalent to obtaining a mutation in a gene of choice at a rate of one in every 200–700 gametes screened. Additionally, it was noted that compared to X-ray-generated deletions, ENU rarely induced mutations in closely linked loci, suggesting that mutations introduced by ENU are subtler. Finally, compared to procarbazine, which is more active in transient post-meiotic cells, ENU preferentially affects spermatogonial stem cells, which are multiplied and replenished during the mouse lifetime, allowing the genetic lesions to be recovered indefinitely.

ENU is an alkylating agent that acts by preferential transfer of its ethyl group to O and N radicals in genomic DNA within mammalian cells [49,50]. Binding of the ethyl to the nucleoradicals creates DNA adducts that provoke mispairing, resulting mainly in base-pair substitutions if not restored by enzymatic DNA repair mechanisms during replication [51,52]. Systematic analysis of the type and frequency of ENU mutations was recently done using whole-exome and whole-genome sequencing [53–55]. Genome-wide, ENU has an average point mutation rate of 1.5 per Mb of genomic DNA [55], with a bias for AT to GC transitions (45%) compared to AT to TA transversions (28%).

The size of a given target gene and its AT density can therefore explain, at least in part, the variable sensitivity to the mutagenic effects of ENU. With a mouse genome size of about 2.7 Mb including 1.5% of protein coding sequence, one can expect about 1,900 new sequence variants per genome of which about 30 are coding.

With a few exceptions (microRNA and cis-elements) [56,57], to date most ENU-induced phenotypes, whose corresponding genotype has been identified, result from nucleotide changes that alter the coding sequencing. A current survey of the Mutagenix database (http://mutagenetix.utsouthwestern.edu/home.cfm) which contains the largest collection of ENU-induced phenotypic mutations (N = 185) [58], revealed that 61% were missense mutations, 19% nonsense alleles, 18% splicing defects, and 2% were frame-shift mutations. Therefore, while targeted mutations producing null alleles are necessary for genetic dissection of phenotypic traits, ENU-induced point mutations can also be used in parallel, revealing the multiple functions of a gene by altering individual protein domains and splicing products. Further, point mutations can produce various types of allelic series: (1) hypermorphic or hypomorphic alleles (increased or reduced activity of the gene product, respectively); (2) antimorphic alleles (the gene product is antagonistic to the wild-type allele); or (3) neomorphic alleles (new molecular function) [59] which can display a broad range of possible phenotypes.

The phenotypes that arise following ENU mutagenesis segregate with different inheritance patterns. Autosomal recessive (68%) is the most commonly observed, followed by dominant or co-dominant segregation (23%); X-linked recessive (4%) or X-linked dominant (1%) are rare, though 4% remain uncharacterized [58]. Once male mice have been treated with ENU, they are crossed to female mice. The resulting large cohorts of offspring are then tested to identify the phenotypically distinct mice most likely to bear a large-effect mutation; this is usually done with dominant or recessive screens. The above data illustrates how recessive screens, which require a three-generation breeding scheme (see below), constitute a more efficient and inclusive design than dominant screens, although the latter are logistically simpler and quicker to conduct since only the first generation offspring are analyzed. Using different breeding schemes, these recessive screens have successfully advanced our understanding of the intricate cellular and molecular cascades involved in immunodeficiency and autoimmunity, as well as in neurological or behavioral disorders, as already reviewed by others (see [39–42]).

Methods for mutagenizing male mice and breeding protocols to recover homozygous mutations have been described previously [60–62]. In our laboratory, we use a recessive screen involving genetically related mouse strains to generate the collections of mutant mice (Figure 2). By using genetically related inbred strains, the number of animals used and the timeline of the experiment can be reduced, as the mice that are screened also serve for mapping of the ENU-induced phenotypes. Moreover, using closely related strains alleviates any possible second-site modifier gene effects that could be present in the mapping strain. Briefly, we use well-validated protocols to induce single nucleotide mutations in 129S1/SvImJ (129S1) and C57BL/6 (B6) mice. This is done using a single intraperitoneal (i.p.) injection of 150 mg/kg of ENU (129S1) or three weekly i.p. injections of 90 mg/kg (B6) [63]. Following treatment, spermatogenesis ceases transiently and fertility is then regained after 11–13 weeks. In a general breeding strategy (Figure 2A), generation 0 (G0) males are then out-crossed with wild-type female mice to produce G1 offspring. These G1 hybrids carry one full set of mutagenized chromosomes and one full set of wild-type chromosomes. Individual G1 males are bred as founders of separate pedigrees, with the aim of bringing B6 or 129S1 sequence variants to homozygosity. To achieve it, G1 males are first crossed with genetically related wild-type females (129X1/SvJ (129X1) females for 129S1 males and C57BL/10 (B10) females for B6 males) to distinguish mutation-bearing chromosomes while preventing the introduction of additional genetic modifiers. The mutations present in the G1 founders are thus propagated in the G2 progeny. Since each G2 offspring should inherit only 50% of sequence variants present in the G1 males, two G2 daughters are backcrossed to their G1 father. This produces G3 progeny, where 12.5% of the G1 sequence variants should come to homozygosity in any given G3 offspring. On average, each G3 offspring is thus expected to be homozygous for about four to five loss-of-function sequence variants of the 30 present in the G1. Therefore, if there is a recessive Mendelian immune variant segregating within a pedigree, researchers can expect to identify 25% of individuals with the same trait or a cluster of two to four deviants by initially screening about 16 G3 offspring in that pedigree. The clustering of heritable variants within a pedigree filters out unavoidable false positives, which occur at a low rate (~5%) in screens for host susceptibility to infection; typically only one individual constitutes a false positive in a given pedigree. Variations of this breeding strategy have been used (Figure 2B) and will be described in the corresponding sections.

Figure 2. Breeding strategies used in our program to produce mice carrying homozygous *N*-ethyl-*N*-nitrosourea (ENU) mutations. (**A**) Treatment with ENU introduces mutations (indicated by a black or gray square) in the germ-line of males of generation 0 (G0). The mutagenized G0 male is out-crossed to a wild-type female to produce first generation (G1) animals. First generation G1 mice are carriers of ENU-induced mutations (indicated by half-filled black squares). G1 males are mated to wild-type females, to produce second generation (G2) animals, which carry about half of the mutation load present in the parental G1. Two G2 daughters are backcrossed to their G1 father to yield third-generation (G3) mice, where the original mutations have been brought to homozygosity (filled black squares). About 25% G3 progeny are expected to present a deviant phenotype in pedigrees that bear a given relevant recessive mutations; (**B**) In this strategy, the G1 progeny of two independent G0 males are intercrossed to produce G2 animals, which in turn are intercrossed to produce G3 mice.

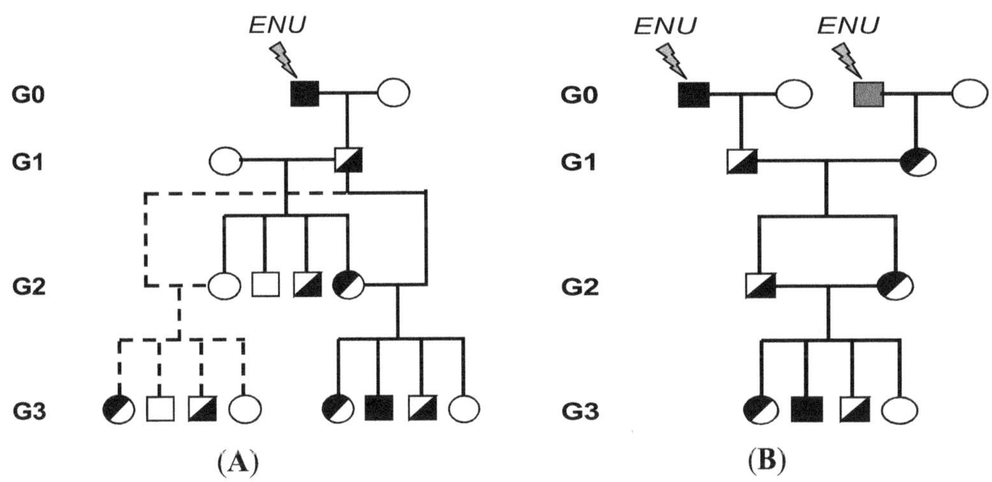

This pedigree structure allows early mapping of heritable variants. At this point, breeding and screening of additional G3 siblings confirm the inheritance of Mendelian recessive infectious traits in one quarter of the offspring. If eight to ten G3 animals displaying a new recessive immune trait are obtained out of 40–50 G3 mice in the pedigree, a genome-wide scan can be performed to establish linkage of the variant to a large initial segment. Before the advent of next generation sequencing (NGS), a time consuming and labor intensive positional cloning approach had to be undertaken to identify candidate genes bearing new genetic variants. The use of NGS techniques has dramatically increased the pace of mutation identification.

4. Gene Identification

The materials and methods underlying phenotype-driven or forward genetics approaches have become considerably more powerful over the years. Traditionally, these approaches required laborious genetic and fine mapping procedures in order to refine regions of interest to large megabase (Mb) chromosomal loci for subsequent PCR amplification and direct sequencing. Nonetheless, they were the methods of choice for the discovery of novel genes and/or novel gene functions in both humans and mice. The introduction of NGS has revolutionized forward genetics approaches, as it allowed the elaboration of robust methods of systematic mutation discovery, thus further closing the gap between phenotype and genotype. However, the sequencing and analysis of whole mammalian genomes remain a substantial bottleneck for many laboratories, both financially and computationally. Instead, inexpensive alternatives have been favored in order to sequence mouse mutations, namely targeting approaches using minimal mapping data. Moreover, targeted sequencing of coding regions of the genomes, or exomes, are particularly relevant for large mutational collections and have become the standard in cases where high-throughput gene mutation discovery methods are needed [64–66]. We describe below some of the standard techniques for sequencing and analysis of *de novo* mutations generated within ENU mouse models in a rapid and unbiased fashion.

Currently, the most widely used commercial mouse exome capture panels (Agilent and NimbleGen) target approximately 37 Mb of the sequences contained within the consensus coding sequence (CCDS) database of the genome, as well as other genomic features (e.g., microRNAs) [53,67] (see Table 1). The protocols contained in each of these kits are very similar. First, labeled DNA (or RNA) baits ranging from 55 to 120 bases are hybridized to fragmented genomic DNA. The baits are pulled down using magnetic beads, and the "captured" genomic fragments are then sequenced using NGS instruments such as SOLiD, Illumina or Roche 454.

Table 1. Comparison between two standard whole exome mouse capture kits.

	Agilent Sureselect Mouse All Exon	Nimblegen SeqCap Ez
Probe size	120 bases	55–105 bases
Target Region size	49.6 Mb	54.3 Mb
Probe Type	RNA	DNA
Number of Targeted Exons	221,784	203,225

Mutation identification and ultimately gene discovery in the context of ENU-designed projects require significant computational analyses, where sequenced DNA fragments are mapped to a mouse reference sequence (C57BL/6J) [68] or to that of a specific mouse strain when available [36], followed by post alignment and variant calling procedures. For a given mouse sample, these procedures typically produce a large amount of single nucleotide variants (SNVs) and insertion/deletions (INDELs), which, depending on the sample's genetic background and coverage, can range from a few thousand to hundreds of thousands in more divergent strains. Further steps are required to filter the strain specific variants if the reference sequence of the mouse background is not used. This can be accomplished, for example, by adding more controls.

Numerous workflows (e.g., Genome Analysis Toolkit (GATK) best practices [69] and McGill University and Genome Quebec Innovation Centre (MUQGIC) [70]) have been designed for mutation discovery. Although each design may vary with regards to the steps and computational programs utilized, the underlying principle of these workflows remains the same. Each one divides the processing and analysis of sequencing data into three key steps: (1) data processing for quality control and filtering of sequenced reads; (2) variant discovery through alignment of filtered reads to known reference genomes; and (3) variant refinement leading to variant calling to identify mutations of interest. A flow diagram similar to GATK best practices [71] but with subdivided steps in file format is shown (Figure 3).

Figure 3. A typical workflow to identify causative mutations in genomic data. The procedures are separated into three general processes: (1) data processing, where raw sequencing data (fastq format) is aligned (sam/bam file format) to a known genome reference followed by alignment improvement steps (*i.e.*, indel realignment, mark duplicates and base recalibration); (2) a variant discovery step in which single nucleotide variants (SNVs) are called from aligned data followed by subsequent filtering (using variant quality thresholds; hard filtering, or Genome Analysis Toolkit (GATK) variant recalibration; and soft filtering); (3) and a variant refinement step to reduce the number of candidate mutations to a manageable number for further validation using Integrative Genomics Viewer (IGV) and/or Sanger sequencing [71].

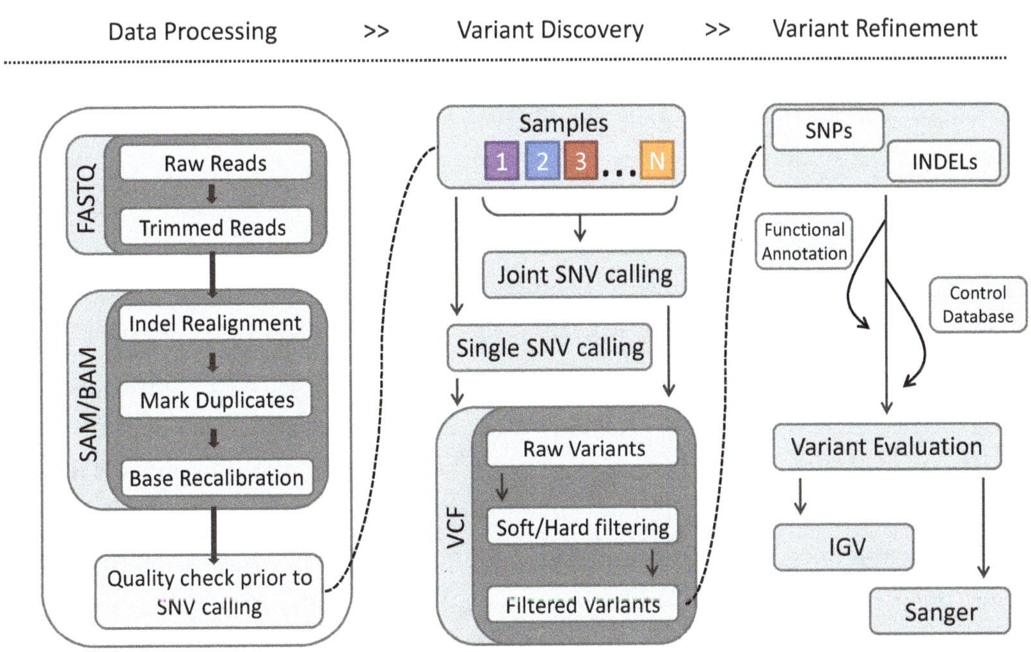

The sequenced reads (in fastq file format) are usually derived from the instrument specific base-calling algorithm (or subsequent steps therein) and contain an identifier for each raw DNA fragment, as well as a phred quality score for each base in the fragment. The raw reads are aligned to a reference genome following a quality control step or "trimmed" to obtain a high quality set of reads for sequence alignment file (sam/bam) generation. The trimming step removes adaptor sequences from the raw reads and optionally removes bases at the 3' end using a specified phred quality threshold, and/or performs a size selection filtering step (e.g., trimmomatic [72]; Figure 3). The trimmed reads are aligned by using either a "hashing" or an effective data compression algorithm called the "Burrows-Wheeler transform" (BWT). Fast, memory-efficient BWT-based aligners, such as BWA [73], are often used in NGS studies. However, these aligners tend to be less sensitive than recent hash-based aligners, such as Novoalign [74], which conversely tend to require more computational resources [75].

Numerous software packages such as GATK [69], samtools [76], and Picard [77] have been developed to attempt to correct for biases incorporated at the sequencing and alignment phases, thus improving variant detection (Figure 3). During library construction and sequencing, duplicated DNA fragments produced by polymerase chain reaction (PCR) amplification and optical duplicates can occur. Software package such as Picard markDup and Samtools rmdup remove or flag potential PCR duplicates if both mates (in the case of paired-end reads) contain the same 5' alignment positions.

At the alignment phase, due in part to the heuristics of the alignment algorithm and the alignment scoring procedure, refinement of mapped reads near indels (GATK indel realigner [69]) and quality scores (GATK base recalibration [69]) are typically required to help reduce false positive variants in downstream analysis. Utilizing these two post-alignment programs, GATK indel realigner transforms regions with misalignments generally introduced by indels into clean reads containing fewer mismatches, whereas base recalibration improves the quality score to better reflect the true base-calling error rates by correcting for variation in quality with respect to machine cycle, sequence context, and other attributes.

To identify the protein-encoding mutations induced by ENU, numerous variant-calling procedures can be employed to convert base calls and quality scores into a set of genotypes on a per sample basis. The most recent variant callers, such as GATK [69], Samtools [75], and FreeBayes [78], use sophisticated statistical models that can be extended to incorporate additional information regarding allele frequencies and/or linkage disequilibrium (LD) patterns. Furthermore, joint analysis of multiple individuals can further improve genotype calling for single samples by taking into account allele frequencies or genotype frequencies [79].

Variant detection programs convert the refined base-calls and quality scores resulting from the post-alignment process and generate variant data containing information regarding the genomic position, SNV quality, *etc.*, of each variant. Generally, thousands of SNVs are generated by the detection protocol. Further annotations and filtering procedures are thus required to identify the expected 50–100 ENU-induced mutations [80]. The use of functional annotation programs such as snpEff [81] and VEP [82], coupled with the exclusion of known variants (for example, on the basis of SNP data from the dbSNP database [83]) and of variants falling below acceptable quality metrics (QUAL, genotype quality (GQ), strand bias, *etc.*), can help to preferentially identify protein coding mutations. However, despite rigorous post-alignment refinement and variant exclusion criteria, recurrent false positive SNVs remain. By comparing a set of ENU samples to unrelated genome or exome sequencing data sets, as well as to mouse genomes data from the Sanger Institute [68] generated using the same analysis workflow, variants commonly shared between related strains or systematic false positives arising from mapping issues related to genome structure (e.g., repetitive or paralogous sequences) or errors (e.g., miss-annotated reference allele) can be flagged for removal. In numerous studies this procedure has proven successful in prioritizing candidate mutations and decreasing their numbers [54,80], and has helped reduce the time requirements and cost of visual inspection (e.g., Integrative Genomics Viewer (IGV) [84]), of Sanger sequencing [85], of validation, and ultimately of novel mutation/gene discovery.

ENU experiments have successfully identified candidate causative mutations residing in protein coding sequences, splice sites or UTRs. However, these causative mutations are not always successfully identified due to either the fact that they may reside in uncaptured regions (*i.e.*, non-coding regions, regulatory regions or un-annotated coding sequences that are not captured by the capture design) or to biases in standard mapping and variant calling procedures. Therefore, further improvements are required in the development of software tools in order to better deal with regions of the genome that are difficult to map (e.g., paralogous sequences and GC-rich regions). The design of exome capture kits must also be improved to extend the set of captured regions. Alternatively, whole genome sequencing may also be a way to identify mutations in regions not captured by whole exome sequencing.

5. Infectious Screens

Establishing an ENU mutagenesis program with the aim of identifying genes involved in the host response to pathogens presents particular challenges. The first is the choice of a pathogen relevant to human health. Mouse models of infection with this pathogen must be available and representative of the corresponding human pathology. Also, the contribution of genetic factors in human and/or mouse response to this pathogen must be proven to support the feasibility of a genetic screen. The second challenge is the choice of the inbred mouse strain to be used for mutagenesis. There is ample evidence that the ENU sensitivity of inbred mice is genetically controlled and thus widely variable across strains [63]. This must be balanced with the varying susceptibility or resistance of inbred strains to infection with specific pathogens. The third challenge is the choice of the screening phenotype. Cell-based phenotypes have been used successfully to identify fundamental mechanisms of innate and acquired immunity [36,37]. The findings, however, require further validation in mouse models to determine a possible role in the infectious process. A clinically relevant, robust, and unequivocal *in vivo* phenotype is also attractive, as it will lead to the identification of the most important molecular determinants for a given infection; it will also minimize the appearance of false positives. Such phenotypes include severe disease (in terms of clinical evaluation or pathogen load) or death, when the mutagenized strain is resistant, or survival, when the mutagenized strain is innately susceptible, following infection. As presented below both screening approaches have led to the identification of key molecules involved in susceptibility or resistance to infectious diseases caused by parasites, bacteria, and viruses.

6. Malaria Parasites

Infecting hundreds of thousands of people every year, malaria is a significant cause of morbidity and mortality in developing countries (www.who.org). Having co-existed with humans for centuries, malaria has exerted a significant selective pressure on the human genome [16,86]. Likely the best-known selection has been the retention of deleterious hemoglobinopathies, such as sickle cell anemia, in malaria endemic regions [87,88]. Other variants associated with reduced susceptibility to malaria infections include those affecting erythrocyte proteins [89–94], the scavanger receptor CD36 [95,96], and elements of the host immune response, including human leukocyte antigen (HLA) [97] and tomor necrosis factor-alpha (TNF-α) [98], among others [99,100]. Despite these clear examples, the genetic component influencing the human response to malarial parasites is complex, multigenic, and influenced by various environmental factors, including parasite virulence [101–103].

Cerebral malaria (CM) is the most severe and lethal complication of *Plasmodium falciparum* infection in humans [104,105]. Prevalent in immunologically naïve children, CM is characterized by high fever and a rapid progression to severe cerebral symptoms including impaired consciousness, seizures, and coma [106,107], resulting in death in about 20% of all cases [16,107]. During CM, parasitized erythrocytes (pRBCs) become trapped within the brain microvasculature [103], triggering a strong pro-inflammatory response [104,105] leading to the activation of the vascular endothelium [106], as well as the recruitment of immune cells and activated platelets [108–110]. This host-directed immune response results in the disruption of blood-brain barrier integrity [111], suggesting that CM pathogenesis is at least partially caused by over-activation of the inflammatory response [16,106,107]. By gaining a more thorough understanding of this disease, including of the host genetic factors affecting differences in susceptibility, novel and more effective prophylactic and therapeutic interventions can be developed.

Mice infected with *Plasmodium berghei* ANKA (PbA) have been used as a model of CM (experimental cerebral malaria, ECM). Mice susceptible to ECM develop neurological symptoms between days five to eight post-infection, including ataxia, hind limb paralysis, coma, and death [112]. ECM-resistant mice survive the cerebral malaria phase, but subsequently succumb to hyperparasitemia and resultanting anemia within three weeks post-infection [16]. Informative crosses between mouse strains of varying degrees of susceptibility to PbA have revealed at least nine quantitative trait loci (QTL) that modulate the host response to ECM [113–118]. However, these methods have failed to identify the causative genes, due in part to the large size of the genomic region and to the high number of positional candidates under the QTL peaks [119]. By introducing random point-mutations and small deletions within a susceptible genetic background, such as B6, B10, or 129S1, ENU-mutagenesis allows for the interrogation and determination of genes that are involved in resistance to ECM.

6.1. Screening for Acquired Resistance to Cerebral Malaria

We have successfully utilized ENU-mutagenesis to identify genes responsible for controlling susceptibility to ECM [119]. Male B6 mice (G0) were mutagenized with the administration of three consecutive i.p. injections of ENU. These G0 males were then bred to wild-type B10, 129S1, or B6 females to establish heterozygous G1 offspring. G1 males were out-crossed a second time to wild-type susceptible females to form the G2 generation. One to two G2 females per pedigree were backcrossed to the paternal G1 to produce G3 offspring, fixing mutations to homozygousity in approximately 25% of all animals (Figure 2A). G3 mice were infected with 10^6 *Plasmodium berghei* ANKA-parasitized RBCs by intravenous injection. The appearance of neurological symptoms and survival time were used as phenotypic markers of ECM disease [119]. Phenodeviant pedigrees were defined as those exhibiting >17% resistant pups in at least three litters or 10 offspring, whichever came first.

Enhanced laboratory resources and technological advances have allowed us to implement three variations of the general protocol outlined above. The first screen out-crossed mutagenized G0 males to the B10 genetic background. G3 animals from this cross were phenotyped for ECM-resistance. To facilitate linkage mapping, G1 males identified as segregating an ECM-resistant phenotype were out-crossed to 129S1 wild-type females. The resulting F1s were intercrossed randomly to generate F2 offspring, which were then phenotyped. Pedigrees identified as resistant were then analyzed for linkage analysis using a genome scan. A total of 6062 G3 mice from 244 G1 males were screened, generating nine phenodeviant pedigrees, with a background survival of approximately 2.8%. From this screen, we have identified an ECM protective mutation in *Jak3* (*Jak3*^{W81R}) [119]. A cytosolic tyrosine kinase that interacts with the common γc chain of cytokine receptors (IL-2, -4, -7, -9, -15, -21), JAK3 is required for STAT family members dependent transcriptional development and activation of inflammatory pathways in NK, T, and B cells [120]. *Jak3*^{W81R} mutants exhibit reduced numbers of NK cells, CD8^{+} T cells, and B cells, as well as severely reduced production levels of IFNγ by CD4^{+} T cells. We also demonstrated that tasocitinib, a JAK3 inhibitor used clinically to treat rheumatoid arthritis (RA) and Crohn's disease (CD), can reduce neuroinflammation and increase survival of *Jak3*$^{-/+}$ heterozygotes in the ECM model [119]. Genetic variants in JAK and STAT family proteins have been identified as causing certain primary immunodeficiencies and are also associated with chronic inflammatory diseases, such as inflammatory bowel disease (IBD), multiple sclerosis (MS), and systemic lupus erythematosus (SLE) in humans [121–123].

With respect to the second screen, we have out-crossed the mutagenized G0 males to the 129S1 genetic background. The 129S1 strain produces larger litters, allowing for the generation of larger numbers of G3 animals. Additionally, out-crossing directly to the 129S1 background eliminated the requirement to complete additional out-crossing of phenodeviant animals. Twenty-eight phenodeviant pedigrees were identified following the screening of 7705 G3 animals in 220 pedigrees, with a background survival of almost 8%. An epistatic interaction between the B6 and 129S1 genetic backgrounds on Chromosomes 4 and 1 was identified in 10 of the 28 phenodeviant pedigrees, potentially masking the effect of ENU-mutagenesis [124]. However, several mutations were identified in this screen, including an abrogated splicing mutation of Exon 6 in the winged-helix transcriptional regulator *Foxn1* gene [125] *Foxn1* mouse mutants are athymic and severely immuno-compromised, while human *FOXN1* mutations cause T-cell immunodeficiency [126]. Heterozygosity for the *Foxn1* mutant allele confers partial protection against ECM, suggesting that FOXN1 transcriptional targets may be relevant to reducing neuroinflammation.

The epistatic interaction between the B6 and 129S1 genetic backgrounds highlights both the limitations and advantages of different variations of the ENU-mutagenesis screen. Both the B6 and 129S1 strains are susceptible to *Plasmodium berghei* ANKA infection, developing neurological symptoms between Days 5 and 10 post-infection. However, in over a third of the phenodeviant pedigrees identified in the mixed background screen, an enrichment of B6 alleles on distal chromosome 4 was associated with resistance to ECM. With such a high percentage of phenodeviant pedigrees mapping to the same locus, we hypothesized that the likelihood of this effect being caused by a single causative ENU-induced mutation was minimal, and that this effect was due to genetic background rearrangements. Additional analysis revealed that ECM resistance on Chromosome 4 (*Berghei* resistance locus 8, named *Berr8*,) was being modulated by a second locus on Chromosome 1 (named *Berr7*). Although we had expected to discover single point mutations due to ENU-mutagenesis, these results highlight the complex nature of cerebral malaria, as well as the difficulty inherent to finding point mutations that are solely responsible for trait modulation.

Due to improvements in technology and the resulting cost reduction, we switched from linkage analysis to exome sequencing analysis for the identification of ENU-induced mutations, removing the requirement for genetic background variations. Hence, the third and final screen was executed on a pure B6 genetic background, wherein the mutagenized G0 males were outcrossed to wild-type B6 females. Switching to the pure genetic background eliminated the likelihood of epistatic interactions between genetic backgrounds, as exhibited in the reduction of background survival rate from almost 8% in the B6x129S1 screen to less than 5% in the pure B6 screen. However, due to smaller litter sizes, almost 40% fewer G3 animals were produced from the 109 screened pedigrees. Even so, eight phenodeviant pedigrees were identified and are currently being investigated.

6.2. Screening for Acquired Resistance to Blood-Stage Malaria

ENU-mutagenesis has also been used to identify genes implicated in host resistance to blood-stage malaria. A dominant ENU-mutagenesis screen for erythrocyte production and maturation defects linked to malaria resistance identified two mutations in the *Ank1* gene: an alternative splice acceptor mutation resulting in a frameshift mutation and premature stop codon was identified in $Mpl^{-/-}$ mice mutagenized on a BALB/c background [127], and a single nonsense mutation was identified in mutagenized SJL/J mice [128]. Both mutations result in early truncation of the ANK1 protein, encoded by *Ank1*. Implicated in hereditary spherocytosis, an inherited form of hemolytic anemia, mouse erythrocytes harboring mutations in *Ank1* exhibit increased resistance to *P. chabaudi*, a model of blood stage malaria, potentially due to parasite maturation impairment [127,128].

6.3. Conclusion

ENU-mutagenesis has enabled the identification of individual genes involved in modulating the host response to both cerebral and blood-stage malaria. We have identified mutations in host inflammatory genes involved in T cell development and/or function (*Jak3* and *Foxn1*), thymus development, and immune cell function [119,125]. These results are consistent with the current understanding of the role of T cells in cerebral malaria pathogenesis [129–132]. Additionally, these genes have been associated with the modulation of other models of acute inflammation as well as of chronic inflammatory conditions [99]. Other labs have identified mutations in the erythrocyte protein ANK1, an important factor in the erythrocyte cytoskeleton [127,128]. Mutations in erythrocytic proteins, including the cell surface Duffy antigen [92] and structural component Band 3 [89–91], have been associated with increased resistance to malaria in humans for several years. Together, these findings advance our understanding of the host response to malaria, and may aid in the discovery of novel drug targets against this devastating disease.

7. *Salmonella* Bacteria Infections

Salmonella enterica infections in humans represent an increasingly significant economic and public health challenge that is associated with high morbidity and mortality in both developing and industrialized countries [133]. In fact, the increase in global population, the emergence of antimicrobial resistance in bacteria, and the prevalence of co-infections (e.g., *Plasmodium*, HIV) exacerbate the burden of this infectious disease [38]. *Salmonella* infection in humans can cause a range of food and waterborne illnesses, from a localized diarrheal disease to the more severe systemic disease, typhoid fever. In fact, nontyphoidal *Salmonella enterica* serovars (e.g., *S. typhimurium*, *S. enteritidis*) are the second leading cause of bacterial food poisoning in the United States. Importantly, about 1%–4% of these *Salmonella*-infected individuals are at an increased risk of developing sepsis, chronic infection or clinical sequelae (ex. chronic arthritis) [134–136]. *Salmonella enterica* Typhi is the etiologic agent of typhoid fever, which is endemic primarily in areas with poor sanitation and a lack of clean drinking water. *Salmonella typhi* causes twenty-one million infections annually, with 220,000 deaths [133]. The outcome of infection depends on the activation of early innate functions, neutrophilic infiltration, phagocytosis by tissue macrophages, and inflammatory cytokine/chemokine secretion (e.g., IFNγ, IL-12, IL-18, TNFα, and IL-6). However, ultimately, the resolution of systemic infection is dependent on both humoral and cell-mediated immune responses [137,138].

In humans, the contribution of host genetics to *Salmonella* infection has been proven by the candidate gene approach and by exome sequencing in patients. Individuals with defects in the IL-12/IL-23 (IL-12β, IL-12Rβ1) and IFNγ (IFNγR1, IFNγR2, STAT1) pathways are in fact predisposed to Mendelian susceptibility to mycobacterial disease (MSMD) and/or disseminated *Salmonella* infection [139–144]. Furthermore, major histocompatabilty complex (MHC) class II and III loci, as well as the TNF haplotype, were significantly associated with typhoid fever in a Vietnamese cohort [145]. Although clinical evidence supports a strong role for host genetics, susceptibility to *Salmonella*-related infections is complex and also influenced by environmental factors and bacterial serotype.

Salmonella typhimurium infection is a recognized experimental model for studying systemic typhoid-like disease in mice [146,147]. Various classical inbred strains of mice demonstrate differential susceptibility/survival following sub-lethal intravenous infection with *S. typhimurium* strain Keller [148]. In particular, the 129 substrains (129S1, 129X1) of mice are highly resistant to virulent infection, compared to DBA/2J mice, which display intermediate mortality, and to the highly susceptible B6 strain. Although the genetic and molecular basis of several mutations important in resistance to *Salmonella* infection in mice have been identified, namely *Nramp1/Slc11a1*, *Tlr4*, and *Pklr*, the low frequency of naturally occurring spontaneous mutations has prompted the use of novel genomic approaches like ENU mutagenesis to identify novel host susceptibility genes to *Salmonella* infection [148–153].

7.1. Screening for Acquired Susceptibility to Salmonella typhimurium

We used ENU mutagenesis to further decipher the host genetic component of susceptibility to *Salmonella* infection *in vivo*. In the screen, G3 ENU-mutagenized mice between 7 and 9 weeks of age were challenged intravenously through the caudal vein with an infectious dose of virulent *Salmonella typhimurium* strain Keller, varying between 1000 to 10,000 colony forming units (CFUs), depending on the background strains used for breeding. Over the course of 14 days, infected mice were monitored for clinical manifestations of illness including a body score index of less than two, muscle wasting, fur ruffling (fever), inactivity, twirling, and shaking. Susceptible mutants were defined as those presenting severe clinical signs between Days 3 to 7 post-infection (prior to background control mice). On average, a minimum of six to eight G3 mice per G2 female were infected with the expectation of identifying two to five heritable deviant pedigrees following the screening of G3 mice derived from roughly 100 G1 males.

Two prototype breeding schemes differing in the genetic contribution of background strains (B6, 129S1, 129X1, and DBA/2J) have been used in five rounds of screening for *Salmonella* susceptibility. Male 129S1 (G0) mice were mutagenized using a single i.p. injection of 150mg/kg of ENU at 8–10 weeks of age. The first breeding scheme involved the generation of G1 mice produced by two independent G0 males (Figure 2B). The G0 males were crossed to B6 females. For each G1 pedigree, four G2 brother-sister pairs were bred to produce G3 progeny. Using this breeding scheme, the *Salmonella* susceptibility allele $Slc11a1^{Asp169}$ from B6 mice was segregated into the G2 population. G2 animals carrying the wild-type *Slc11a1* alleles were then selected for further breeding. As the introduction of susceptibility to the B6 background was interfering with our capacity to capture recessive alleles acting in later infection stages (past Day 4), we subsequently modified the breeding scheme as in Figure 2A. Hence in the second round of screening, G0 males were out-crossed to wild-type 129X1 females to generate G1 heterozygote offspring. G1 males were further backcrossed to 129X1 females to generate G2 mice. G2 females were then backcrossed to the G1 male to give rise to G3 progeny, which were then used for primary phenotyping of susceptibility to infection using survival analysis with 10,000 CFUs. Using the following scheme, 643 G3 mice derived from 39 G1 males were screened and two deviant pedigrees were identified: *Oxie & Celie* (*Ity14*) (*Immunity to Typhimurium* locus *14*) and *Jody & Cloe* (*Ity15*). In this particular case, we used a strain that was closely related to the mutagenized males to prevent or minimize the impact of the genetic background on the expressivity of the phenotype while allowing mapping in the G3 animals. We identified 105 SNPs between 129S1 and 129X1. However, their clustering in the genome did not allow the mapping of some pedigrees. Variations of these protocols (Figure 2) were used to facilitate mapping resolution using SNPs between 129S1 and DBA/2J directly in the G3 population. In the third round of screening, G1 males were out-crossed to DBA/2J, and the resulting G2 mice were randomly intercrossed to generate G3 progeny. G3 mice were then screened with an infectious dose of 5000 CFUs. Using this scheme, 1570 G3 mice derived from 65 G1 males were screened, and one deviant pedigree, *Ity16*, was identified, validated, and cloned [154]. In the fourth round of screening, G0 males were out-crossed directly to DBA/2J in order to introduce genetic variability as early as possible in the breeding scheme, thus facilitating mapping (Figure 2B). In this round, 3,348 G3 mice derived from 208 G1 males were screened and four deviant pedigrees were identified: *Cherrie & Walter* (*Ity17*), *Jeanine & Harman*

(Ity18), *Lexie & Leona*, and *Philippe & Desiree*. Lastly, with the onset of whole-exome sequencing as an alternative to mapping using genetic variation between parental strains, the breeding scheme shown in Figure 2B was carried out on an 129S1 background. From the following screen we have infected 580 G3 mice derived from 41 G1 males, and two deviant pedigrees, *Rakeem & Athena* and *Lessie & Virgie*, were identified.

In summary, 8,389 G3 mice derived from 491 G1 males were screened for increased susceptibility to *Salmonella typhimurium* infection as measured by survival analysis. A total of 10 deviant pedigrees have been identified (Table 2). From this screen, we have to date identified, cloned, and characterized *Salmonella* susceptible mutations in *Usp18* ($Usp18^{L361F}$), *Ank1* ($Ank1^{Gln1357Ter}$), and *Stat4* ($Stat4^{G418\_E445}$) [154–156]. USP18 (Ubiquitin Specific Peptidase 18) both regulates type I IFN signaling and functions as a protease to remove ISG15 adducts from substrate proteins [157,158]. We have reported that decreased survival in mice that carry the $Usp18^{L361F}$ mutation results from increased bacterial loads in the spleen and liver, as well as increased inflammatory response leading to septic shock [156,159]. In more recent studies, we have shown that regulation of type I IFN signaling is the predominant mechanism affecting the susceptibility of $Usp18^{L361F}$ mice to bacterial infection. Also, we have found that hyperactivation of type I IFN signaling leads to increased ISGylation and IL-10 production, as well as decreased expression of markers of autophagy [160]. Additionally, we have shown that $Usp18^{L361F}$ mice are more susceptible to infection with *Mycobacterium tuberculosis* (same as above).

Table 2. Summary of the three ENU-mutagenesis screens for experimental cerebral malaria, *Salmonella*, and herpes simplex virus (HSV)-1.

	Malaria	*Salmonella*	HSV-1
G1 males	573	491	265
G3 mice	16,411	8,415	7,802
Deviant pedigrees (in progress)	45	16	11
Confirmed pedigrees	5	3	2

The transcription factor STAT4 (Signal Transducer and Activator of Transcription Factor 4) is a critical mediator of IL-12 signaling. It plays an important role in both innate and adaptive immunity by regulating the transcription of target genes such as *Ifng* and those mediating NK cell cytotoxicity, T helper 1 cell differentiation, and immunoglobulin isotype switching to IgG1. The $Stat4^{G418\_E445}$ mutation results in impaired innate IFNγ secretion, primarily from splenic NK and NKT cells, contributing to increased hepatosplenic bacterial loads. These findings support the importance of the IL-12/IFNγ axis in resistance to *Salmonella* infection.

ANK1 is a structural protein of the erythrocyte membrane, which plays an important role in membrane stability by mediating the attachment of band 3 (SLC4A1) and protein 4.2 (EPB4.2) to the spectrin-based membrane cytoskeleton [161]. Mice homozygous for the $Ank1^{Gln1357Ter}$ mutation develop hemolytic anemia and present clinicopathological features of human hereditary spherocytosis, the most common form of congenital chronic hemolysis in Europe and North America [162]. On one hand, as observed with other mutations affecting red blood cell turnover [163], *Ank1* deficits protect mice against malaria [128]. On the other hand, normal ANK1 function is critical for an effective host response against infection with *Salmonella. Salmonella* susceptibility in *Ank1*^{Gln1357Ter} mutant mice is the result of a combination of factors, namely the concomitant deposition of iron in tissues, which favors bacterial growth, and low levels of the iron regulatory hormone hepcidin [154]. In addition, the strong induction of heme oxygenase 1 (*Hmox1*) expression observed during malaria infection and in *Ank1*^{Gln1357Ter} mutant results in impaired oxidative burst function, which favors the intracellular replication of bacteria [154,164].

7.2. Ex Vivo and in Vivo ENU Screens for Susceptibility to Bacteria Infections

Additional ENU initiatives have uncovered novel genetic determinants of resistance to bacterial infections. Different primary screens in G3 offspring were used, including: (1) measurement of TNF bioactivity after *ex vivo* challenge of thioglycolate-induced peritoneal macrophages with various pathogen-associated molecular patterns (PAMPs) (*Cd36*, *Tnf*, *Map3k8*) [165–167]; (2) measurement of type I IFN bioactivity after *ex vivo* challenge of thioglycolate-induced peritoneal macrophages with *Listeria monocytogenes* (*Tmem173/Sting*) [168]; (3) *in vivo* screen for other classes of pathogens (*Slfn2*) [169]; (4) mutations affecting hematopoetic cell development (Genista-*Gfi1*) [170]; and (5) visible phenodeviants presenting inflammatory lesions of the skin (*Scd1*) [171] or of the feet (*Ptpn6/Shp1*) [172]. For example, a TLR2 agonist screen in macrophages identified the *Oblivious* pedigree, which possesses a mutation in *Cd36* resulting in increased susceptibility to infection with Gram positive bacterium *Staphylococcus aureus* [165]. In addition, the *Sluggish* pedigree, which carries a mutation in the *Map3k8* kinase, has impaired type I IFN production downstream of TLR7 and TLR9 signaling, rendering it susceptible to Group B streptococcus infection *in vivo* [166]. Another example is the ENU-induced mutation in *Gfi1* within the *Genista* pedigree, wherein depletion of PMNs confers resistance to *Brucella abortus* infection [173,174] and increased susceptibility to oral infection with *Salmonella typhimurium sfiA*$^{-}$ [170]. Moreover, the *ex vivo* ENU screen using *Listeria monocytogenes* identified the *Goldenticket* pedigree as carrying a mutation in *Tmem173/Sting,* further demonstrating the importance of type I IFN signaling during bacterial infection [168].

7.3. Conclusion

ENU-mutagenesis identified single gene effects (novel allele and novel function) within critical pathways involved in immunity to bacterial infection that could potentially be translatable to infection with other classes of pathogens and/or to chronic inflammatory diseases. The findings have emphasized the importance of IFN signaling (*Usp18*, *Stat4*, *Sting*, *Map3k8*) during bacterial infections [155,156,166,168], as well as erythropoeisis and iron metabolism, (*Ank1*) in the case of *Salmonella* pathogenesis [159].

8. Herpes Viruses

The *Herpesviridae* family is a large ancient family with a long history of coevolution with their hosts probably predating the origin of the primate lineage. Altogether the nine human herpesviruses infect 90% of the world population causing different types of pathologies that vary considerably according to the immune status of the infected individual. These ubiquitous viruses constitute a striking example of the intricate interplay that can be gradually established between host and pathogen, and show that important information can be gleaned from the study of host-pathogen interactions, namely the contribution of both viral immune evasion and host resistance genes to the outcome of infection.

8.1. Cytomegaloviruses

Human cytomegaly virus (HCMV) is the most frequent congenital viral infection in developing countries, potentially leading to blindness, deafness or mental retardation in affected infants. Primary infection or reactivation of the virus can result in severe morbidity and mortality, especially in immune-compromised individuals such as transplant recipients, leukemia or lymphoma patients and AIDS patients. Fortunately, HCMV is closely related to its murine homologue, mouse cytomegalovirus and both cause death in immunocompromised individuals [175,176]. Thus, infection of mice with MCMV represents an excellent model for the study of HCMV pathology and indeed it is an important tool for virologists, immunologists, and geneticists, all of whom have benefited from the well-developed state of the model. Forward genetic studies in inbred mouse strains identified major epistatic (*Klra16*/*H2*^{k}) or single gene effects (*Klra7*, *Klra8*) demonstrating the crucial role that natural killer (NK) cell specific activating (Ly49H, Ly49P) and inhibitory (Ly49G) receptors play in response to virus infections (reviewed in [177]).

8.2. Screening for Altered Immune Responses to MCMV

Beutler and colleagues were the ones to initiate the ENU screen for MCMV susceptibility (for the latest review, see [178]). With this strategy, over 20,000 G3 B6 mice carrying ENU mutations were infected i.p. with 10^5 plaque forming units (pfu) of MCMV. This viral dose was chosen because wild-type B6 mice are uniformly resistant in this infectious experimental situation. However, the pheno-deviant offspring that exhibited clinical signs of disease or/and high viral titers in the spleen were considered susceptible. Several mice with immunodeficiency phenotypes identified from other screens made by Beulter's group, such as defects in toll-like receptor (TLR) signaling or adaptive immunity, were also tested for their potential MCMV susceptibility. Here, we highlight some of the most important findings that have been made using ENU-mutagenesis to test susceptibility to MCMV infection.

Dendritic cells (DCs) are specialized cells of the hematopoietic system that alert the immune system to the presence of infection. Therefore, they generally represent the first line of defense against pathogens. In the context of MCMV infection, DCs recognize the virus through TLR3 and TLR9, which are able to respectively detect double-stranded RNA (an intermediate product of viral replication) and viral double stranded DNA. Following MCMV recognition, DCs and plasmacytoid DCs (pDCs) in particular, produce large amounts of antiviral type I IFN cytokines (IFN-α/β), which are essential mediators of the innate and adaptive immune responses. Thus, loss-of-function mutations in genes that encode components necessary for the expression of IFN-α/β (such as *Tlr9*, *Tlr3*, *Myd88*, *Trif*, and *Unc93b1*), or that are involved in the IFN-α/β signaling pathway (ie, downstream of IFN-α/β receptor), like *Stat1*, have been shown to increase susceptibility to MCMV infection [179–182]. It should be noted that among all of these ENU mutations, only the one in *Stat1* was initially identified from the MCMV screen, the others being deduced from immune screens. The NF-κB signaling pathway is also essential for survival to MCMV infection. This is attested by the identification of a loss-of-function mutation in the *Ikbkg* gene encoding NEMO, a regulatory subunit of the IKK complex responsible for the nuclear translocation of NF-κB [183]. *Ex vivo* screens for increased susceptibility to MCMV infection have been performed on peritoneal macrophages isolated from ENU-mutagenized mice, and revealed a missense mutation in the *Eif2ak4* gene encoding GCN2 [184]. This protein is related to PKR, an effector known to inhibit viral replication via phosphorylation of the alpha subunit of eukaryotic initiation factor 2 (eIF2α). The loss-of-function mutation identified in *Eif2ak4* affects the phosphorylation of eIF2α in response to MCMV infection and was therefore associated with an increased susceptibility to MCMV. The MCMV screen, together with the immune screens, led to the identification of several phenodeviants with mutations in genes that contribute to the establishment of an efficient immune response against pathogens, as they act at different levels of IFN-α/β production (TLR9, TRIF and UNC93B1), of IFN-α/β signaling (STAT1), and of the antiviral response (GCN2).

DCs are not the only sites of MCMV recognition. Natural killer (NK) cells are also important responders to MCMV infection, playing a crucial role in containing it at early times post-infection [185,186]. This was initially demonstrated by *in vivo* depletion studies, in which specific antibodies were used to transiently eliminate NK cells before infection with the virus [186–188]. Then, the differential susceptibility of the BALB/c and B6 strains was shown to be due to the presence of the NK-activating receptor Ly49H in the latter [189,190]. This receptor engages the MCMV viral protein m157 [191,192], leading to NK cell proliferation and target cell killing [193]. ENU studies allowed the initial discovery of mutations in the *Gimap5* and *Unc13d* genes, in the context of two screens that had been designed to detect *in vivo* defective NK cells and cytotoxic T lymphocyte (CTL) responses [194] and MCMV susceptibility [195], respectively. In both cases, *Gimap5*^{G38C} and *Unc13d*$^{jinx/jinx}$ were shown to be associated with defects in NK cell activity and impaired resistance to MCMV infection, which are consistent with the crucial function of NK cells in the early control of MCMV replication. *Gimap5*^{G38C} affects NK cell development, whereas *Unc13d*$^{jinx/jinx}$ NK cells fail to degranulate, a deficit also observed in activated CD8$^+$ T cells. Individuals carrying another deleterious mutation, this time in the *Itgb2* gene encoding the integrin β2 CD18, which partially affects NK cell development, are, however, fully resistant to MCMV [196]. In this case, it suggests that even if the β2 integrins are required for optimal NK cell maturation, their partial deficiency could be overcome during MCMV infection, highlighting the robustness of antiviral protective responses.

Other ENU mutations revealed from the screen for host survival against MCMV infection were independently identified in the *Flt3* [197] and *Slfn2* [169] genes. *Flt3*$^{wmfl/wmfl}$ mice have been shown to have impaired DC development, making these cells incapable of supporting the effector function of NK cells [197]. In contrast to *Flt3*$^{wmfl/wmfl}$, neither DCs, nor NK cells are impaired in *Slfn2*^{I135N} mice [169]. However, both bacterial and viral infections trigger death by apoptosis of peripheral T cells and inflammatory monocytes in *Slfn2*^{I135N} mice, indicating the crucial role of Slfn2 in maintaining quiescence in some immune cells. In addition to these ENU mutants recovered from the MCMV screen, four unrelated mutants, called *Mayday*, *Solitaire*, *Goodnight*, and *Slumber*, were shown to die very early post-infection (*i.e.*, D2-D3 p.i.) before high viral titers could be observed in the spleen and the liver [198]. Their abrupt death was probably not caused by the direct lytic effects of the virus, but mostly by collateral damage, such as the accompanying inflammatory reaction in response to MCMV infection, since this phenotype was also observed after lipopolysaccharides (LPS) or CpG administration. Based on the comparative sequence analysis of these four mutants, their MCMV susceptibility has been shown to be due to a genetic rearrangement of the *Kcnj8* locus that is likely to have occurred in B6 mice prior to ENU treatment. *Kcnj8* encodes the potassium channel Kir6.1, which maintains the host homeostatic state during the innate immune response. Altogether, these mutations highlight genes that are directly involved in the immune system, but also show the importance of other non-immune signaling pathways, such as homoestasis, in host survival.

8.3. Herpes Simplex Virus 1

HSV-1 is the causative agent of herpes simplex encephalitis (HSE), a lethal neurological disease. It is acknowledged that environmental factors have no effect on the pathogenesis of HSE, and no geographical or seasonal patterns in the distribution of the disease have been observed [199,200]. Despite the high seroprevalence of HSV-1 (up to 90%) [201], HSE pathology is rare and affects only a small proportion of otherwise healthy individuals. Therefore, in addition to HSV-1 infection, the second major cause of the disease is the presence of rare host genetic factors, which play a large part in determining the susceptibility of an individual to HSE. Loss-of-function mutations in the *UNC93B1*, *TLR3*, *TRIF*, *TRAF3*, and *TBK1* genes have been associated with a human genetic predisposition to HSE [202–207], illustrating the critical role of the UNC93B-TLR3-type I IFN pathway in protection against HSV-1. However, these mutations exhibit incomplete penetrance and represent only a minority of HSE cases. This indicates the likely existence of other anti-HSE pathways and may reflect the effects of additional host genetics factors.

8.4. Screening for Acquired Susceptibility to HSE

Two breeding schemes have been used in the mutagenesis screen to identify host susceptibility genes to HSV-1 infection. We started with the B6/B10 screen, where mutagenized B6 G0 males were out-crossed to B10. This allowed linkage mapping with the use of a panel of 255 B6/B10 polymorphic markers (SNPs) distributed across the genome [208]. We then switched to a pure B6 genetic background to eliminate the likelihood of epistatic interactions between the B6 and B10 genetic backgrounds. In total (Table 2), 7,802 G3 B6 mice carrying ENU mutations derived from 265 G1 males were infected i.p. with 10^4 pfu of HSV-1 strain 17. This dose led to lethal encephalitis in susceptible A/J mice, whereas wild-type B6 mice remained unaffected. Following infection, the ENU-mutagenized mice were monitored for two weeks. The phenodeviant offsprings that exhibited clinical signs of disease or succumbed to the infection were considered susceptible. Using this strategy, we revealed eleven deviant pedigrees. One of these led to the identification of a premature stop codon (L3X) in the Ptprc gene, which encodes the leukocyte common antigen CD45. $Ptprc^{L3X}$ mutant mice showed reduced numbers of $CD3^+$ T and mature follicular B cells, suggesting defects in T and B cell development [209]. In this report, we also demonstrated that $CD4^+$ Th1 cells, by producing IFNγ, help $CD8^+$ T cell recruitment to prevent the dissemination of HSV-1 into the central nervous system, thus protecting mice from lethal HSV-1 infection. Altogether, our data point to CD45 as the first host component involved in the adaptive immune response that directly contributes to susceptibility to HSV-1 and HSE pathology. We are currently investigating the 10 other deviant pedigrees, which have, once again, shown the crucial role of T cells in host survival, but have also revealed that anti-inflammatory factors are critical to protection against HSV-1-induced encephalitis [210].

9. Conclusions and Perspectives

ENU-mutagenesis constitutes an inherently unbiased and powerful approach to the production of new alleles. Technological improvements in high-throughput DNA sequencing, combined with the completion of the mouse genome project [68], have greatly facilitated their identification. The recent introduction of NGS has led to a faster and more efficient identification of ENU mutations, which is particularly helpful for analyzing large mutant collections, especially when mapping data are not available to guide an analysis. New variants generated by ENU-mutagenesis mirror those existing in the human population and also represent a natural complement to null alleles being produced by gene targeting. Finding new ENU-induced alleles will also benefit from the new CRISPR/Cas9 technology. ENU variants, although easier to pinpoint by sequencing, need to be validated experimentally as in any forward genetic approach of gene identification. The CRISPR/Cas9 system appears to be an excellent complement to ENU mutagenesis, allowing candidate point mutations identified by NGS to be efficiently confirmed as causative mutations. The ENU mutagenesis approach has proven to be extremely useful in dissecting the genetic architecture of host defenses against infectious diseases. The approach promises to remain current in the field, being constantly renewed by technological advances such as NGS or genome editing.

As summarized in Table 2, over 30,000 G3 mice were screened by our group for either resistance to *Plasmodium berghei* or susceptibility to *Salmonella typhimurium* and HSV-1 infection. In total, 72 deviant pedigrees have been identified and we have to date confirmed ENU-induced mutations for 10 pedigrees. These mutations highlight gene functions that are directly involved in the immune system (*Foxn1*, *Jak3*, *Stat4*, *Usp18* and *Ptprc*), but also show the importance of other non-immune pathways, such as erythropoeisis and iron metabolism (*Ank1*), in host survival (Table 3). Beutler and colleagues also used the ENU mutagenesis approach, and over 20,000 G3 mice were screened for their susceptibility to MCMV. In parallel, they also developed several "immune" ENU screens, where some phenodeviant pedigrees, characterized by defects in the TLR signaling pathway and/or in T/NK cells functions, were then tested for their potential MCMV susceptibility. Of these, it should be noted that among the ENU mutations identified by the group of Beutler, only few were initially revealed by the MCMV *in vivo* screen (*Stat1*, *Unc13d*, *Flt3* and *Slfn2*), the others being deduced from other screens (*Tlr9*, *Trif*, *Unc93b1*, *Ikbkg*, *Eif2ak4*, *Gimap5*) [178]. This observation can be explained by the fact that *in vivo* models are more complex than *in vitro* systems. Indeed, deficiencies in one particular immune cell or signaling pathway can be compensated by the presence of other competent immune cells, making the identification of defective alleles more difficult *in vivo*.

The ENU mutations identified in *Jak3* ($Jak3^{W81R}$) and *Ptprc* ($Ptprc^{L3X}$) highlighted the critical nature of T cell function for CM pathogenesis and protection against HSV1 infection, respectively. The robustness of these mouse models of neuroinflammation and their ability to detect genetic effects regulating common pathways critical for neuroinflammation are highlighted by the complementary observations that the $Jak3^{W81R}$ mutant allele (protective in the ECM screen) confers susceptibility to HSV encephalitis (HSE), while the $Ptprc^{L3X}$ (causing susceptibility to HSE screen) is protective in the ECM model [211]. This approach could be generalized to other interesting pedigrees, where the role of the ENU mutations could be assessed in these different mouse models of infectious diseases. By cross-testing these mutant pedigrees, it should be possible to reveal common and specific pathways, as well as cells and proteins, that are crucial in the protection against malaria and *Salmonella* or viral infections. Moreover, the role of ENU mutations identified in the neuroinflammatory models of ECM and HSE could also be tested in other models of inflammation, such as the model of experimental encephalitis (EAE) that mimics MS, or DSS colitis that models IBD. Preliminary experiments using the EAE model have already suggested that $Ptprc^{L3X}$ mice are more resistant to EAE symptoms than wild-type and heterozygous littermate controls [210]. Thus, the cross-testing of these mutant pedigrees in different models of inflammation may provide additional information on the gene function, including its role in the pro- and anti-inflammatory balance. It can also provide novel targets for the development of new drugs that could be used in therapy for acute and chronic inflammatory diseases. As an example, a JAK3 inhibitor, currently in clinical use for the treatment of RA and CD (tasocitinib; Pfizer, New York, NY, USA), has been shown to reduce neuroinflammation and increase survival of $Jak3^{-/+}$ heterozygotes in our ECM model [119]. Therefore, pharmacological modulation of JAK3 mimics the effect of its genetic inactivation, indicating that the ECM screen can identify novel pharmacological targets for drug discovery.

Table 3. Genes and pathways identified in ENU screens described in this review.

Pathway	Gene	Screen	Phenotype	Reference
TLR signaling	*Cd36*	Immunity→*S. aureus*	Susceptible	[165]
	Map3k8	Immunity→Group B streptococcus	Susceptible	[166]
	Ptpn6	Autoimmunity→*L. monocytogenes*	Susceptible	[172]
	Tlr9	Immunity→MCMV	Susceptible	[179]
	Trif	Immunity→MCMV	Susceptible	[180]
	Unc93b1	Immunity→MCMV	Susceptible	[181]
	Ikbkg	Immunity→MCMV	Susceptible	[183]
Type I IFN signal	*Usp18*	*S.* Typhimurium	Susceptible	[156,159]
	Stat1	MCMV	Susceptible	[182]
Effector	*Eif2ak4*	MCMV	Susceptible	[184]
Cellular immunity	*Jak3*	*P. Berghei*	Resistant	[119]
	Foxn1	*P. Berghei*	Resistant	[125]
	Stat4	*S.* Typhimurium	Susceptible	[155,156]
	Tnf	Immunity→L. *monocytogenes*	Resistant	[167]
	Gfi1	Immunity→*S.* Typhimurium	Susceptible	[170]
	Gimap5	Immunity→MCMV	Susceptible	[194]
	Unc13d	MCMV	Susceptible	[195]
	Flt3	MCMV	Susceptible	[197]
	Slfn2	MCMV	Susceptible	[169]
	Ptprc	HSV-1	Susceptible	[209]
Red cell cytoskeleton	*Ank1*	*S.* Typhimurium	Susceptible	[154]
	Ank1	*P. Chabaudi*	Resistant	[127,128]
Homeostasis	*Kcnj8*	MCMV	Susceptible	[198]
Lipid metabolism	*Scd1*	Immunity→*S. Pyogenes*	Susceptible	[171]

One objective of the ENU-mutagenesis approach is to translate and validate knowledge obtained in the mouse infectious context to an improved understanding of human immunity and susceptibility to infection. As a starting point, mouse studies are fundamental for exploring host-pathogen interactions, especially when orthologous human genes exist. One striking example came from the discovery of the ENU-induced mutation in the mouse *Unc93b1* gene that causes susceptibility to MCMV [181]. Based on this finding, the group of JL Casanova identified an autosomal recessive UNC93B deficiency in two human patients with HSE [202]. Furthermore, a survey of the literature has shown that human variants identified in our ECM and HSE screens are risk factors for inflammatory diseases. For example, genetic variants in *JAK* and *STAT* family members have been associated with IBD, MS, RA, and SLE [121,122]. *PTPRC* polymorphisms are associated with autoimmune and inflammatory conditions including MS, SLE, and myasthenia gravis [212]. Thus, the ENU-mutagenesis approach should be continued in combination with GWAS studies, thus providing important insights into the pathways, cells, and proteins that directly impact susceptibility to pathogens, as it constitutes an invaluable resource for identifying novel therapeutic treatments.

Acknowledgments

This work was support by a Team grant (Danielle Malo, Philippe Gros, Silvia M. Vidal) from the Canadian Institutes of Health Research (CTP-87520). Grégory Caignard is supported by the Canadian Institutes of Health Research. Megan M. Eva is supported by the McGill University Health Centre. Silvia M. Vidal is supported by the Canada Research Chairs Program. We thank Eve-Marie Gendron-Pontbriand for proofreading the manuscript.

Author Contributions

Robert Eveleigh and Guillaume Bourque contributed to the section on gene identification (section 4); Rebekah van Bruggen and Philippe Gros contributed to the section about Malarial parasites infection and screen (section 6); Megan M. Eva and Danielle Malo contributed to the section about Salmonella infection and secreen (section 7), Grégory Caignard and Silvia M. Vidal contributed to the sections about herpes virus infection and screen (section 8) as well as the introductory sections (sections 1, 2, 3 and 5) and the conclusions (section 9).

References

1. Brussow, H. Europe, the bull and the Minotaur: The biological legacy of a Neolithic love story. *Environ. Microbiol.* **2009**, *11*, 2778–2788.
2. McMichael, A.J. Environmental and social influences on emerging infectious diseases: Past, present and future. *Philos. Trans. R. Soc. Lond. B. Biol. Sci.* **2004**, *359*, 1049–1058.
3. Casanova, J.L.; Abel, L. Inborn errors of immunity to infection: The rule rather than the exception. *J. Exp. Med.* **2005**, *202*, 197–201.
4. Fauci, A.S.; Morens, D.M. The perpetual challenge of infectious diseases. *N. Engl. J. Med.* **2012**, *366*, 454–461.
5. Chapman, S.J.; Hill, A.V. Human genetic susceptibility to infectious disease. *Nat. Rev. Genet.* **2012**, *13*, 175–188.
6. Cobat, A.; Orlova, M.; Barrera, L.F.; Schurr, E. Host genomics and control of tuberculosis infection. *Publ. Health Genet.* **2013**, *16*, 44–49.

7. Plantinga, T.S.; Johnson, M.D.; Scott, W.K.; Joosten, L.A.; van der Meer, J.W.; Perfect, J.R.; Kullberg, B.J.; Netea, M.G. Human genetic susceptibility to Candida infections. *Med. Mycol.* **2012**, *50*, 785–794.
8. Keynan, Y.; Malik, S.; Fowke, K.R. The role of polymorphisms in host immune genes in determining the severity of respiratory illness caused by pandemic H1N1 influenza. *Publ. Health Genet.* **2013**, *16*, 9–16.
9. Min-Oo, G.; Gros, P. Erythrocyte variants and the nature of their malaria protective effect. *Cell Microbiol.* **2005**, *7*, 753–763.
10. Lederman, M.M.; Penn-Nicholson, A.; Cho, M.; Mosier, D. Biology of CCR5 and its role in HIV infection and treatment. *JAMA* **2006**, *296*, 815–826.
11. Lindesmith, L.; Moe, C.; Marionneau, S.; Ruvoen, N.; Jiang, X.; Lindblad, L.; Stewart, P.; LePendu, J.; Baric, R. Human susceptibility and resistance to Norwalk virus infection. *Nat. Med.* **2003**, *9*, 548–553.
12. Von Bernuth, H.; Picard, C.; Puel, A.; Casanova, J.L. Experimental and natural infections in MyD88- and IRAK-4-deficient mice and humans. *Eur. J. Immunol.* **2012**, *42*, 3126–3135.
13. International Human Genome Sequencing. Finishing the euchromatic sequence of the human genome. *Nature* **2004**, *431*, 931–945.
14. The International HapMap Consortium. A haplotype map of the human genome. *Nature* **2005**, *437*, 1299–1320.
15. Hohl, T.M. Overview of vertebrate animal models of fungal infection. *J. Immunol. Methods* **2014**, doi:10.1016/j.jim.2014.03.022.
16. Longley, R.; Smith, C.; Fortin, A.; Berghout, J.; McMorran, B.; Burgio, G.; Foote, S.; Gros, P. Host resistance to malaria: Using mouse models to explore the host response. *Mamm. Genet.* **2011**, *22*, 32–42.
17. Sancho-Shimizu, V.; Zhang, S.Y.; Abel, L.; Tardieu, M.; Rozenberg, F.; Jouanguy, E.; Casanova, J.L. Genetic susceptibility to herpes simplex virus 1 encephalitis in mice and humans. *Curr. Opin. Allergy Clin. Immunol.* **2007**, *7*, 495–505.
18. Wick, M.J. Innate immune control of *Salmonella enterica* serovar Typhimurium: Mechanisms contributing to combating systemic *Salmonella* infection. *J. Innate Immun.* **2011**, *3*, 543–549.
19. Waterston, R.H.; Lindblad-Toh, K.; Birney, E.; Rogers, J.; Abril, J.F.; Agarwal, P.; Agarwala, R.; Ainscough, R.; Alexandersson, M.; An, P.; *et al.* Initial sequencing and comparative analysis of the mouse genome. *Nature* **2002**, *420*, 520–562.
20. Church, D.M.; Goodstadt, L.; Hillier, L.W.; Zody, M.C.; Goldstein, S.; She, X.; Bult, C.J.; Agarwala, R.; Cherry, J.L.; DiCuccio, M.; *et al.* Lineage-specific biology revealed by a finished genome assembly of the mouse. *PLoS Biol.* **2009**, *7*, e1000112.
21. Guenet, J.L. Animal models of human genetic diseases: Do they need to be faithful to be useful? *Mol. Genet. Genet.* **2011**, *286*, 1–20.
22. Guan, C.; Ye, C.; Yang, X.; Gao, J. A review of current large-scale mouse knockout efforts. *Genesis* **2010**, *48*, 73–85.
23. Ayadi, A.; Birling, M.C.; Bottomley, J.; Bussell, J.; Fuchs, H.; Fray, M.; Gailus-Durner, V.; Greenaway, S.; Houghton, R.; Karp, N.; *et al.* Mouse large-scale phenotyping initiatives: Overview of the European Mouse Disease Clinic (EUMODIC) and of the Wellcome Trust

Sanger Institute Mouse Genetics Project. *Mamm. Genome* **2012**, *23*, 600–610.

24. Gaj, T.; Gersbach, C.A.; Barbas, C.F., 3rd. ZFN, TALEN, and CRISPR/Cas-based methods for genome engineering. *Trends Biotechnol.* **2013**, *31*, 397–405.
25. Wang, H.; Yang, H.; Shivalila, C.S.; Dawlaty, M.M.; Cheng, A.W.; Zhang, F.; Jaenisch, R. One-step generation of mice carrying mutations in multiple genes by CRISPR/Cas-mediated genome engineering. *Cell* **2013**, *153*, 910–918.
26. Blake, J.A.; Bult, C.J.; Eppig, J.T.; Kadin, J.A.; Richardson, J.E. The Mouse Genome Database: Integration of and access to knowledge about the laboratory mouse. *Nucl. Acids Res.* **2014**, *42*, D810–D817.
27. Smith, C.M.; Finger, J.H.; Hayamizu, T.F.; McCright, I.J.; Xu, J.; Berghout, J.; Campbell, J.; Corbani, L.E.; Forthofer, K.L.; Frost, J,P. The mouse Gene Expression Database (GXD): 2014 update. *Nucl. Acids Res.* **2014**, *42*, D818–D824.
28. Begley, D.A.; Krupke, D.M.; Neuhauser, S.B.; Richardson, J.E.; Bult, C.J.; Eppig, J.T.; Sundberg, J.P. The Mouse Tumor Biology Database (MTB): A central electronic resource for locating and integrating mouse tumor pathology data. *Vet. Pathol.* **2012**, *49*, 218–223.
29. Wiltshire, S.A.; Leiva-Torres, G.A.; Vidal, S.M. Quantitative trait locus analysis, pathway analysis, and consomic mapping show genetic variants of Tnni3k, Fpgt, or H28 control susceptibility to viral myocarditis. *J. Immunol.* **2011**, *186*, 6398–6405.
30. Di Pietrantonio, T.; Hernandez, C.; Girard, M.; Verville, A.; Orlova, M.; Belley, A.; Behr, M.A.; Loredo-Osti, J.C.; Schurr, E. Strain-specific differences in the genetic control of two closely related mycobacteria. *PLoS Pathog.* **2010**, *6*, e1001169.
31. Toth, L.A.; Trammell, R.A.; Williams, R.W. Mapping complex traits using families of recombinant inbred strains: An overview and example of mapping susceptibility to Candida albicans induced illness phenotypes. *Pathog. Dis.* **2014**, *71*, 232–246.
32. Boivin, G.A.; Pothlichet, J.; Skamene, E.; Brown, E.G.; Loredo-Osti, J.C.; Sladek, R.; Vidal, S.M. Mapping of clinical and expression quantitative trait loci in a sex-dependent effect of host susceptibility to mouse-adapted influenza H3N2/HK/1/68. *J. Immunol.* **2012**, *188*, 3949–3960.
33. Ferris, M.T.; Aylor, D.L.; Bottomly, D.; Whitmore, A.C.; Aicher, L.D.; Bell, T.A.; Bradel-Tretheway, B.; Bryan, J.T.; Buus, R.J.; Gralinski, L.E.; *et al.* Modeling host genetic regulation of influenza pathogenesis in the collaborative cross. *PLoS Pathog.* **2013**, *9*, e1003196.
34. Guenet, J.L.; Bonhomme, F. Wild mice: An ever-increasing contribution to a popular mammalian model. *Trends Genet.* **2003**, *19*, 24–31.
35. Chia, R.; Achilli, F.; Festing, M.F.; Fisher, E.M. The origins and uses of mouse outbred stocks. *Nat. Genet.* **2005**, *37*, 1181–1186.
36. Keane, T.M.; Goodstadt, L.; Danecek, P.; White, M.A.; Wong, K.; Yalcin, B.; Heger, A.; Agam, A.; Slater, G.; Goodson, M.; *et al.* Mouse genomic variation and its effect on phenotypes and gene regulation. *Nature* **2011**, *477*, 289–294.
37. Simon, M.M.; Greenaway, S.; White, J.K.; Fuchs, H.; Gailus-Durner, V.; Wells, S.; Sorg, T.; Wong, K.; Bedu, E.; Cartwright, E.J.; *et al.* A comparative phenotypic and genomic analysis of C57BL/6J and C57BL/6N mouse strains. *Genet. Biol.* **2013**, *14*, R82.
38. Vidal, S.M.; Malo, D.; Marquis, J.F.; Gros, P. Forward genetic dissection of immunity to infection in the mouse. *Annu. Rev. Immunol.* **2008**, *26*, 81–132.

39. Cook, M.C.; Vinuesa, C.G.; Goodnow, C.C. ENU-mutagenesis: Insight into immune function and pathology. *Curr. Opin. Immunol.* **2006**, *18*, 627–633.
40. Hoebe, K.; Beutler, B. Forward genetic analysis of TLR-signaling pathways: An evaluation. *Adv. Drug Deliv. Rev.* **2008**, *60*, 824–829.
41. Hoyne, G.F.; Goodnow, C.C. The use of genomewide ENU mutagenesis screens to unravel complex mammalian traits: Identifying genes that regulate organ-specific and systemic autoimmunity. *Immunol. Rev.* **2006**, *210*, 27–39.
42. Oliver, P.L.; Davies, K.E. New insights into behaviour using mouse ENU mutagenesis. *Hum. Mol. Genet.* **2012**, *21*, R72–R81.
43. Russell, W.L. Radiation and chemical mutagenesis and repair in mice. *Johns Hopkins Med. J. Suppl.* **1972**, *1*, 239–247.
44. Jaenisch, R. Germ line integration and Mendelian transmission of the exogenous Moloney leukemia virus. *Proc. Natl. Acad. Sci. USA* **1976**, *73*, 1260–1264.
45. Russell, L.B.; Russell, W.L. Frequency and nature of specific-locus mutations induced in female mice by radiations and chemicals: A review. *Mutat. Res.* **1992**, *296*, 107–127.
46. Russell, W.L.; Kelly, E.M.; Hunsicker, P.R.; Bangham, J.W.; Maddux, S.C.; Phipps, E.L. Specific-locus test shows ethylnitrosourea to be the most potent mutagen in the mouse. *Proc. Natl. Acad. Sci. USA* **1979**, *76*, 5818–5819.
47. Russell, W.L.; Hunsicker, P.R.; Carpenter, D.A.; Cornett, C.V.; Guinn, G.M. Effect of dose fractionation on the ethylnitrosourea induction of specific-locus mutations in mouse spermatogonia. *Proc. Natl. Acad. Sci. USA* **1982**, *79*, 3592–3593.
48. Hitotsumachi, S.; Carpenter, D.A.; Russell, W.L. Dose-repetition increases the mutagenic effectiveness of N-ethyl-N-nitrosourea in mouse spermatogonia. *Proc. Natl. Acad. Sci. USA* **1985**, *82*, 6619–6621.
49. Singer, B. All oxygens in nucleic acids react with carcinogenic ethylating agents. *Nature* **1976**, *264*, 333–339.
50. Bignami, M.; Vitelli, A.; di Muccio, A.; Terlizzese, M.; Calcagnile, A.; Zapponi, G.A.; Lohman, P.H.M.; den Engelse, L.; Dogliotti1, E. Relationship between specific alkylated bases and mutations at two gene loci induced by ethylnitrosourea and diethyl sulfate in CHO cells. *Mutat. Res.* **1988**, *193*, 43–51.
51. Van Zeeland, A.A. Molecular dosimetry of alkylating agents: Quantitative comparison of genetic effects on the basis of DNA adduct formation. *Mutagenesis* **1988**, *3*, 179–191.
52. Vogel, E.W.; Natarajan, A.T. DNA damage and repair in somatic and germ cells *in vivo*. *Mutat. Res.* **1995**, *330*, 183–208.
53. Fairfield, H.; Gilbert, G.J.; Barter, M.; Corrigan, R.R.; Curtain, M.; Ding, Y.M.; Ascenzo, M.D.; Gerhardt, D.J.; He, C.; Huang, W.H.; *et al.* Mutation discovery in mice by whole exome sequencing. *Genet. Biol.* **2011**, *12*, R86.
54. Andrews, T.D.; Whittle, B.; Field, M.A.; Balakishnan, B.; Zhang, Y.; Shao, Y.; Cho, V.; Kirk, M.; Singh, M.; Xia, Y.; *et al.* Massively parallel sequencing of the mouse exome to accurately identify rare, induced mutations: An immediate source for thousands of new mouse models. *Open Biol.* **2012**, *2*, 120061.
55. Bull, K.R.; Rimmer, A.J.; Siggs, O.M.; Miosge, L.A.; Roots, C.M.; Enders, A.; Bertram, E.M.;

Crockford, T.L.; Whittle, B.; Potter, P.K.; *et al.* Unlocking the bottleneck in forward genetics using whole-genome sequencing and identity by descent to isolate causative mutations. *PLoS Genet.* **2013**, *9*, e1003219.

56. Lewis, M.A.; Quint, E.; Glazier, A.M.; Fuchs, H.; de Angelis, M.H.; Langford, C.; van Dongen, S.; Abreu-Goodger, C.; Piipari, M.; Redshaw, N.; *et al.* An ENU-induced mutation of miR-96 associated with progressive hearing loss in mice. *Nat. Genet.* **2009**, *41*, 614–618.
57. Masuya, H.; Sezutsu, H.; Sakuraba, Y.; Sagai, T.; Hosoya, M.; Kanedaa, H.; Miuraa, I.; Kobayashia, K.; Sumiyamad, K.; Shimizu, A.; *et al.* A series of ENU-induced single-base substitutions in a long-range cis-element altering sonic hedgehog expression in the developing mouse limb bud. *Genomics* **2007**, *89*, 207–214.
58. Arnold, C.N.; Barnes, M.J.; Berger, M.; Blasius, A.L.; Brandl, K.; Croker, B.; Crozat, K.; Du, X.; Eidenschenk, C.; Georgel, P.; *et al.* ENU-induced phenovariance in mice: Inferences from 587 mutations. *BMC Res. Notes* **2012**, *5*, 577.
59. Puk, O.; Moller, G.; Geerlof, A.; Krowiorz, K.; Ahmad, N.; Wagner, S.; Adamski, J.; de Angelis, M.H.; Graw, J. The pathologic effect of a novel neomorphic Fgf9(Y162C) allele is restricted to decreased vision and retarded lens growth. *PLoS ONE* **2011**, *6*, e23678.
60. Caspary, T. Phenotype-driven mouse ENU mutagenesis screens. *Methods Enzymol.* **2010**, *477*, 313–327.
61. Probst, F.J.; Justice, M.J. Mouse mutagenesis with the chemical supermutagen ENU. *Methods Enzymol.* **2010**, *477*, 297–312.
62. Georgel, P.; Du, X.; Hoebe, K.; Beutler, B. ENU mutagenesis in mice. *Methods Mol. Biol.* **2008**, *415*, 1–16.
63. Justice, M.J.; Carpenter, D.A.; Favor, J.; Neuhauser-Klaus, A.; de Angelis, M.H.; Soewarto, D.; Moser, A.; Cordes, S.; Miller, D.; Chapman, V.; *et al.* Effects of ENU dosage on mouse strains. *Mamm. Genet.* **2000**, *11*, 484–488.
64. Bainbridge, M.N.; Wang, M.; Burgess, D.L.; Kovar, C.; Rodesch, M.J.; D'Ascenzo, M.; Kitzman, J.; Wu, Y.-Q.; Newsham, I.; Richmond, T.A.; *et al.* Whole exome capture in solution with 3 Gbp of data. *Genet. Biol.* **2010**, *11*, R62.
65. Choi, M.; Scholl, U.I.; Ji, W.; Liu, T.; Tikhonova, I.R.; Zumbob, P.; Nayirc, A.; Bakkaloğlud, A.; Özend, S.; Sanjad, S.; *et al.* Genetic diagnosis by whole exome capture and massively parallel DNA sequencing. *Proc. Natl. Acad. Sci. USA* **2009**, *106*, 19096–19101.
66. Ng, S.B.; Buckingham, K.J.; Lee, C.; Bigham, A.W.; Tabor, H.K.; Dent, K.M.; Huff, C.D.; Shannon, P.T.; Jabs, E.W.; Nickerson, D.A.; *et al.* Exome sequencing identifies the cause of a mendelian disorder. *Nat. Genet.* **2010**, *42*, 30–35.
67. Pruitt, K.D.; Harrow, J.; Harte, R.A.; Wallin, C.; Diekhans, M.; Maglott, D.R.; Searle, S.; Farrell, C.M.; Loveland, J.E.; Ruef, B.J.; *et al.* The consensus coding sequence (CCDS) project: Identifying a common protein-coding gene set for the human and mouse genomes. *Genet. Res.* **2009**, *19*, 1316–1323.
68. Mouse Genome Project. Available online: http://www.sanger.ac.uk/resources/mouse/genomes/ (accessed on 28 August 2014).
69. DePristo, M.A.; Banks, E.; Poplin, R.; Garimella, K.V.; Maguire, J.R.; Hartl, C.; Philippakis, A.A.;

del Angel, G.; Rivas, M.A.; Hanna, M.; *et al.* A framework for variation discovery and genotyping using next-generation DNA sequencing data. *Nat. Genet.* **2011**, *43*, 491–498.

70. Pipeline Space Home. Available online: https://biowiki.atlassian.net/wiki/display/PS/Pipeline+Space+Home (accessed on 28 August 2014).
71. GATK Best Practices. Available online: http://www.broadinstitute.org/gatk/guide/best-practices (accessed on 8 August 2014).
72. Lohse, M.; Bolger, A.M.; Nagel, A.; Fernie, A.R.; Lunn, J.E.; Stitt, M.; Usadel, B. RobiNA: A user-friendly, integrated software solution for RNA-Seq-based transcriptomics. *Nucl. Acids Res.* **2012**, *40*, W622–W627.
73. Li, H.; Durbin, R. Fast and accurate long-read alignment with Burrows—Wheeler transform. *Bioinformatics* **2010**, *26*, 589–595.
74. Li, H.; Homer, N. A survey of sequence alignment algorithms for next-generation sequencing. *Br. Bioinform.* **2010**, *11*, 473–483.
75. Li, H.; Handsaker, B.; Wysoker, A.; Fennell, T.; Ruan, J.; Homer, N.; Marth, G.; Abecasis, G.; Durbin, R. The Sequence Alignment/Map format and SAMtools. *Bioinformatics* **2009**, *25*, 2078–2079.
76. Skliris, G.P.; Rowan, B.G.; Al-Dhaheri, M.; Williams, C.; Troup, S.; Begic, S.; Parisien, M.; Watson, P.H.; Murphy, L.C. Immunohistochemical validation of multiple phospho-specific epitopes for estrogen receptor alpha (ERalpha) in tissue microarrays of ERalpha positive human breast carcinomas. *Breast Cancer Res. Treat* **2009**, *118*, 443–453.
77. Picard. Available online: http://picard.sourceforge.net (accessed on 28 August 2014).
78. Garrison, E.; Marth, G. Haplotype-based variant detection from short-read sequencing. Available online: http://arxiv.org/abs/1207.3907 (accessed on 28 August 2014).
79. Nielsen, R.; Paul, J.S.; Albrechtsen, A.; Song, Y.S. Genotype and SNP calling from next-generation sequencing data. *Nat. Rev. Genet.* **2011**, *12*, 443–451.
80. Moresco, E.M.; Li, X.; Beutler, B. Going forward with genetics: Recent technological advances and forward genetics in mice. *Am. J. Pathol.* **2013**, *182*, 1462–1473.
81. Cingolani, P.; Platts, A.; Wang le, L.; Coon, M.; Nguyen, T.; Land, S.J.; Lu, X.; Ruden, D.M. A program for annotating and predicting the effects of single nucleotide polymorphisms, SnpEff: SNPs in the genome of Drosophila melanogaster strain w1118; iso-2; iso-3. *Fly (Austin)* **2012**, *6*, 80–92.
82. McLaren, W.; Pritchard, B.; Rios, D.; Chen, Y.; Flicek, P.; Cunningham, F. Deriving the consequences of genomic variants with the Ensembl API and SNP Effect Predictor. *Bioinformatics* **2010**, *26*, 2069–2070.
83. Sherry, S.T.; Ward, M.H.; Kholodov, M.; Baker, J.; Phan, L.; Smigielski, E.M.; Sirotkin, K. DbSNP: The NCBI database of genetic variation. *Nucl. Acids Res.* **2001**, *29*, 308–311.
84. Robinson, J.T.; Thorvaldsdottir, H.; Winckler, W.; Guttman, M.; Lander, E.S.; Getz, G.; Mesirov, J.P. Integrative genomics viewer. *Nat. Biotechnol.* **2011**, *29*, 24–26.
85. Sanger, F.; Nicklen, S.; Coulson, A.R. DNA sequencing with chain-terminating inhibitors. *Proc. Natl. Acad. Sci. USA* **1977**, *74*, 5463–5467.
86. Kwiatkowski, D.P. How malaria has affected the human genome and what human genetics can teach us about malaria. *Am. J. Hum. Genet.* **2005**, *77*, 171–192.

87. Ayi, K.; Turrini, F.; Piga, A.; Arese, P. Enhanced phagocytosis of ring-parasitized mutant erythrocytes: A common mechanism that may explain protection against falciparum malaria in sickle trait and beta-thalassemia trait. *Blood* **2004**, *104*, 3364–3371.
88. Friedman, M.J. Erythrocytic mechanism of sickle cell resistance to malaria. *Proc. Natl. Acad. Sci. USA* **1978**, *75*, 1994–1997.
89. Allen, S.J.; O'Donnell, A.; Alexander, N.D.; Mgone, C.S.; Peto, T.E.; Clegg, J.B.; Alpers, M.P.; Weatherall, D.J. Prevention of cerebral malaria in children in Papua New Guinea by southeast Asian ovalocytosis band 3. *Am. J. Trop. Med. Hyg.* **1999**, *60*, 1056–1060.
90. Foo, L.C.; Rekhraj, V.; Chiang, G.L.; Mak, J.W. Ovalocytosis protects against severe malaria parasitemia in the Malayan aborigines. *Am. J. Trop. Med. Hyg.* **1992**, *47*, 271–275.
91. Genton, B.; al-Yaman, F.; Mgone, C.S.; Alexander, N.; Paniu, M.M.; Alpers, M.P.; Mokela, D. Ovalocytosis and cerebral malaria. *Nature* **1995**, *378*, 564–565.
92. Miller, L.H.; Mason, S.J.; Clyde, D.F.; McGinniss, M.H. The resistance factor to Plasmodium vivax in blacks. The Duffy-blood-group genotype, FyFy. *N. Engl. J. Med.* **1976**, *295*, 302–304.
93. Ruwende, C.; Khoo, S.C.; Snow, R.W.; Yates, S.N.; Kwiatkowski, D.; Gupta, S.; Warn, P.; Allsopp, C.E.; Gilbert, S.C.; Peschu, N.; *et al.* Natural selection of hemi- and heterozygotes for G6PD deficiency in Africa by resistance to severe malaria. *Nature* **1995**, *376*, 246–249.
94. Tishkoff, S.A.; Varkonyi, R.; Cahinhinan, N.; Abbes, S.; Argyropoulos, G.; Destro-Bisol, G.; Drousiotou, A.; Dangerfield, B.; Lefranc, G.; Loiselet, J.; *et al.* Haplotype diversity and linkage disequilibrium at human G6PD: Recent origin of alleles that confer malarial resistance. *Science* **2001**, *293*, 455–462.
95. Aitman, T.J.; Cooper, L.D.; Norsworthy, P.J.; Wahid, F.N.; Gray, J.K.; Curtis, B.R.; McKeigue, P.M.; Kwiatkowski, D.; Greenwood, B.M.; Snow, R.W.; *et al.* Malaria susceptibility and CD36 mutation. *Nature* **2000**, *405*, 1015–1016.
96. Omi, K.; Ohashi, J.; Patarapotikul, J.; Hananantachai, H.; Naka, I.; Pottere, S.; Medanad, I.M.; Miua, J.; Ball, H.J. CD36 polymorphism is associated with protection from cerebral malaria. *Am. J. Hum. Genet.* **2003**, *72*, 364–374.
97. Hill, A.V.; Allsopp, C.E.; Kwiatkowski, D.; Anstey, N.M.; Twumasi, P.; Rowe, P.A.; Bennett, S.; Brewster, D.; McMichael, A.J.; Greenwood, B.M. Common west African HLA antigens are associated with protection from severe malaria. *Nature* **1991**, *352*, 595–600.
98. McGuire, W.; Hill, A.V.; Allsopp, C.E.; Greenwood, B.M.; Kwiatkowski, D. Variation in the TNF-alpha promoter region associated with susceptibility to cerebral malaria. *Nature* **1994**, *371*, 508–510.
99. Bongfen, S.E.; Laroque, A.; Berghout, J.; Gros, P. Genetic and genomic analyses of host-pathogen interactions in malaria. *Trends Parasitol.* **2009**, *25*, 417–422.
100. Verra, F.; Mangano, V.D.; Modiano, D. Genetics of susceptibility to Plasmodium falciparum: From classical malaria resistance genes towards genome-wide association studies. *Parasit. Immunol.* **2009**, *31*, 234–253.
101. Fortin, A.; Stevenson, M.M.; Gros, P. Complex genetic control of susceptibility to malaria in mice. *Genes Immun.* **2002**, *3*, 177–186.
102. Hunt, N.H.; Golenser, J.; Chan-Ling, T.; Parekh, S.; Rae, C.; Pottere, S.; Medanad, I.M.; Miua, J.; Ball, H.J. Immunopathogenesis of cerebral malaria. *Int. J. Parasitol.* **2006**, *36*, 569–582.

103. Miller, L.H.; Baruch, D.I.; Marsh, K.; Doumbo, O.K. The pathogenic basis of malaria. *Nature* **2002**, *415*, 673–679.
104. Newton, C.R.; Hien, T.T.; White, N. Cerebral malaria. *J. Neurol. Neurosurg. Psychiatry* **2000**, *69*, 433–441.
105. Tripathi, A.K.; Sha, W.; Shulaev, V.; Stins, M.F.; Sullivan, D.J., Jr. Plasmodium falciparum-infected erythrocytes induce NF-kappaB regulated inflammatory pathways in human cerebral endothelium. *Blood* **2009**, *114*, 4243–4252.
106. Brown, H.; Hien, T.T.; Day, N.; Mai, N.T.; Chuong, L.V.; Chau, T.T.; Loc, P.P.; Phu, N.H.; Bethell, D.; Farrar, J.; *et al.* Evidence of blood-brain barrier dysfunction in human cerebral malaria. *Neuropathol. Appl. Neurobiol.* **1999**, *25*, 331–340.
107. Mishra, S.K.; Newton, C.R. Diagnosis and management of the neurological complications of falciparum malaria. *Nat. Rev. Neurol.* **2009**, *5*, 189–198.
108. Hafalla, J.C.; Claser, C.; Couper, K.N.; Grau, G.E.; Renia, L.; de Souza, J.B.; Riley, E.M. The CTLA-4 and PD-1/PD-L1 inhibitory pathways independently regulate host resistance to Plasmodium-induced acute immune pathology. *PLoS Pathog.* **2012**, *8*, e1002504.
109. Lacerda-Queiroz, N.; Rodrigues, D.H.; Vilela, M.C.; Rachid, M.A.; Soriani, F.M.; Sousa, L.P.; Campos, R.D.; Quesniaux, V.F.; Teixeira, M.M.; Teixeira, A.L. Platelet-activating factor receptor is essential for the development of experimental cerebral malaria. *Am. J. Pathol.* **2012**, *180*, 246–255.
110. Lou, J.; Lucas, R.; Grau, G.E. Pathogenesis of cerebral malaria: Recent experimental data and possible applications for humans. *Clin. Microbiol. Rev.* **2001**, *14*, 810–820.
111. De Souza, J.B.; Riley, E.M. Cerebral malaria: The contribution of studies in animal models to our understanding of immunopathogenesis. *Microbes Infect.* **2002**, *4*, 291–300.
112. Senaldi, G.; Vesin, C.; Chang, R.; Grau, G.E.; Piguet, P.F. Role of polymorphonuclear neutrophil leukocytes and their integrin CD11a (LFA-1) in the pathogenesis of severe murine malaria. *Infect. Immun.* **1994**, *62*, 1144–1149.
113. Bagot, S.; Idrissa Boubou, M.; Campinom, S.; Behrschmidt, C.; Gorgette, O.; Guénet, J.L.; Penha-Gonçalves, C.; Mazier, D.; Pied, S.; Cazenave, P.A. Susceptibility to experimental cerebral malaria induced by Plasmodium berghei ANKA in inbred mouse strains recently derived from wild stock. *Infect. Immun.* **2002**, *70*, 2049–2056.
114. Berghout, J.; Min-Oo, G.; Tam, M.; Gauthier, S.; Stevenson, M.M.; Gros, P. Identification of a novel cerebral malaria susceptibility locus (Berr5) on mouse chromosome 19. *Genes Immun.* **2010**, *11*, 310–318.
115. Bopp, S.E.; Rodrigo, E.; Gonzalez-Paez, G.E.; Frazer, M.; Barnes, S.W.; Valim, C.; Watson, J.; Walker, J.R.; Schmedt, C.; Winzeler, E.A. Identification of the Plasmodium berghei resistance locus 9 linked to survival on chromosome 9. *Malar. J.* **2013**, *12*, 316.
116. Campino, S.; Bagot, S.; Bergman, M.L.; Almeida, P.; Sepulveda, N.; Pied, S.; Penha-Gonçalves, C.; Holmberg, D.; Cazenave, P.-A. Genetic control of parasite clearance leads to resistance to Plasmodium berghei ANKA infection and confers immunity. *Genes Immun.* **2005**, *6*, 416–421.
117. Ohno, T.; Nishimura, M. Detection of a new cerebral malaria susceptibility locus, using CBA mice. *Immunogenetics* **2004**, *56*, 675–678.
118. Bopp, S.E.; Ramachandran, V.; Henson, K.; Luzader, A.; Lindstrom, M.; Spooner, M.; Steffy, B.M.;

Suzuki, O.; Janse, C.; Waters, A.P.; *et al.* Genome wide analysis of inbred mouse lines identifies a locus containing Ppar-gamma as contributing to enhanced malaria survival. *PLoS ONE* **2010**, *5*, e10903.

119. Bongfen, S.E.; Rodrigue-Gervais, I.G.; Berghout, J.; Torre, S.; Cingolani, P.; Wiltshire, S.A.; Leiva-Torres, G.A.; Letourneau, L.; Sladek, R.; Blanchette, M.; *et al.* An N-ethyl-N-nitrosourea (ENU)-induced dominant negative mutation in the JAK3 kinase protects against cerebral malaria. *PLoS ONE* **2012**, *7*, e31012.
120. Nosaka, T.; van Deursen, J.M.; Tripp, R.A.; Thierfelder, W.E.; Witthuhn, B.A.; McMickle, A.P.; Doherty, P.C.; Grosveld, G.C.; Ihle, J.N. Defective lymphoid development in mice lacking Jak3. *Science* **1995**, *270*, 800–802.
121. Jostins, L.; Ripke, S.; Weersma, R.K.; Duerr, R.H.; McGovern, D.P.; Hui, K.Y.; Lee, J.C.; Schumm, L.P.; Sharma, Y.; Anderson, C.A.; *et al.* Host-microbe interactions have shaped the genetic architecture of inflammatory bowel disease. *Nature* **2012**, *491*, 119–124.
122. International Multiple Sclerosis Genetics Consortium. The genetic association of variants in CD6, TNFRSF1A and IRF8 to multiple sclerosis: A multicenter case-control study. *PLoS ONE* **2011**, *6*, e18813.
123. Jakkula, E.; Leppa, V.; Sulonen, A.M.; Varilo, T.; Kallio, S.; Kemppinen, A.; Purcell, S.; Koivisto, K.; Tienari, P.; Sumelahti, M.; *et al.* Genome-wide association study in a high-risk isolate for multiple sclerosis reveals associated variants in STAT3 gene. *Am. J. Hum. Genet.* **2010**, *86*, 285–291.
124. Torre, S.; van Bruggen, R.; Kennedy, J.M.; Berghout, J.; Bongfen, S.E.; Langat, P.; Lathrop, M.; Vidal, S.M.; Gros, P. Susceptibility to lethal cerebral malaria is regulated by epistatic interaction between chromosome 4 (Berr6) and chromosome 1 (Berr7) loci in mice. *Genes Immun.* **2013**, *14*, 470.
125. Torre, S.; Gros, P. Department of Human Genetics, Department of Biochemistry, and Complex Traits Group, McGill University, Montréal, QC, Canada. Unpublished data, 2014.
126. Pignata, C.; Fusco, A.; Amorosi, S. Human clinical phenotype associated with FOXN1 mutations. *Adv. Exp. Med. Biol.* **2009**, *665*, 195–206.
127. Rank, G.; Sutton, R.; Marshall, V.; Lundie, R.J.; Caddy, J.; Romeo, T.; Fernandez, K.; Mccormack, M.P.; Cooke, B.M.; Foote, S.J.; *et al.* Novel roles for erythroid Ankyrin-1 revealed through an ENU-induced null mouse mutant. *Blood* **2009**, *113*, 3352–3362.
128. Greth, A.; Lampkin, S.; Mayura-Guru, P.; Rodda, F.; Drysdale, K.; Roberts-Thomson, M.; McMorran, B.J.; Foote, S.J.; Burgio, G. A novel ENU-mutation in ankyrin-1 disrupts malaria parasite maturation in red blood cells of mice. *PLoS One* **2012**, *7*, e38999.
129. Belnoue, E.; Kayibanda, M.; Vigario, A.M.; Deschemin, J.C.; van Rooijen, N.; Viguier, M.; Snounou, G.; Rénia, L. On the pathogenic role of brain-sequestered alphabeta CD8+ T cells in experimental cerebral malaria. *J. Immunol.* **2002**, *169*, 6369–6375.
130. Finley, R.W.; Mackey, L.J.; Lambert, P.H.; Virulent, P. berghei malaria: Prolonged survival and decreased cerebral pathology in cell-dependent nude mice. *J. Immunol.* **1982**, *129*, 2213–2218.
131. Renia, L.; Potter, S.M.; Mauduit, M.; Rosa, D.S.; Kayibanda, M.; Deschemina, J.; Snounoub, G.; Grüner, A.C. Pathogenic T cells in cerebral malaria. *Int. J. Parasitol.* **2006**, *36*, 547–554.

132. Grau, G.E.; Piguet, P.F.; Engers, H.D.; Louis, J.A.; Vassalli, P.; Lambert, P.H. L3T4+ T lymphocytes play a major role in the pathogenesis of murine cerebral malaria. *J. Immunol.* **1986**, *137*, 2348–2354.
133. Crump, J.A.; Luby, S.P.; Mintz, E.D. The global burden of typhoid fever. *Bull. World Health Organ.* **2004**, *82*, 346–353.
134. Mastroeni, P.; Maskell, D. *Salmonella Infections: Clinical, Immunological, and Molecular Aspects*; Cambridge University Press: Cambridge, UK, 2006; Volume 13, p. 381.
135. Gordon, M.A.; Banda, H.T.; Gondwe, M.; Gordon, S.B.; Boeree, M.J.; Walsh, A.L.; Corkill, J.E.; Hart, C.A.; Gilks, C.F.; Molyneux, M.E. Non-typhoidal *Salmonella* bacteraemia among HIV-infected Malawian adults: High mortality and frequent recrudescence. *AIDS* **2002**, *16*, 1633–1641.
136. Majowicz, S.E.; Musto, J.; Scallan, E.; Angulo, F.J.; Kirk, M.; O'Brien, S.J.; Jones, T.F.; Fazil, A.; Hoekstra, R.M. The global burden of nontyphoidal *Salmonella* gastroenteritis. *Clin. Infect. Dis.* **2010**, *50*, 882–889.
137. Mittrucker, H.W.; Kaufmann, S.H. Immune response to infection with *Salmonella typhimurium* in mice. *J. Leukoc. Biol.* **2000**, *67*, 457–463.
138. Dougan, G.; John, V.; Palmer, S.; Mastroeni, P. Immunity to salmonellosis. *Immunol. Rev.* **2011**, *240*, 196–210.
139. Alcais, A.; Abel, L.; Casanova, J.L. Human genetics of infectious diseases: Between proof of principle and paradigm. *J. Clin. Invest.* **2009**, *119*, 2506–2514.
140. Bustamante, J.; Zhang, S.Y.; von Bernuth, H.; Abel, L.; Casanova, J.L. From infectious diseases to primary immunodeficiencies. *Immunol. Allergy Clin. North Am.* **2008**, *28*, 235–258.
141. Casanova, J.L.; Fieschi, C.; Zhang, S.Y.; Abel, L. Revisiting human primary immunodeficiencies. *J. Intern. Med.* **2008**, *264*, 115–127.
142. De Beaucoudrey, L.; Samarina, A.; Bustamante, J.; Cobat, A.; Boisson-Dupuis, S.; Feinberg, J.; Al-Muhsen, S.; Jannière, L.; Rose, Y.; Desurenaim, M.; *et al.* Revisiting human IL-12Rbeta1 deficiency: A survey of 141 patients from 30 countries. *Med. (Baltim.)* **2010**, *89*, 381–402.
143. Gordon, M. *Salmonella* infections in immunocompromised adults. *J. Infect.* **2008**, *56*, 413–422.
144. Lammas, D.A.; Casanova, J.L.; Kumararatne, D.S. Clinical consequences of defects in the IL-12-dependent interferon-gamma (IFN-gamma) pathway. *Clin. Exp. Immunol.* **2000**, *121*, 417–425.
145. Dunstan, S.J.; Stephens, H.A.; Blackwell, J.M.; Duc, C.M.; Lanh, M.N.; Dudbridge, F.; Phuong, C.X.; Luxemburger, C.; Wain, J.; Ho, V.A.; *et al.* Genes of the class II and class III major histocompatibility complex are associated with typhoid fever in Vietnam. *J. Infect. Dis.* **2001**, *183*, 261–268.
146. House, D.; Bishop, A.; Parry, C.; Dougan, G.; Wain, J. Typhoid fever: Pathogenesis and disease. *Curr. Opin. Infect. Dis. 2001*, *14*, 573–578.
147. Santos, R.L.; Zhang, S.; Tsolis, R.M.; Kingsley, R.A.; Adams, L.G.; Bäumler, A.J. Animal models of *Salmonella* infections: Enteritis *versus* typhoid fever. *Microbes Infect.* **2001**, *3*, 1335–1344.
148. Roy, M.F.; Malo, D. Genetic regulation of host responses to *Salmonella* infection in mice. *Genes Immun.* **2002**, *3*, 381–393.

149. Malo, D.; Vogan, K.; Vidal, S.; Hu, J.; Cellier, M.; Schurr, E.; Fuks, A.; Bumstead, N.; Morgan, K.; Gros, P. Haplotype mapping and sequence analysis of the mouse Nramp gene predict susceptibility to infection with intracellular parasites. *Genomics* **1994**, *23*, 51–61.
150. Qureshi, S.T.; Lariviere, L.; Leveque, G.; Clermont, S.; Moore, K.J.; Gros, P.; Malo, D. Endotoxin-tolerant mice have mutations in Toll-like receptor 4 (Tlr4). *J. Exp. Med.* **1999**, *189*, 615–625.
151. Poltorak, A.; He, X.; Smirnova, I.; Liu, M.Y.; van Huffel, C.; Du, X.; Birdwell, D.; Alejos, E.; Silva, M.; Galanos, C.; *et al.* Defective LPS signaling in C3H/HeJ and C57BL/10ScCr mice: Mutations in Tlr4 gene. *Science* **1998**, *282*, 2085–2088.
152. Vidal, S.M.; Malo, D.; Vogan, K.; Skamene, E.; Gros, P. Natural resistance to infection with intracellular parasites: Isolation of a candidate for Bcg. *Cell* **1993**, *73*, 469–485.
153. Roy, M.F.; Riendeau, N.; Bedard, C.; Helie, P.; Min-Oo, G.; Turcotte, K.; Gros, P.; Canonne-Hergaux, F.; Malo, D. Pyruvate kinase deficiency confers susceptibility to *Salmonella typhimurium* infection in mice. *J. Exp. Med.* **2007**, *204*, 2949–2961.
154. Yuki, K.E.; Eva, M.M.; Richer, E.; Chung, D.; Paquet, M.; Cellier, M.; Canonne-Hergaux, F.; Vaulont, S.; Vidal, S.M.; Malo, D.; *et al.* Suppression of hepcidin expression and iron overload mediate *Salmonella* susceptibility in ankyrin 1 ENU-induced mutant. *PLoS One* **2013**, *8*, e55331.
155. Eva, M.M.; Yuki, K.E.; Dauphinee, S.M.; Schwartzentruber, J.A.; Pyzik, M.; Paquet, M.; Lathrop, M.; Majewski, J.; Vidal, S.M.; Malo, D.; *et al.* Altered IFN-gamma-mediated immunity and transcriptional expression patterns in N-Ethyl-N-nitrosourea-induced STAT4 mutants confer susceptibility to acute typhoid-like disease. *J. Immunol.* **2014**, *192*, 259–270.
156. Richer, E.; Prendergast, C.; Zhang, D.E.; Qureshi, S.T.; Vidal, S.M.; Malo, D. N-ethyl-N-nitrosourea-induced mutation in ubiquitin-specific peptidase 18 causes hyperactivation of IFN-alphass signaling and suppresses STAT4-induced IFN-gamma production, resulting in increased susceptibility to *Salmonella typhimurium*. *J. Immunol.* **2010**, *185*, 3593–3601.
157. Malakhova, O.; Malakhov, M.; Hetherington, C.; Zhang, D.E. Lipopolysaccharide activates the expression of ISG15-specific protease UBP43 via interferon regulatory factor 3. *J. Biol. Chem.* **2002**, *277*, 14703–14711.
158. Kim, K.I.; Yan, M.; Malakhova, O.; Luo, J.K.; Shen, M.F.; Zou, W.; de la Torre, J.C.; Zhang, D. Ube1L and protein ISGylation are not essential for alpha/beta interferon signaling. *Mol. Cell. Biol.* **2006**, *26*, 472–479.
159. Richer, E.; Yuki, K.E.; Dauphinee, S.M.; Lariviere, L.; Paquet, M.; Malo, D. Impact of Usp18 and IFN signaling in *Salmonella*-induced typhlitis. *Genes Immun.* **2011**, *12*, 531–543.
160. Dauphinee, S.M.; Richer, E.; Eva, M.M.; McIntosh, F.; Paquet, M.; Dangoor, D.; Burkart, C.; Zhang, D.E.; Gruenheid, S.; Gros, P.; *et al.* Contribution of increased ISG15, ISGylation and deregulated type I IFN signaling in Usp18 mutant mice during the course of bacterial infections. *Genes Immun.* **2014**, *15*, 282–292.
161. Perrotta, S.; Gallagher, P.G.; Mohandas, N. Hereditary spherocytosis. *Lancet* **2008**, *372*, 1411–1426.
162. Eber, S.W.; Gonzalez, J.M.; Lux, M.L.; Scarpa, A.L.; Tse, W.T.; Dornwell, M.; Herbers, J.; Kugler, W.; Ozcan, R.; Pekrun, A.; *et al.* Ankyrin-1 mutations are a major cause of dominant and recessive hereditary spherocytosis. *Nat. Genet.* **1996**, *13*, 214–218.

163. Min-Oo, G.; Fortin, A.; Tam, M.F.; Nantel, A.; Stevenson, M.M.; Gros, P. Pyruvate kinase deficiency in mice protects against malaria. *Nat. Genet.* **2003**, *35*, 357–362.

164. Cunnington, A.J.; de Souza, J.B.; Walther, M.; Riley, E.M. Malaria impairs resistance to *Salmonella* through heme- and heme oxygenase-dependent dysfunctional granulocyte mobilization. *Nat. Med.* **2012**, *18*, 120–127.

165. Hoebe, K.; Georgel, P.; Rutschmann, S.; Du, X.; Mudd, S.; Crozat, K.; Sovath, S.; Shame, L.; Hartung, T.; Zähringer, U.; *et al.* CD36 is a sensor of diacylglycerides. *Nature* **2005**, *433*, 523–527.

166. Xiao, N.; Eidenschenk, C.; Krebs, P.; Brandl, K.; Blasius, A.L.; Xia, Y.; Khovananth, K.; Smart, N.G.; Beutler, B. The Tpl2 mutation Sluggish impairs type I IFN production and increases susceptibility to group B streptococcal disease. *J. Immunol.* **2009**, *183*, 7975–7983.

167. Rutschmann, S.; Hoebe, K.; Zalevsky, J.; Du, X.; Mann, N.; Dahiyat, B.I.; Steed, P.; Beutler, B. PanR1, a dominant negative missense allele of the gene encoding TNF-alpha (Tnf), does not impair lymphoid development. *J. Immunol.* **2006**, *176*, 7525–7532.

168. Sauer, J.D.; Sotelo-Troha, K.; von Moltke, J.; Monroe, K.M.; Rae, C.S.; Brubaker, S.W.; Hyodo, M.; Hayakawa, Y.; Woodward, J.J.; Portnoy, D.A.; *et al.* The N-ethyl-N-nitrosourea-induced Goldenticket mouse mutant reveals an essential function of Sting in the *in vivo* interferon response to Listeria monocytogenes and cyclic dinucleotides. *Infect. Immun.* **2011**, *79*, 688–694.

169. Berger, M.; Krebs, P.; Crozat, K.; Li, X.; Croker, B.A.; Siggs, O.M.; Popkin, D.; Du, X.; Lawson, B.R.; Theofilopoulos, A.N.; *et al.* An Slfn2 mutation causes lymphoid and myeloid immunodeficiency due to loss of immune cell quiescence. *Nat. Immunol.* **2010**, *11*, 335–343.

170. Ordonez-Rueda, D.; Jonsson, F.; Mancardi, D.A.; Zhao, W.; Malzac, A.; Liang, Y.; Bertosio, E.; Grenot, P.; Blanquet, V.; Sabrautzki, S.; *et al.* A hypomorphic mutation in the Gfi1 transcriptional repressor results in a novel form of neutropenia. *Eur. J. Immunol.* **2012**, *42*, 2395–2408.

171. Georgel, P.; Crozat, K.; Lauth, X.; Makrantonaki, E.; Seltmann, H.; Sovath, S.; Hoebe, K.; Du, X.; Rutschmann, S.; Jiang, Z.; *et al.* A toll-like receptor 2-responsive lipid effector pathway protects mammals against skin infections with gram-positive bacteria. *Infect. Immun.* **2005**, *73*, 4512–4521.

172. Croker, B.A.; Lawson, B.R.; Rutschmann, S.; Berger, M.; Eidenschenk, C.; Blasius, A.L.; Moresco, E.M.Y.; Sovath, S.; Cengia, L.; Shultz, L.D.; *et al.* Inflammation and autoimmunity caused by a SHP1 mutation depend on IL-1, MyD88, and a microbial trigger. *Proc. Natl. Acad. Sci. USA* **2008**, *105*, 15028–15033.

173. Jaeger, B.N.; Donadieu, J.; Cognet, C.; Bernat, C.; Ordonez-Rueda, D.; Barlogis, V.; Mahlaoui, N.; Fenis, A.; Narni-Mancinelli, E.; Beaupain, B.; *et al.* Neutrophil depletion impairs natural killer cell maturation, function, and homeostasis. *J. Exp. Med.* **2012**, *209*, 565–580.

174. Barquero-Calvo, E.; Martirosyan, A.; Ordonez-Rueda, D.; Arce-Gorvel, V.; Alfaro-Alarcon, A.; Lepidi, H.; Malissen, B.; Malissen, M.; Gorvel, J.; Moreno, E. Neutrophils exert a suppressive effect on Th1 responses to intracellular pathogen Brucella abortus. *PLoS Pathog.* **2013**, *9*, e1003167.

175. Krmpotic, A.; Bubic, I.; Polic, B.; Lucin, P.; Jonjic, S. Pathogenesis of murine cytomegalovirus infection. *Microbes Infect.* **2003**, *5*, 1263–1277.

176. Rawlinson, W.D.; Farrell, H.E.; Barrell, B.G. Analysis of the complete DNA sequence of murine cytomegalovirus. *J. Virol.* **1996**, *70*, 8833–8849.

177. Pyzik, M.; Gendron-Pontbriand, E.M.; Vidal, S.M. The impact of Ly49-NK cell-dependent

recognition of MCMV infection on innate and adaptive immune responses. *J. Biomed. Biotechnol.* **2011**, *2011*, 641702.

178. Moresco, E.M.; Beutler, B. Resisting viral infection: The gene by gene approach. *Curr. Opin. Virol.* **2011**, *1*, 513–518.
179. Tabeta, K.; Georgel, P.; Janssen, E.; Du, X.; Hoebe, K.; Crozat, K.; Mudd, S.; Shamel, L.; Sovath, S.; Goode, J.; *et al.* Toll-like receptors 9 and 3 as essential components of innate immune defense against mouse cytomegalovirus infection. *Proc. Natl. Acad. Sci. USA* **2004**, *101*, 3516–3521.
180. Hoebe, K.; Du, X.; Georgel, P.; Janssen, E.; Tabeta, K.; Kim, S.O.; Goode, J.; Lin, P.; Mann, N.; Mudd, S.; *et al.* Identification of Lps2 as a key transducer of MyD88-independent TIR signalling. *Nature* **2003**, *424*, 743–748.
181. Tabeta, K.; Hoebe, K.; Janssen, E.M.; Du, X.; George, P.; Crozat, K.; Mudd, S.; Mann, N.; Sovath, S.; Goode, J.; *et al.* The Unc93b1 mutation 3d disrupts exogenous antigen presentation and signaling via Toll-like receptors 3, 7 and 9. *Nat. Immunol.* **2006**, *7*, 156–164.
182. Crozat, K.; Georgel, P.; Rutschmann, S.; Mann, N.; Du, X.; Hoebe, K.; Beutlerm, B. Analysis of the MCMV resistome by ENU mutagenesis. *Mamm. Genome* **2006**, *17*, 398–406.
183. Siggs, O.M.; Berger, M.; Krebs, P.; Arnold, C.N.; Eidenschenk, C.; Huberb, C.; Piriea, E.; Smarta, N.G.; Khovanantha, K.; Xia, Y.; *et al.* A mutation of Ikbkg causes immune deficiency without impairing degradation of IkappaB alpha. *Proc. Natl. Acad. Sci. USA* **2010**, *107*, 3046–3051.
184. Won, S.; Eidenschenk, C.; Arnold, C.N.; Siggs, O.M.; Sun, L.; Brandla, K.; Mullenb, T.; Nemerowb, G.R.; Morescoa, E.M.Y.; Beutler, B. Increased susceptibility to DNA virus infection in mice with a GCN2 mutation. *J. Virol.* **2012**, *86*, 1802–1808.
185. Biron, C.A. Initial and innate responses to viral infections—Pattern setting in immunity or disease. *Curr. Opin. Microbiol.* **1999**, *2*, 374–381.
186. Bukowski, J.F.; Woda, B.A.; Habu, S.; Okumura, K.; Welsh, R.M. Natural killer cell depletion enhances virus synthesis and virus-induced hepatitis *in vivo*. *J. Immunol.* **1983**, *131*, 1531–1538.
187. Bukowski, J.F.; Woda, B.A.; Welsh, R.M. Pathogenesis of murine cytomegalovirus infection in natural killer cell-depleted mice. *J. Virol.* **1984**, *52*, 119–128.
188. Welsh, R.M.; Dundon, P.L.; Eynon, E.E.; Brubaker, J.O.; Koo, G.C.; O'Donnell, C.L. Demonstration of the antiviral role of natural killer cells *in vivo* with a natural killer cell-specific monoclonal antibody (NK 1.1). *Nat. Immun. Cell Growth Regul.* **1990**, *9*, 112–120.
189. Brown, M.G.; Dokun, A.O.; Heusel, J.W.; Smith, H.R.; Beckman, D.L.; Blattenberger, E.A.; Dubbelde, C.E.; Stone, L.R.; Scalzo, A.A.; Yokoyama, W.M. Vital involvement of a natural killer cell activation receptor in resistance to viral infection. *Science* **2001**, *292*, 934–937.
190. Lee, S.H.; Girard, S.; Macina, D.; Busa, M.; Zafer, A.; Belouchi, A.; Gros, P.; Vidal, S.M. Susceptibility to mouse cytomegalovirus is associated with deletion of an activating natural killer cell receptor of the C-type lectin superfamily. *Nat. Genet.* **2001**, *28*, 42–45.
191. Arase, H.; Mocarski, E.S.; Campbell, A.E.; Hill, A.B.; Lanier, L.L. Direct recognition of cytomegalovirus by activating and inhibitory NK cell receptors. *Science* **2002**, *296*, 1323–1326.
192. Smith, H.R.; Heusel, J.W.; Mehta, I.K.; Kim, S.; Dorner, B.G.; Naidenko, O.V.; Iizuka, K.; Furukawa, H.; Beckman, D.L.; Pingel, J.T.; *et al.* Recognition of a virus-encoded ligand by a natural killer cell activation receptor. *Proc. Natl. Acad. Sci. USA* **2002**, *99*, 8826–8831.
193. Dokun, A.O.; Kim, S.; Smith, H.R.; Kang, H.S.; Chu, D.T.; Yokoyama, W.M. Specific and

nonspecific NK cell activation during virus infection. *Nat. Immunol.* **2001**, *2*, 951–956.

194. Barnes, M.J.; Aksoylar, H.; Krebs, P.; Bourdeau, T.; Arnold, C.N.; Xia, Y.; Khovananth, K.; Engel, I.; Sovath, S.; Lampe, K.; *et al.* Loss of T cell and B cell quiescence precedes the onset of microbial flora-dependent wasting disease and intestinal inflammation in Gimap5-deficient mice. *J. Immunol.* **2010**, *184*, 3743–3754.
195. Crozat, K.; Hoebe, K.; Ugolini, S.; Hong, N.A.; Janssen, E.; Rutschmann, S.; Mudd, S.; Sovath, S.; Vivier, E.; Beutler, B. Jinx, an MCMV susceptibility phenotype caused by disruption of Unc13d: A mouse model of type 3 familial hemophagocytic lymphohistiocytosis. *J. Exp. Med.* **2007**, *204*, 853–863.
196. Crozat, K.; Eidenschenk, C.; Jaeger, B.N.; Krebs, P.; Guia, S.; Beutler, B.; Vivier, E.; Ugolini, S. Impact of beta2 integrin deficiency on mouse natural killer cell development and function. *Blood* **2011**, *117*, 2874–2882.
197. Eidenschenk, C.; Crozat, K.; Krebs, P.; Arens, R.; Popkin, D.; Arnolda, C.N.; Blasiusa, A.L.; Benedictb, C.A.; Morescoa, E.M.Y.; Xiaa, Y.; *et al.* Flt3 permits survival during infection by rendering dendritic cells competent to activate NK cells. *Proc. Natl. Acad. Sci. USA* **2010**, *107*, 9759–9764.
198. Croker, B.; Crozat, K.; Berger, M.; Xia, Y.; Sovath, S.; Schaffer, L.; Eleftherianos, I.; Imler, J.; Beutler, B. ATP-sensitive potassium channels mediate survival during infection in mammals and insects. *Nat. Genet.* **2007**, *39*, 1453–1460.
199. Abel, L.; Plancoulaine, S.; Jouanguy, E.; Zhang, S.Y.; Mahfoufi, N.; Nicolas, N.; Sancho-Shimizu, V.; Alcaïs, A.; Guo, Y.; Cardon, A.; *et al.* Age-dependent Mendelian predisposition to herpes simplex virus type 1 encephalitis in childhood. *J. Pediatr.* **2010**, *157*, 623–629.
200. Yao, H.W.; Ling, P.; Chen, S.H.; Tung, Y.Y.; Chen, S.H. Factors affecting herpes simplex virus reactivation from the explanted mouse brain. *Virology* **2012**, *433*, 116–123.
201. Kennedy, P.G.; Chaudhuri, A. Herpes simplex encephalitis. *J. Neurol. Neurosurg. Psychiatry* **2002**, *73*, 237–238.
202. Casrouge, A.; Zhang, S.Y.; Eidenschenk, C.; Jouanguy, E.; Puel, A.; Yang, K.; Alcais, A.; Picard, C.; Mahfoufi, N.; Nicolas, N.; *et al.* Herpes simplex virus encephalitis in human UNC-93B deficiency. *Science* **2006**, *314*, 308–312.
203. Zhang, S.Y.; Jouanguy, E.; Ugolini, S.; Smahi, A.; Elain, G.; Romero, P.; Segal, D.; Sancho-Shimizu, V.; Lorenzo, L.; Puel, A.; *et al.* TLR3 deficiency in patients with herpes simplex encephalitis. *Science* **2007**, *317*, 1522–1527.
204. Sancho-Shimizu, V.; Perez de Diego, R.; Lorenzo, L.; Halwani, R.; Alangari, A.; Israelsson, E.; Fabrega, S.; Cardon, A.; Maluenda, J.; Tatematsu, M.; *et al.* Herpes simplex encephalitis in children with autosomal recessive and dominant TRIF deficiency. *J. Clin. Invest.* **2011**, *121*, 4889–4902.
205. Perez de Diego, R.; Sancho-Shimizu, V.; Lorenzo, L.; Puel, A.; Plancoulaine, S.; Picard, C.; Herman, M.; Cardon, A.; Durandy, A.; Bustamante, J.; *et al.* Human TRAF3 adaptor molecule deficiency leads to impaired Toll-like receptor 3 response and susceptibility to herpes simplex encephalitis. *Immunity* **2010**, *33*, 400–411.

206. Guo, Y.; Audry, M.; Ciancanelli, M.; Alsina, L.; Azevedo, J.; Herman, M.; Anguiano, E.; Sancho-Shimizu, V.; Lorenzo, L.; Pauwels, E.; *et al.* Herpes simplex virus encephalitis in a patient with complete TLR3 deficiency: TLR3 is otherwise redundant in protective immunity. *J. Exp. Med.* **2011**, *208*, 2083–2098.
207. Herman, M.; Ciancanelli, M.; Ou, Y.H.; Lorenzo, L.; Klaudel-Dreszler, M.; Pauwels, E.; Sancho-Shimizu, V.; de Diego, R.P.; Abhyankar, A.; Israelsson, E.; *et al.* Heterozygous TBK1 mutations impair TLR3 immunity and underlie herpes simplex encephalitis of childhood. *J. Exp. Med.* **2012**, *209*, 1567–1582.
208. Xia, Y.; Won, S.; Du, X.; Lin, P.; Ross, C.; la Vine, D.; Wiltshire, S.; Leiva, G.; Vidal, S.M.; Whittle, B.; *et al.* Bulk segregation mapping of mutations in closely related strains of mice. *Genetics* **2010**, *186*, 1139–1146.
209. Caignard, G.; Leiva-Torres, G.A.; Leney-Greene, M.; Charbonneau, B.; Dumaine, A.; Fodil-Cornu, N.; Pyzik, M.; Cingolani, P.; Schwartzentruber, J.; Dupaul-Chicoine, J.; *et al.* Genome-wide mouse mutagenesis reveals CD45-mediated T cell function as critical in protective immunity to HSV-1. *PLoS Pathog.* **2013**, *9*, e1003637.
210. Caignard, G.; Vidal, S.M. Department of Human Genetics and Complex Traits Group, McGill University, Montréal, QC, Canada. Unpublished data, 2014.
211. Caignard, G.; Gros, P.; Vidal, S.M. Department of Human Genetics, Department of Biochemistry, and Complex Traits Group, McGill University, Montréal, QC, Canada. Unpublished data, 2014.
212. Tchilian, E.Z.; Beverley, P.C. Altered CD45 expression and disease. *Trends Immunol.* **2006**, *27*, 146–153.

8

An Efficient Estimator of the Mutation Parameter and Analysis of Polymorphism from the 1000 Genomes Project

Yunxin Fu [1,2]

[1] Division of Biostatistics and Human Genetics Center, The University of Texas Health Science Center at Houston, 1200 Herman Pressler, Houston, TX 77025, USA; E-Mail: yunxin.fu@uth.tmc.edu;

[2] Laboratory for Conservation and Utilization of Bio-Resources, Yunnan University, Kunming 650091, China

---

Abstract: The mutation parameter θ is fundamental and ubiquitous in the analysis of population samples of DNA sequences. This paper presents a new highly efficient estimator of θ by utilizing the phylogenetic information among distinct alleles in a sample of DNA sequences. The new estimator, called Allelic BLUE, is derived from a generalized linear model about the mutations in the allelic genealogy. This estimator is not only highly accurate, but also computational efficient, which makes it particularly useful for estimating θ for large samples, as well as for a large number of cases, such as the situation of analyzing sequence data from a large genome project, such as the 1000 Genomes Project. Simulation shows that Allelic BLUE is nearly unbiased, with variance nearly as small as the minimum achievable variance, and in many situations, it can be hundreds- or thousands-fold more efficient than a previous method, which was already quite efficient compared to other approaches. One useful feature of the new estimator is its applicability to collections of distinct alleles without detailed frequencies. The utility of the new estimator is demonstrated by analyzing the pattern of θ in the data from the 1000 Genomes Project.

Keywords: mutation parameter; coalescent; allelic genealogy; 1000 Genomes Project

---

1. Introduction

The pattern of genetic polymorphism in a population can be influenced by a number of factors, among which the mutation parameter (commonly denoted by θ) plays a central role. θ is defined as $4Nu$ and

$2Nu$ for diploid and haploid genomes, respectively, where N is the effective population size and u is the mutation rate per sequence per generation. Almost all existing summary statistics for polymorphism are related to θ. Well-known examples include the number of alleles in a sample [1] the number (K) of segregating sites (or polymorphic sites) [2], mean number (Π) of nucleotide differences between two sequences [3] and the number of mutations of various sizes [4].

Due to the fundamental nature of this parameter for understanding both population dynamics, as well as the mechanism of evolution, it is important that it can be estimated as accurately as possible. Classical estimators include Watterson's estimator [2], Tajima's estimator [3], Ewens' estimator based on the number of alleles [5] in the sample and the heterozygosity estimator [6]. Under the assumption of a single random mating population evolving according to the Wright–Fisher model with constant population size and neutral mutations, these estimators are all either unbiased or nearly unbiased. However, their variances, which are the primary measure of accuracy of an estimator, can differ considerably and, furthermore, are substantially larger than the minimum achievable variance [7]:

$$V_{min} = \theta \left[\sum_{i=1}^{n-1} \frac{1}{\theta + i} \right]^{-1} \tag{1}$$

where n is the sample size. Realizing the limitations of these classical estimators, several new approaches were developed in the 1990s, all utilizing the fine structural result of coalescent theory [3,8,9]. Representative are Griffiths and Tavare's Markov Chain Monte Carlo (MCMC) estimator [10,11] based on recurrent equations for the probability of the polymorphism configuration, Knuher and Felsenstein's MCMC method [12] based on Metropolitan-Hasting sampling and Fu's BLUE estimators [13,14] based on linear regression taking advantage of the linear relationship between mutations in the genealogy of a sample and the mutation parameter. These new groups of estimators can all achieve substantially smaller variances and may even reach the minimum variance [13]. One common feature of these estimators is that they are all computationally intensive and, as a result, are suitable for only relatively smaller samples. Such limitations are particularly serious for the MCMC-based approach.

The potential for genetic research based on population samples has been greatly enhanced by the steady reduction in the cost of sequencing. As a result, sample sizes in these studies are substantially larger than before, and the trend will continue with the arrival of next generation sequencers. Already, it is commonplace to see sequenced samples of many hundreds of individuals and even thousands (such as the sample in the 1000 Genomes Project [15]). The reduction of sequencing cost also leads to a larger region of the genome or even the entire genome being sequenced (e.g., 1000 Genomes Project). Consequently, new approaches that are both highly accurate and efficient in computation are desirable. This paper presents one such method and demonstrates its utility by analyzing polymorphism from the 1000 Genomes Project.

2. Theory and Method

2.1. The Theory

Assume that a sample of n DNA sequences at a locus without recombination is taken from a single population evolving according to the Wright–Fisher model and all mutations are selectively neutral.

The sample genealogy thus consists of $2(n-1)$ branches, each spanning at least one coalescent time (Figure 1). The number of mutations that occurred in a branch is thus the sum of the numbers of mutations in the coalescent time it spans. Consider one branch, and without loss of generality, assume it spans the i-th coalescent time. Then, during the i-th coalescent time, the number of mutations occurred in the branch has expectation and variance equal to:

$$\frac{\theta}{i(i-1)} \quad \text{and} \quad \frac{\theta}{i(i-1)} + \frac{1}{i^2(i-1)^2}\theta^2 \tag{2}$$

respectively. These are consequences of the coalescent time being exponentially distributed and the number of mutations in a given number of generations following a Poisson distribution. Consider the number of mutations in another branch that spans the j-th coalescent time. Then, the covariance between the two numbers of mutations is equal to:

$$\frac{\theta^2}{i^2(i-1)^2} \tag{3}$$

if $i = j$, and zero otherwise.

Figure 1. A sample genealogy with different coalescent times separated by dashed lines. Branch 1 spans the third to the sixth coalescent times, $\chi_1(2) = 0,\ \chi_1(i) = 1,$ for $i = 3, ..., 6$, while Branch 2 spans the fourth to the sixth coalescent times, $\chi_2(2) = \chi_2(3) = 0,\ \chi_2(4) = \chi_2(5) = \chi_2(6) = 1$. Combining Branches 1 and 2 results in $\phi(2) = 0, \phi(3) = 1, \phi(4) = \phi(5) = \phi(6) = 2$.

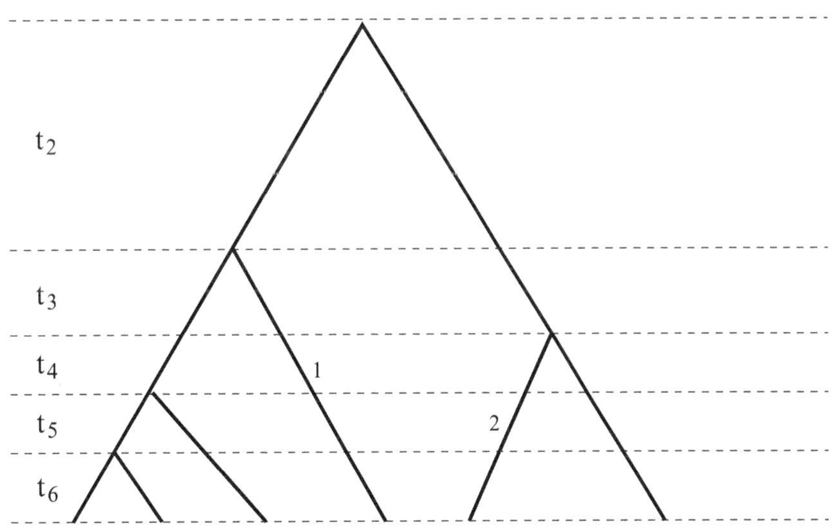

For the branch $k (k = 1, ..., 2(n-1))$ in the genealogy, define an index $\chi_k(i)$, such that it takes value one if the branch spans the i-th coalescent time and zero otherwise. Then, m_k, the number of mutations, has its expectation and variance equal to:

$$E(m_k) = \theta \sum_{i=2}^{n} \frac{\chi_k(i)}{i(i-1)} \tag{4}$$

$$V(m_k) = E(m_k) + \theta^2 \sum_{i=2}^{n} \frac{\chi_k^2(i)}{i^2(i-1)^2} \tag{5}$$

respectively, and for two different branches a and b:

$$Cov(m_a, m_b) = \theta^2 \sum_{i=2}^{n} \frac{\chi_a(i)\chi_b(i)}{i^2(i-1)^2} \tag{6}$$

The previous results are readily generalized. Instead of considering the mutations in different branches separately, one can combine mutations in several branches. Suppose branches $(k_1, ..., k_t)$ are combined. Define for the combined branches a variable ϕ as:

$$\phi(i) = \sum_{j=1}^{t} \chi_{k_j}(i) \tag{7}$$

Consider a population dynamics model in which the effective population size can change only at the time a coalescent event occurs. Although such a model does not stem from biological reality, its laddered changes in population sizes allow a reasonable approximation of reality and makes the mathematics simpler. Let θ_i represent the θ during the i-th coalescent period. Suppose the combined branches is denoted by branch (group) k, then m_k, the number of mutations in branch k has expectation and variance equal to:

$$E(m_k) = \sum_{i=2}^{n} \frac{\phi_k(i)\theta_i}{i(i-1)} \tag{8}$$

$$V(m_k) = E(m_k) + \sum_{i=2}^{n} \frac{\phi_k^2(i)\theta_i^2}{i^2(i-1)^2} \tag{9}$$

respectively, and for two such branches a and b, we have:

$$Cov(m_a, m_b) = \sum_{i=2}^{n} \frac{\phi_a(i)\phi_b(i)\theta_i^2}{i^2(i-1)^2} \tag{10}$$

Suppose that the $2(n-1)$ branches of the sample genealogy are divided into M $(\leq 2(n-1))$ disjoint groups (*i.e.*, each branch belongs to one and only one group). Let m_k represent the number of mutations in branch group k and $\boldsymbol{m} = (m_1, ..., m_k)^T$. Then, similar to the previous result [13], the relationship between $\boldsymbol{\theta} = (\theta_2, ..., \theta_n)^T$ and $\boldsymbol{m}$ can be expressed by a generalized linear model:

$$\boldsymbol{m} = \boldsymbol{\alpha\theta} + \boldsymbol{e} \tag{11}$$

where $\boldsymbol{\alpha}$ is a matrix of dimension $M \times n$ with: $\alpha_{ij} = \frac{\phi_i(j)}{j(j-1)}$ and $\boldsymbol{e}$ a vector of length M representing error terms. Let $\Gamma(\boldsymbol{\theta}) = Var(\boldsymbol{m})$. Then:

$$\Gamma(\boldsymbol{\theta}) = \boldsymbol{\gamma}(\boldsymbol{\theta}) + \boldsymbol{\beta}(\boldsymbol{\theta}) \tag{12}$$

where $\boldsymbol{\gamma}$ and $\boldsymbol{\beta}$ are both $M \times M$ matrices defined as:

$$\boldsymbol{\gamma}(\boldsymbol{\theta}) = Diag\{\boldsymbol{\alpha}_1\boldsymbol{\theta}, ..., \boldsymbol{\alpha}_M\boldsymbol{\theta}\} \tag{13}$$

$$\boldsymbol{\beta}(\boldsymbol{\theta}) = \left\{ \sum_{k=2}^{n} \frac{\phi_i(k)\phi_j(k)\theta_k^2}{k^2(k-1)^2} \right\} \tag{14}$$

where $\boldsymbol{\alpha}_k$ represents the k-th row vector of $\boldsymbol{\alpha}$. Following the previous approach [13,14], a best linear unbiased estimator of $\boldsymbol{\theta}$ can be obtained as the limit of the series:

$$\boldsymbol{\theta}^{(k+1)} = [\boldsymbol{\alpha}^T\Gamma(\boldsymbol{\theta}^{(k)})^{-1}\boldsymbol{\alpha}]^{-1}\boldsymbol{\alpha}^T\Gamma(\boldsymbol{\theta}^{(k)})^{-1}\boldsymbol{m} \tag{15}$$

with $\boldsymbol{\theta}^{(0)}$ being the initial estimate of $\boldsymbol{\theta}$ (for example, setting all θ_i equal to Watterson's estimate of θ).

The above formulation allows maximal $n-1$ different values of θ corresponding to the $n-1$ coalescent periods. Although very flexible, such an extreme model may lead to reduced accuracy of estimation for individual parameters, so some compromise is likely to be useful. When two or more consecutive θ values are assumed to be the same, the length of the $\boldsymbol{\theta}$ vector will be reduced. At the extreme, if all of the θs are the same, $\boldsymbol{\theta}$ is reduced to a single quantity, and when $M = 2(n-1)$, it further defaults to BLUE [13]. Since efficient estimators for a single value of θ will continue to be useful in the analysis of the whole genome polymorphism of large samples, we will focus on developing one such scheme in this paper.

2.2. Allelic BLUE estimator

In order to take advantage of the BLUE estimator, sample genealogy needs to be inferred. Furthermore, the key to developing a highly efficient BLUE estimator is to define the M groups of branches, which not only retains the detail mutational information, but also satisfies the relationship $M << 2(n-1)$. Fu's UPBLUE [13] corresponds to the extreme in which $M = 2(n-1)$, *i.e.*, each branch belongs to its own group. While this may retain the maximal mutational information, it leads to computational inefficiency. Fu's [14] approach is more or less equivalent to $M = n - 1$, with groups defined by the size of mutations. This achieves computational efficiency with the expense of reduced accuracy due to over condensation of mutational information. Thus, the goal here is to strive for a balance.

We recognize that much of the phylogenetic information in a sample resides in the pattern of differences between distinct alleles. The phylogenetic method, UPGMA (e.g., [16]), which was found to be appropriate in Fu [13], will continue to be used in our new method. Since UPGMA is a sequential method, which at each step joins two sequences (or two groups of sequences) that differ the least. As a result, copies of sequences of the same allele will be joined together before any pair of sequences of distinct alleles. The resulting sample genealogy can be roughly divided into two portions (see Figure 2), with the bottom portion corresponding to the coalescent within allelic class and the upper portion the coalescent among allelic classes. Combine all the branches (or segments of branches) underneath the dashed line into one group, which will be referred to as the within allele branch. Suppose there are L distinct alleles in the sample; then we have for the within allele branch:

$$\phi(i) = \begin{cases} 0, & i \leq L \\ i, & i > L \end{cases} \tag{16}$$

Then, the number of mutations in the within allele branch have expectation and variance equal respectively to:

$$(a_n - a_L)\theta \qquad \text{and} \qquad (a_n - a_L)\theta + (b_n - b_L)\theta^2 \,. \tag{17}$$

Figure 2. Schematic relationship between sample genealogy and allelic genealogy. The dark portion is the genealogy of distinct alleles, while the light portion (which is below the horizontal line) is the coalescent within alleles and contains no mutation or only a few in the dashed segments of the branches.

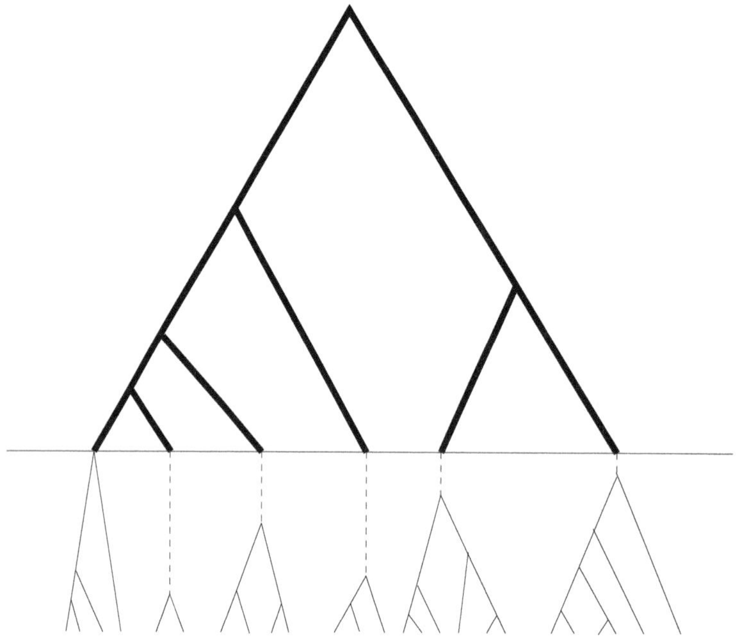

Furthermore, we assume that there is no mutation in the within allele branch (which should be a good approximation, although technically, the assertion may not be true). Since the within allele branch does not span any coalescent time that overlaps with those of branches in the allelic genealogy, we have (assumed that the last branch group represents the within allele subtree) that:

$$\boldsymbol{\alpha} = \begin{pmatrix} \boldsymbol{\alpha}^* \\ a_n - a_L \end{pmatrix} \tag{18}$$

where $\boldsymbol{\alpha}^*$ a vector of length $2(L-1)$ with the k-th element equal to $\sum_{i=2}^{L} \frac{\phi_k(i)}{i(i-1)}$. The inverse of the covariance matrix of $\boldsymbol{m}$ is:

$$\begin{pmatrix} \boldsymbol{\Gamma}^*(\theta)^{-1} & \mathbf{0} \\ 0 & [(a_n - a_L)\theta + (b_n - b_L)\theta^2]^{-1} \end{pmatrix} \tag{19}$$

where $\boldsymbol{\Gamma}^*$ is defined for branch groups of the allelic genealogy. Let $\boldsymbol{m}^*$ be the vector of mutations in branches of the allelic genealogy (the dark portion in the genealogy of Figure 2). Then, Equation (15) becomes:

$$\theta^{(k+1)} = \frac{(\alpha^*)^T \boldsymbol{\Gamma}^*(\theta^{(k)})^{-1}}{(a_n - a_L)^2[(a_n - a_L) + (b_n - b_L)\theta^{(k)}]^{-1} + (\alpha^*)^T \boldsymbol{\Gamma}^*(\theta^{(k)})^{-1}\boldsymbol{\alpha}^*} \boldsymbol{m}^* \tag{20}$$

This limit will be referred to as the Allelic BLUE estimator denoted by θ_{ab}. Since, for large samples, the number of distinct alleles is typically much smaller than the sample size; thus, the new estimator is expected to be highly efficient computationally.

To determine if it is indeed true that merging those branches representing within allele coalescent does not lead to significant loss of information and, thus, would not reduce the accuracy of estimation, we compared Allelic BLUE with the original BLUE using simulated samples for a number of combinations of θ and n. The correlation between the two estimates is around 0.99. Therefore, Allelic BLUE is expected to be as accurate as BLUE without merging branches.

2.3. Bias-Corrected Allelic BLUE Estimator

Since UPGMA systematically introduces bias in the inferred sample genealogy, the resulting Allelic BLUE estimate is expected to be biased similar to the BLUE estimator [13]. Therefore, it is necessary to correct the bias. Similar to Fu [13], we used simulated samples to derive understanding of the pattern of biases. A total of 550 combinations of θ and n were examined with 25 different θ values: $0.5, 0.75, 1, 1.5, 2(1)5, 6(2)12, 15(5), 50, 60(10)100$ and 150, and 25 different sample sizes n: $10(5)25, 30(10)60, 80, 100(25)200, 250,\ 300, 400, 500, 750, 1000(1000)5000$. For each combination of the parameters, 1000 samples were simulated, and for each simulated sample, θ_{ab} was obtained and their mean value computed over all simulated samples. Similar to those in Fu [13], the estimates in almost all situations are underestimates of the true θ. In general, the underestimate is the result of systematic bias of the UPGMA algorithm used to construct the genealogy, because UPGMA leads to early coalescent for more similar sequences and, thus, has a tendency to place more mutations in branches that are deeper into the tree. In the current situation, it is further compounded by our simplification that, up to the $i + 1$ coalescent, there are no mutations.

Using regression analysis, Fu [13] showed that the relationship:

$$\sqrt{\theta_u} = -0.0336\sqrt{n-2} + 1.002\sqrt{\theta} \tag{21}$$

summarizes well the BLUE estimate (with $M = 2(n-1)$), θ and sample size n, which is not larger than 100. When larger sample sizes were examined, the above equation is not adequate, and log transformation, rather than square-root transformation, can lead to a better regression [17]. Therefore, log-transformation was chosen in our regression analysis. Table 1 showed that $ln(\theta_{ab})$ can be summarized very well by the following equation:

$$\begin{aligned} ln(\theta_{ab}) = &- 0.1981 + 0.0080ln(n) + 1.0043ln(\theta) \\ &- 0.0019ln(n)ln(\theta) + 0.0108ln^2(\theta) - 0.0012ln(n)ln^2(\theta) \end{aligned} \tag{22}$$

which leads to an estimate of θ from the solution for the above quadratic equation with regard to $ln(\theta_a)$:

$$\hat{\theta}_a = \exp\left[\frac{-b + \sqrt{b^2 - 4ac}}{2a}\right] \tag{23}$$

where:

$$a = 0.0108 - 0.0012ln(n) \tag{24}$$

$$b = 1.0043 - 0.0019ln(n) \tag{25}$$

$$c = -ln(\theta_{ab}) - 0.1981 + 0.008ln(n) \tag{26}$$

Although this estimator in most situations is excellent, we found that regression equations for a narrower range of sample sizes provides estimates that are more robust in some situations (particularly when θ is large). As a result, we derive our final estimator $\hat{\theta}_a$ of θ using Equation (23) with values of a, b and c, as provided in Table 2.

Table 1. Summary of regression analysis between θ_{ab}, θ and n.

Source	Sum of Squares	Degrees of freedom	Mean Square
Model	1,715.227	5	343.0454
Residual	0.074	644	0.0001
Total	1,715.301	649	2.6430

Term	Coefficient	Standard Error	t test	$P > \|t\|$
$ln(n)$	0.0080	0.0005	16.73	0.000
$ln(\theta)$	1.0043	0.0025	398.16	0.000
$ln(n)ln(\theta)$	-0.0019	0.0004	-4.19	0.000
$ln^2(\theta)$	0.0108	0.0005	19.80	0.000
$ln(n)ln^2(\theta)$	-0.0012	0.0001	-12.74	0.000
constant	-0.1981	0.0027	-72.99	0.000

Note: Number of points for regression = 650, $R^2 = 1.000$ and MSE = 0.0107

Table 2. Coefficients for estimating θ using Equation (23).

	Coefficients ($n' = ln(n)$)		
Sample Size	**a**	**b**	**c**
$n < 50$	$0.0112 - 0.0012n'$	$1.0076 - 0.0026n'$	$-ln(\theta_a) - 0.2101 + 0.0107n'$
$50 \leq n < 500$	$0.0131 - 0.0017n'$	$1.0009 - 0.0016n'$	$-ln(\theta_a) - 0.1980 + 0.0088n'$
$n \geq 500$	$0.0069 - 0.0007n'$	$0.9850 - 0.0008n'$	$-ln(\theta_a) - 0.1581 + 0.0025n'$

Figure 3 plots the relationship between θ, sample size (n) and the allelic BLUE estimates (θ_{ab}) for a subset of these parameter combinations. It is easy to see that the match between prediction and the mean value of θ_a is excellent.

Figure 4 shows the distributions of the estimate $\hat{\theta}_a$ from simulated samples in the case of $n = 500$ with $\theta = 5$, and $n = 2000$ with $\theta = 50$, respectively. It appears in both cases that the normalities are sufficiently accurate approximations, which is indeed expected from the theory of least squares estimators.

The ultimate measure of the quality of an estimator is its bias and standard deviation for samples independent of those used to derive the estimator. Therefore, we simulated another set of samples for a number of combinations of θ and n and applied $\hat{\theta}_a$, as well as UPBLUE to these samples. Table 3 presents the summary of these simulations, particularly the efficiency of the new approach.

Figure 3. The relationship between Allelic Blue estimate θ_a, θ and sample size. Solid lines represents the prediction of θ_a based on Equation (23) with the coefficients given in Table 2.

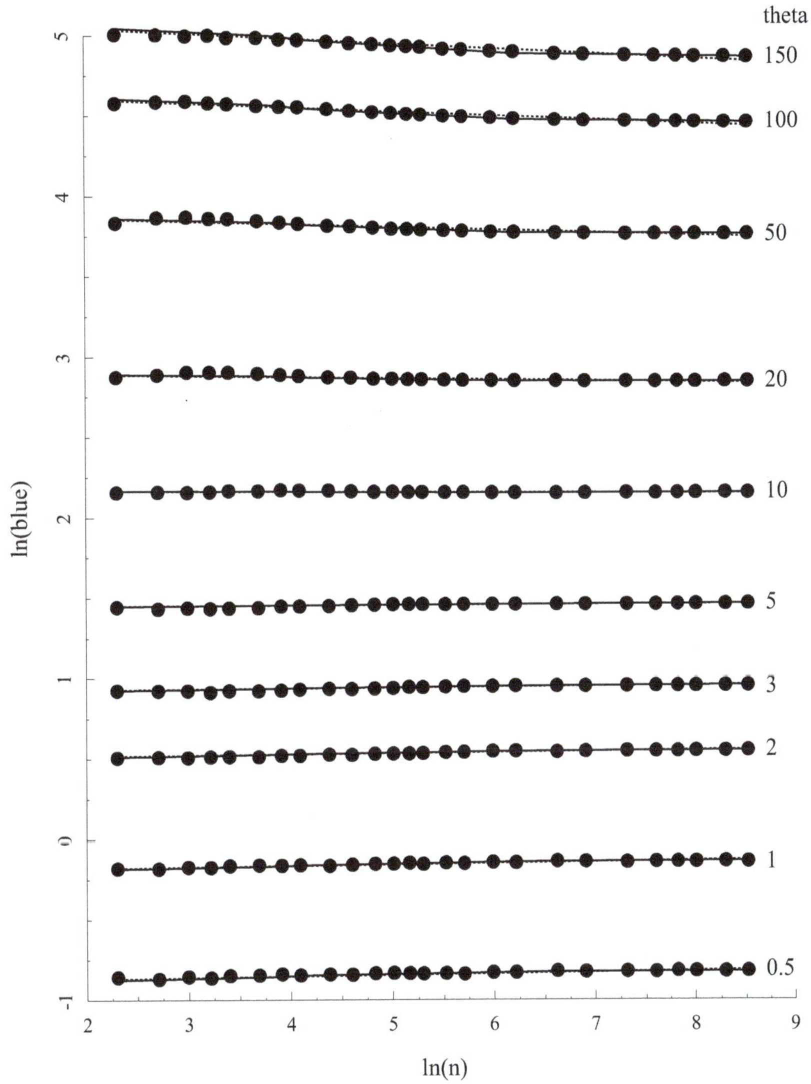

Figure 4. Distribution of $\hat{\theta}_a$ based on 1000 simulated samples. (**Left**) $n = 500$ and $\theta = 5$; (**right**) $n = 2000$ and $\theta = 50$.

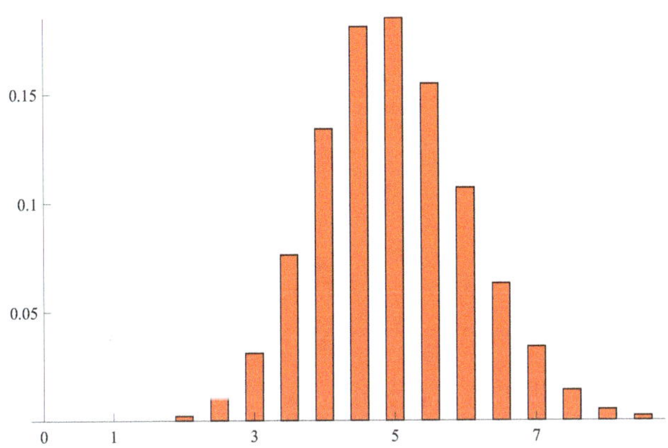

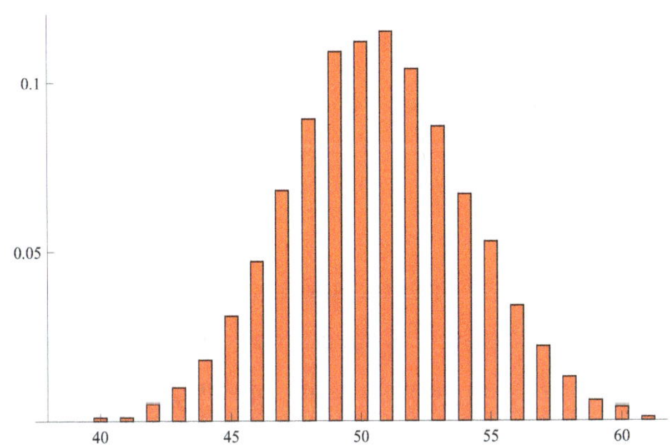

Table 3. Performance of $\hat{\theta}_a$.

θ	n	mean $\hat{\theta}_a$	SE	SD_{min}	Speed	Ratio
2	20	1.97	1.00	0.97	0.00	10
	50	1.98	0.84	0.81	0.00	59
	100	1.98	0.75	0.74	0.00	280
	500	2.00	0.63	0.61	0.02	1,569
	1,000	2.00	0.58	0.58	0.27	1,615
	2,500	2.00	0.54	0.54	2.59	3,104
5	20	4.93	1.90	1.83	0.00	4.5
	50	4.96	1.52	1.48	0.00	23
	100	4.99	1.34	1.30	0.00	78
	500	5.01	1.08	1.05	0.04	970
	1,000	5.00	1.00	0.98	0.47	1,306
	2,500	4.91	0.87	0.90	4.90	1,565
20	20	20.33	5.84	5.52	0.00	2.0
	50	20.27	4.20	4.05	0.01	4.4
	100	20.06	3.47	3.37	0.06	8.7
	500	20.04	2.56	2.49	0.18	359
	1,000	19.99	2.32	2.26	0.91	727
	2,500	19.96	2.07	2.04	16	842
50	20	50.92	12.58	12.51	0.01	1.6
	50	50.45	8.68	8.59	0.02	2.5
	100	50.09	6.99	6.79	0.07	3.8
	500	50.06	4.70	4.58	0.97	72
	1,000	49.90	4.15	4.06	3.6	206
	2,500	49.80	3.94	3.57	42	356
100	20	100.25	23.47	24.03	0.01	1.5
	50	100.20	15.52	15.87	0.15	1.7
	100	99.97	12.24	12.08	0.20	2
	500	100.38	7.86	7.48	4	22
	1,000	99.89	6.59	6.47	16	49
	2,500	99.69	5.58	5.54	75	202

Note: Speed is the average CPU time (in seconds) for obtaining $\hat{\theta}_a$ for a simulated sample in a Linux machine with a 2.3-Ghz CPU. SD_{min} is the minimum standard deviation computed as the square root of Equation (28) in [18]. Ratio is the ratio of speed for UPBLUE [13] and speed of $\hat{\theta}_a$.

Table 3 shows that the speed of $\hat{\theta}_a$ increases with the sample size slowly, while it increases faster with θ. This is because $\hat{\theta}_a$'s speed is dependent on the number of alleles in the sample, which is more closely related to θ than sample size. In comparison, UPBLUE is considerably less efficient, particularly with increasing sample size. Take the case of $\theta = 100$ and $n = 5000$, it takes on average about one minute for $\hat{\theta}_a$ to complete the estimation, while it takes about 6 h for UPBLUE to do the same.

3. Exploring θ in Data from the 1000 Genomes Project

The 1000 Genomes Project generated a very valuable set of genome-wide polymorphism data [15] for which the newly developed Allelic Blue estimator is applicable. Phase I (May, 2012, release) contains polymorphism, as well as inferred phases compiled from 1092 individuals from 14 different populations. The rich information captured by the genome-wide polymorphism deserves to be analyzed from various angles [20], and our main purpose here is to illustrate that our efficient estimator of θ provides additional insight into the pattern of polymorphism in addition to the conventional estimates. We chose to report the analysis for a subset of samples, which consists of the three African samples (YRI, LWKand ASW) with 246 individuals (thus, a sample size of $n = 492$).

In our analysis, each of the 22 autosomal chromosomes was divided into non-overlapping consecutive windows of 2000 bps (within which the average impact of recombination should be negligible), and θ was estimated for each window. Since the phases of SNPs for each individual were the result of inference, there are some segments in which the quality of inference appears to be poor due to an unreasonably larger number of inferred alleles than the number of SNPs. We thus removed all of the segments in which there is evidence of either recombination or a poor quality of inference, that is, when the number of alleles is larger than the number of SNPs plus one. It should be noted that the SNPs reported in the 1000 Genomes Project data set are those that passed various quality controls and filtering. In our analysis, no further filtering is applied, except for the aforementioned exclusion of segments that are suspected to be the result of poor phrase inference. A total of 648,903 windows were analyzed. This analysis required about 3 h to complete the estimates of θ in a desktop computer equipped with an Intel Xeon CPUs at 3.33 Ghz. In comparison, UBPLUE ran a couple days without being able to finish the same task.

Since both Watterson's estimator $\hat{\theta}_W$ and Tajima's estimator ($\hat{\theta}_T$) are widely used, we present our results in terms of the relative values with regard to $\hat{\theta}_W$ and contrast them with the relative values of $\hat{\theta}_W$ to $\hat{\theta}_T$. Since testing neutrality is not the purpose, we do not employ testing statistics, such as Tajima's test [19] or Fu and Li's tests [7]. Figure 5 plots the proportional difference between $\hat{\theta}_W = K/a_n$ (K is the number of segregating sites and a_n is a constant depending on the sample size) and $\hat{\theta}_T$. The overwhelming characteristic of the plot is that θ_W is, in general, larger than $\hat{\theta}_T$, with an average of 1.96-times the value of $\hat{\theta}_T$; similar patterns were observed previously (for example, [20]). In general, an estimated θ can be viewed as a weighted average of SNPs of various sizes. $\hat{\theta}_T$ gives on average more weight to SNPs that occurred long ago than those arisen recently, while $\hat{\theta}_W$ gives equal weight to every SNP. Therefore, elevated $\hat{\theta}_W$ values across the whole genome compared to $\hat{\theta}_T$ were considered as evidence of recent population growth. It should be noted that there is no obvious extended regions with smaller or larger values for θ_W or θ_T. In comparison, Figure 6 plots the proportional difference between $\hat{\theta}_a$ and $\hat{\theta}_W$.

Figure 5. Plot of $(\hat{\theta}_W - \hat{\theta}_T)/\hat{\theta}_T$ along 22 autosomes with windows of a size of 2000 bps (each dot represents a mean over 10 consecutive windows). Chromosomes 1 to 22 are presented from left to right separated by vertical lines. The overall mean value is 0.96.

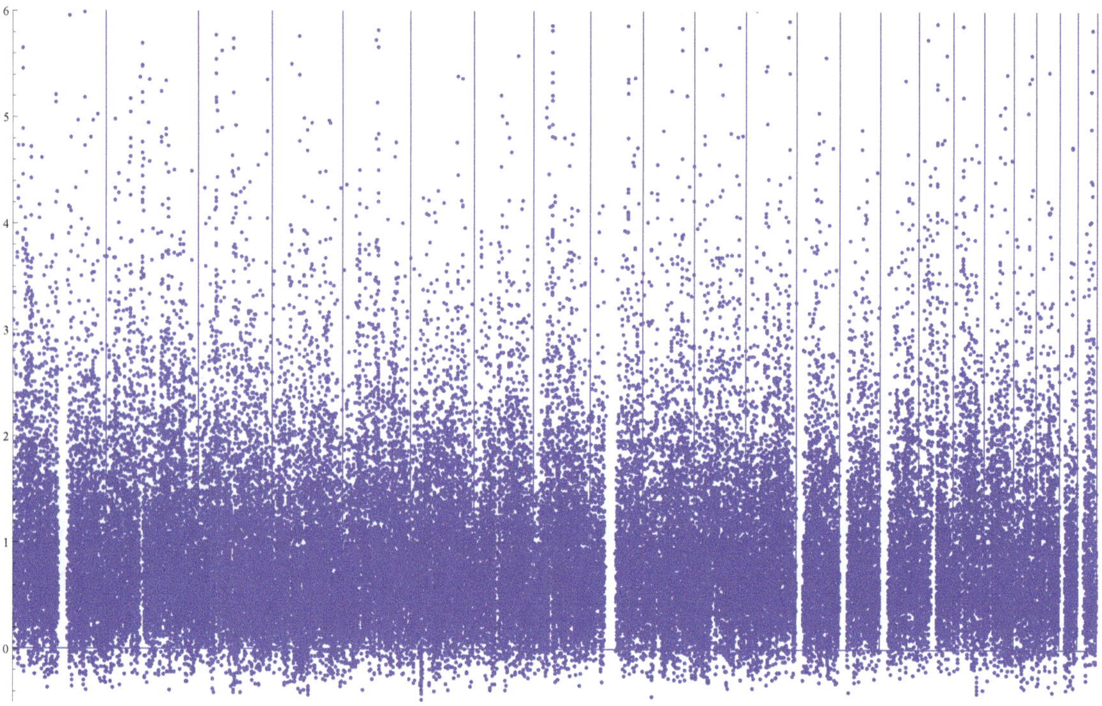

Figure 6. Plot of $(\hat{\theta}_a - \hat{\theta}_W)/\hat{\theta}_W$ along 22 autosomes with windows of a size of 2000 bps (each dot represents a mean over 10 consecutive windows). Chromosomes 1 to 22 are presented from left to right separated by vertical lines. The overall mean value is 0.20.

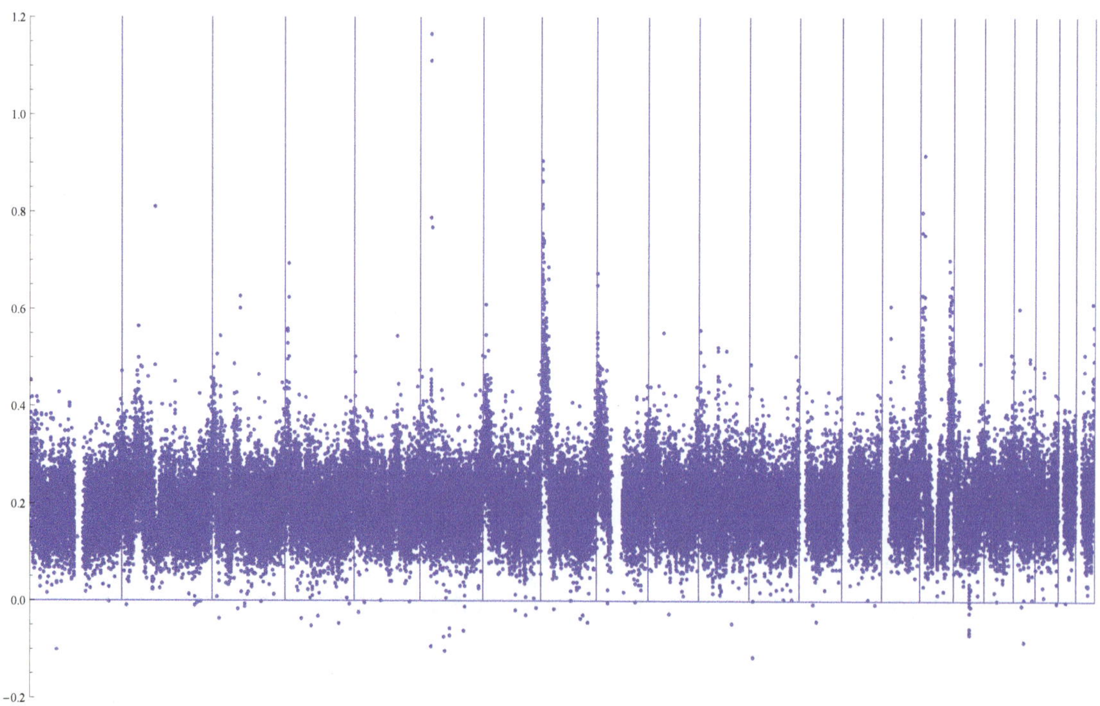

The overwhelming pattern shown in Figure 6 is again that $\hat{\theta}_a$ in general is larger than $\hat{\theta}_W$, which means that the difference to $\hat{\theta}_T$ will be more pronounced than that of $\hat{\theta}_T$. This is the result that more weight is given to recent mutations than the old ones in $\hat{\theta}_a$. Beside some sporadic large values, there are regions at either the beginning or the end of some chromosomes that yield considerably elevated values of $\hat{\theta}_a$ (for example, for chromosomes 7, 8, 9, 16 and 22). We are not sure how to interpret these patterns, but suspect that they may partially suggest the decreased quality of phase inference at the beginning and end of chromosomes.

4. Discussion and Conclusions

The Allelic BLUE estimator of θ presented in this paper is a high quality estimator with little bias and its variance nearly as small as the minimum achievable variance. Furthermore, it is highly efficient computationally, because its speed depends on the number of distinct alleles in a sample rather than the sample size. This later characteristic makes it very useful for estimating θ for large samples and for situations in which a large region (or the whole genome) is sequenced, while θ needs to be estimated for successive windows of a genome, such as the case of 1000 Genomes Projects. Since θ_a and UPBLUE are both based on the same idea of utilizing phylogenetic information in a sample with generalized linear regression, their estimates are highly correlated, which are seen in both the simulation and in real data. However, since θ_a is computationally much more efficient, it is thus superior to UPBLUE [13] and, thus, can replace UPBLUE for general use. The analysis of the polymorphism from the 1000 Genomes Project shows that although UPBLUE is a relatively efficient estimator among sophisticated estimators; it has nearly reached its limit for exploratory data analysis for large genome projects. Therefore, the new Allelic BLUE estimator arrival is timely.

One additional utility of the new estimator $\hat{\theta}_a$ is for estimating θ from a collection of distinct alleles, which are collected without recording the multiplicity of each allele, as long as the number of alleles examined is roughly known. Such situations sometime arise when the collection of data is focused on identifying distinct alleles, such as in the survey of infectious pathogens or when data are collected over years and pooled from multiple sources. To illustrate this utility, we simulated samples of size 200 with $\theta = 10$. If only the distinct alleles are recorded (which implicitly assumes that the sample size is the same as the number of distinct alleles), then θ_a yields an average value of 22.8, which is more than twice as large as the true value. However, if the sample size used in the estimation is 20% smaller or larger than the actual value (thus, 160 and 240, respectively), the corresponding $\hat{\theta}_a$ are 10.6 and 9.5, respectively, both of which are quite close to the true θ value.

The Allelic BLUE estimator is developed under the assumption of one single random mating population evolving according to the Wright–Fisher model with a constant effective population size. The restriction to constant population size results in an estimator of average θ since the MRCAof a sample, which is comparable to some classical estimators and, thus, provides an informative contrast to other estimators (and may be used to construct hypothesis tests in the future). However, the restriction does make the method unsuitable for exploring historical changes in effective population sizes. On the other hand, the theoretical foundation for exploring changes in population sizes in the linear regression framework has been established in this paper, and we will study its statistical properties, as well as its application elsewhere.

The Java programs for performing the Allelic Blue estimation can be downloaded from the author's web page [21].

Acknowledgments

This work was supported in part by grants from NIH (5U01HG005728 Fu, 2U54 HG003273 Gibbs), Betty Wheless Trotter Endowment Fund from The University of Texas Health Science Center at Houston (Fu) and the fund from Yunnan University, Kunming, China.

References

1. Ewens, W.J. The sampling theory of selectively neutral alleles. *Theor. Popul. Biol.* **1972**, *3*, 87–112.
2. Watterson, G.A. On the number of segregation sites. *Theor. Popul. Biol.* **1975**, *7*, 256–276.
3. Tajima, F. Evolutionary relationship of DNA sequences in finite populations. *Genetics* **1983**, *105*, 437–460.
4. Fu, Y.X. Statistical properties of segregating sites. *Theor. Popul. Biol.* **1995**, *48*, 172–197.
5. Ewens, W.J. *Mathematical Population Genetics*; Springer-Verlag: New York, NY, USA, 2004.
6. Xu, H.; Fu, Y.X. Estimating effective population size or mutation rate with microsatellites. *Genetics* **2004**, *166*, 555–563.
7. Fu, Y.X.; Li, W.H. Statistical tests of neutrality of mutations. *Genetics* **1993**, *133*, 693–709.
8. Kingman, J.F.C. The coalescent. *Stoch. Process. Their Appl.* **1982**, *13*, 235–248.
9. Hudson, R.R. Properties of a neutral allele model with intragenic recombination. *Theor. Popul. Biol.* **1983**, *23*, 183–201.
10. Griffiths, R.C.; Tavaré, S. Ancestral inference in population genetics. *Statist. Sci.* **1994**, *9*, 307–319.
11. Griffiths, R.C.; Tavaré, S. Monte Carlo inference methods in population genetics. *Math. Comput. Model.* **1996**, *23*, 141–158.
12. Kuhner, M.K.; Yamato, Y.; Felsenstein, J. Estimating effective population size and mutation rate from sequence data using Metropolis-Hastings sampling. *Genetics* **1995**, *140*, 1421–1430.
13. Fu, Y.X. A phylogenetic estimator of effective population size or mutation rate. *Genetics* **1994**, *136*, 685–692.
14. Fu, Y.X. Estimating effective population size or mutation rate using the frequencies of mutations of various classes in a sample of DNA sequences. *Genetics* **1994**, *138*, 1375–1386.
15. 1000 Genomes Project Consortium. Website for the data generated by 1000 Genomes Project: http://www.1000genomes.org/ (accessed on 1 July 2014).
16. Nei, M. *Molecular Evolutionary Genetics*; Columbia University Press: New York, NY, USA, 1987.
17. Zhang, F. Statistical Methods for Estimating Mutation Rate and Effective Population Size from Samples of DNA Sequences. Ph.D. Dissertation, The University of Texas Health Science Center at Houston, Houston, TX, 2003.

18. Fu, Y.X.; Li, W.H. Maximum likelihood estimation of population parameters. *Genetics* **1993**, *134*, 1261–1270.
19. Tajima, F. Statistical method for testing the neutral mutation hypothesis by DNA polymorphism. *Genetics* **1989**, *123*, 585–595.
20. 1000 Genomes Project Consortium. An integrated map of genetic variation from 1092 human genomes. *Nature* **2012**, *49*, 56–65.
21. Fu, Y.X. The Java programs for performing the Allelic Blue estimation. Availiable online: https://sph.uth.edu/yfu/ (accessed on 1 July 2014).

9

The Little Fly that Could: Wizardry and Artistry of *Drosophila* Genomics

Radoslaw K. Ejsmont [1,2,*] and Bassem A. Hassan [1,2,*]

[1] VIB Center for the Biology of Disease, VIB, 3000 Leuven, Belgium
[2] Center for Human Genetics, University of Leuven School of Medicine, 3000 Leuven, Belgium

* Authors to whom correspondence should be addressed;
E-Mails: radoslaw.ejsmont@cme.vib-kuleuven.be (R.K.E.); bh@kuleuven.be (B.A.H.);

Abstract: For more than 100 years now, the fruit fly *Drosophila melanogaster* has been at the forefront of our endeavors to unlock the secrets of the genome. From the pioneering studies of chromosomes and heredity by Morgan and his colleagues, to the generation of fly models for human disease, *Drosophila* research has been at the forefront of genetics and genomics. We present a broad overview of some of the most powerful genomics tools that keep *Drosophila* research at the cutting edge of modern biomedical research.

Keywords: *Drosophila*; genetics; genomics

---

1. Introduction

The Human Genome Project, on its way to producing an assembled genome of *Homo sapiens*, has gone through several test runs yielding sequenced genomes of other organisms of high relevance for research into human development and disease. The first published genome of a free-living organism was that of the proteobaterium *Haemophilus influenzae* [1], followed by sequencing of the genome of *Saccharomyces cerevisiae* yeast, the first eukaryotic genome sequenced [2], and the genome of *Caenorhabditis elegans*, the first genome of a multicellular organism and the first animal genome [3].

The second animal genome sequenced was that of the fruit fly *Drosophila melanogaster* [4]. In this review, we discuss the significance of the sequencing of the *Drosophila* genome as well as the technical advances and new research avenues that have accompanied it.

2. Drosophila as a Model

2.1. In Development

The fruit fly has been studied for over a century and the lessons learned from fly research makes it almost impossible to enumerate but a few of the most notable cases. The pioneering studies that identified genes involved in *Drosophila* embryo segmentation [5,6] and establishment of segment polarity [6] were seminal for understanding conserved developmental strategies in the animal kingdom. The discovery of homeotic genes is one of the best-known examples of genes discovered in the fruit fly, and these were found to be conserved and play analogous roles in humans [7–9]. *Drosophila* has played a seminal role in sensory organ development research. The discovery of the *eyeless* gene [10], a fly homolog of human and mouse PAX6 [11,12], and determination of its targets [13] shed light on vertebrate eye development and led to discovery of novel disease related genes in humans [14]. The proneural gene atonal plays a crucial role in the development of *Drosophila* photoreceptor neurons [15] and chordotonal organs [16]. Its function is conserved in mammals, where its homologs Math5 and Math1 were shown to be involved in regulating formation of retinal ganglion cells [17] and inner ear mechanosensory hair cells [18].

2.2. In Signaling

Drosophila has been extensively used for studies of signaling pathways. In Hedgehog signaling, both the Hedgehog ligand itself [6,19,20] and its receptor Patched [6,21,22] were first identified in the fly, though the link between the two was first established in mammals [23,24]. The ligand of the Wnt signaling pathway turned out to be a well-known *Drosophila* segment polarity protein, *i.e.*, Wingless. The Wnt receptor, Frizzled [25], and several other signal transduction cascade members were identified in the fly as members of the Wnt pathway [26–28]. The planar cell polarity (PCP) pathway is yet another example of a signaling cascade in which key players and mechanisms of action have been, to a large extend, identified in *Drosophila* [29,30]. The Notch signaling pathway, associated with cell fate control, lateral inhibition, and signal integration during development, has been discovered and extensively studied in fruit flies [31–33]. Finally, major components and mechanisms of action of the Hippo signaling pathway have been described in *Drosophila* [34–36]. All these pathways play major roles in human development and disease.

2.3. In Disease

Over the past two decades the fruit fly became an increasingly popular model organism for the study of human disease, with focus on neurodegenerative [37] and neuromuscular [38] diseases as well as cancer [39]. Neurological discases that havc bccn modclcd in *Drosophila* include trinucleotide repeat disorders [40–42], Alzheimer's disease [43–46], Parkinson's disease [47,48], amyotrophic lateral sclerosis [49,50], and dystrophy [51]. Other examples that include use of the fruit fly model are studies of alcohol abuse [52,53], cocaine addiction [54], obesity [55] and diabetes [56], cardiac diseases [57], and asthma [58]. *Drosophila* has been demonstrated to be a great model to identify tumor suppressor genes [59] or genes involved in metastasis [60]. Thanks to the conservation of major signaling pathways, tumor suppressors and oncogenes, various fly cancer models have been established. Understanding how signal transduction pathways like Hippo, Notch, Dpp or JAK-STAT affect tumor formation was aided by research in fruit flies [61–63]. *Drosophila* has been used as a model for tumor invasion and metastasis [64], and as a platform to identify novel therapeutic targets [65].

3. Meet the Drosophila Genome

The *Drosophila* genome is estimated to be approximately 200 Mb, with one third of it forming pericentric heterochromatin [66]. It is organized on three autosomes (numbered 2, 3 and 4) and sex chromosomes, X (also referred to as the first chromosome) and Y. The initial assembly of the fruit fly genome was published in March 2000, after almost a year of whole genome shotgun sequencing. The first published assembly, referred to as Release 1 of the genome, included 13,991 genes encoding for 14,080 peptides. Over two thirds of annotated genes were assigned gene ontology (GO) terms upon annotation. The initial assembly contained ~1300 gaps in mapped sequences [4] that were filled with subsequent releases.

The third release of the genome was the first that included pericentric heterochromatin sequences [67]. The mutations indicated in the sequenced strain's genotype, as well as several other identified mutations, have been corrected with wild-type sequence [68]. With that release, a comprehensive set of resources were published, including a library of full-length cDNAs for 40% of genes [69] and an atlas of gene expression patterns during embryogenesis [70]. Sequence analysis provided insights into transposable elements within the genome [71], core promoter structures [72], and largely improved annotation of gene models [68].

The current, fifth assembly of the genome has closed all but 9 gaps in the main assembly. The sequenced genome covers over 120 Mb of euchromatin, and over 9 Mb of mapped and over 10 Mb of unmapped heterochromatin. The current annotation revision contains 13,942 protein coding genes and over 2354 non-coding RNA genes, including ribosomal (rRNAs), transport (tRNAs), micro- (miRNAs), and small nuclear (snRNA) and small nucleolar (snoRNA) RNAs [73]. Through genome analysis, fruit flies have been found to contain complex gene structures. Approximately 7.5% of all genes, including non-coding RNAs, are located within the introns of other genes. Messenger RNAs for about 15% of genes overlap with mRNAs of genes on the opposite strands. Over 30 genes have been identified as dicistronic, *i.e.*, producing single mRNA encoding for two separate protein products through independent translation initiation events [68]. Over 30% of *Drosophila melanogaster* genes were found to be alternatively spliced [74], yielding a diverse set of almost 30,000 protein-coding transcripts [73]. The next release of the genome assembly (Release 6) is expected this year (2014).

Improved assembly and annotation of the fruit fly genome was possible not only due to new sequencing data, but also thanks to advances in bioinformatics tools. An integrated computational pipeline and a tailored database schema have been developed to facilitate genomic data storage and automated sequence annotation [75]. Computed annotations have been manually curated by experts and to aid in this task, a dedicated annotation editor was developed [76]. Finally, automated genome annotation in general requires the use of computational tools, some of which were first applied in the *Drosophila* genome project [72,77].

4. Genomes by the Dozen

Analysis of coding parts of the genome can be facilitated by comparison of genomic sequences with sequences of cDNAs originating from the same species. Most of the DNA in the majority of species, however, is non-coding. One approach to identify functional non-coding DNA segments, such as *cis*-regulatory elements, relies on finding conserved regions or motifs across related species. This naturally requires having more than one genome sequenced and was a driving force behind sequencing of the genomes of *Schizosaccharomyces pombe* [78] and *Caenorhabditis briggsae* [79] in the yeast and worm research communities, respectively. In the *Drosophila* genus, the comparative genomics era began with sequencing of the *Drosophila pseudoobscura* genome [80]. The two genomes were found to be very similar, despite 25–55 million years of evolutionary divergence. Synteny is preserved in blocks containing 10.7 genes on average, which corresponds to ~83 kb. The vast majority of synteny breaks were caused by intrachromosomal rearrangements. On average, ~48% of the base pairs are conserved between these two species.

The next advance in *Drosophila* comparative genomics came with sequencing of ten further species, *Drosophila sechellia, simulans, yakuba, erecta, ananassae, persimilis, willistoni, mojavensis, virilis*, and *grimshawi*. These species span a broad spectrum of morphologies, ecologies, and behaviors, yet have identical body plans and very similar life cycles [81]. Furthermore, these species share approximately 70% of their genes. Genome sizes estimated by flow cytometry vary between 130 Mb in *D. mojavensis* to 364 Mb in *D. virilis* [66]. The synteny conservation between sequenced species varies with an average of 122 genes per block between *D. melanogaster* and *simulans* down to 8 genes per block between *D. melanogaster* and *grimshawi*. Overall genome size, number of genes, distribution of transposable element classes, and patterns of codon usage are all very similar across the 12 sequenced genomes. At a finer scale, however, the number of structural changes and rearrangements is larger, including rearrangements of genes within the Hox cluster or highly dynamic sizes and content of multigene families [81].

Together, the 12 *Drosophila* genomes provide a solid platform for annotation and analysis of both coding and non-coding DNA. This unprecedented dataset enabled the use of evolutionary signatures—specific patterns of change in DNA elements upon selection—for *de novo* prediction and correction of previously annotated protein-coding gene models [82], non-coding RNAs, and transcription factor (TF) binding sites [83]. Identification of TF binding motifs has traditionally been based on DNA alignments. Alignment-based methods can also be used for the identification of *cis*-regulatory modules (CRMs), which are comprised of a number of TF binding motifs [84]. In many cases, however, the number and order of individual motifs varies between species, especially when these are distant, while preserving regulatory outcome. To address such cases, alignment-free approaches have also been developed [85,86].

5. Genomes by Population

Drosophila provides an unmatched set of resources for studying quantitative traits [87]. In the post genomic era, genome-wide association studies (GWAS) have become a preferred method for analyzing complex traits. The GWAS methodology is now routinely and successfully applied in the identification of human disease-associated genes [88]. Two fruit fly resources, the *Drosophila* Genetic Reference Panel (DGRP) [89] and the *Drosophila* Synthetic Population Resource (DSPR) [90], offer large sets of sequenced and mapped fly lines tailored for GWAS and quantitative trait loci (QTL) mapping.

The DGRP is a collection of more than 200 fully sequenced recombinant inbred lines (RILs) that were established from mated females collected from a market in Raleigh, North Carolina, USA. The genomic sequences of these lines contain over 4.5 million single nucleotide polymorphisms (SNPs), over one hundred thousand polymorphic microsatellites, and over 36 thousand transposable elements [89]. The DGRP has been extensively characterized and in addition to detailed genomic sequence analysis includes microarray [91] and RNA-seq [92] datasets for selected lines. To date, numerous genome-wide association studies have been published on various traits using DGRP, including oxidative stress [93], mitochondrial function [94], viral infection resistance [95], and sleep [96].

The *Drosophila* Synthetic Population Resource uses a different approach. Over 1,700 DSPR RILs were established from 15 isogenic founder lines created from geographically distinct *Drosophila* populations. The founder lines were split in two groups of eight (with one line in both groups) and mixed for 50 generations to create two synthetic populations, from which two sets of RILs were established. The founder lines were fully sequenced and each RIL was mapped using restriction-site associated DNA (RAD) markers onto the founders' sequence with 17 kb median resolution. The number of SNPs in the founder lines exceeds 1.6 million [90]. The DSPR is complementary to DGRP and both resources can be used together for cross-validation and to increase the mapping power [97].

6. Decoding the Genome's Secrets

6.1. The modENCODE Project

The information encoded by genomes goes far beyond a simple trinucleotide code used to translate nucleic acid sequence into protein. A plethora of information is hidden within introns, UTRs, non-coding RNAs, *cis*-regulatory elements, and chromatin marks. These elements are known to regulate where and when a gene product is expressed. The human ENCODE (ENCyclopedia Of DNA Elements) project [98] aims to identify and understand the information carried by the human genome. The modENCODE project is the model organism counterpart of ENCODE with focus on two species *C. elegans* [99] and *D. melanogaster* [100]. The fruit fly modENCODE data includes high-throughput transcriptome sequencing (RNA-seq), chromatin immunoprecipitation followed by sequencing (ChIP–seq) for transcription factor binding sites and histone modifications, DNA replication patterns, and nucleosome occupancy. The samples have been collected from 12–30 developmental time points of the sequenced *D. melanogaster* strain and from several cell lines [100].

6.2. The Transcriptome

Comprehensive transcriptomics data has redefined gene models for 75% of fly genes by adding new exons or splice variants. The majority of annotation changes were supported by direct cDNA evidence. Analysis of transcription start sites (TSSs) for over half of *Drosophila* genes resulted in identification of over 1500 novel promoters. The structural analysis of RNA-seq-identified transcripts that did not seem to encode proteins revealed that a majority of them has no thermodynamically stable secondary structure, suggesting structure-independent functions. Among structural non-coding RNAs, several hundred novel small regulatory RNAs (miRNAs, siRNAs, and piRNA) have been identified. Additionally, transcription start sites for both protein coding and non-coding RNAs have been derived from the presence of chromatin marks characteristic of transcriptionally active regions, such as H3K4me3 enrichment, H3K9ac, and presence of RNA polymerase II in TSS-proximal regions [100].

6.3. Chromatin Landscape

Eukaryotic genomes are organized into large domains that exhibit distinct chromatin properties [100]. Analysis of large-scale organization of the chromatin landscape has revealed unexpected complexity and plasticity among different cell types. Some regions in the usually silent pericentric heterochromatin exhibited surprisingly high gene expression activity. Conversely, large regions of normally transcriptionally active euchromatin harbored histone marks (H3K9me2) typical for heterochromatin [100,101]. Chromatin signatures characteristic of various functional elements have been identified by ChIP-chip for 18 histone modifications (both activating, such as H3K4me, H3K9/18/27ac, H2B ubiquitination and repressive, such as H3K9me2/3 or Polycomb associated H3K27me3) and variants (H1, H4) from several cell lines and developmental stages. Correlating chromatin signatures with transcriptome and protein binding data (replication factors, insulator-binding proteins, and transcription factors) helped identify marks specific for promoters, actively transcribed regions, introns, insulators, and origins of replication [100]. The presence of specific chromatin marks was found to correlate with the physical properties of chromatin, where transcriptionally active chromatin exhibited high solubility and high nucleosome-turnover rates [100]. Computational analysis of combinatorial patterns of histone modifications revealed distinct chromatin states associated with active TSSs, exons, introns, and other open chromatin as well as closed chromatin states [100,102,103].

6.4. Transcriptional Regulation

The modENCODE project has identified binding sites for almost 40 transcription factors through both ChIP-chip and ChIP-seq. The analysis has revealed that out of nearly 40,000 identified unique binding sites found, 5% are bound by 8 different transcription factors or more and are considered High Occupancy Target (HOT) regions. Furthermore, almost 40% of the sites can be bound by more than two factors [103]. The HOT regions exhibit decreased nucleosome density, increased nucleosome turnover and often colocalize with TSS and ORC (origin recognition complex) binding sites, suggesting interplay between chromatin regulation, TF binding, and DNA replication [100,103]. In total, modENCODE ChIP experiments revealed over 500 silencers, 2300 new promoters, over 14 candidate CBP-bound *cis*-regulatory elements, and over 7500 putative insulators [103]. Pairwise analysis of binding site co-occurrence has revealed over 800 known and putative transcription factor co-binding interactions. Binding sites for transcription factors regulating biologically opposing roles exhibited negative associations. The modENCODE TF binding data sets combined with external data were used to construct a network covering over 80 transcription factors and characterizing over 800, largely novel, regulatory interactions. Binding site co-occurrence among various analyzed promoters corresponded to temporal co-expression of the respective target genes, supporting the existence of combinatorial transcription factor codes [103].

7. The Fruit Fly Toolkit

7.1. Getting Constructs in

Drosophila is famous for its extensive range of forward and reverse genetics tools. The powerful toolkit was primed by the discovery of P transposable element-based germline transformation [104]. This revolutionary development allowed, for the first time, efficient delivery of foreign DNA into the genome. Development of the vast majority of *Drosophila* tools required at some stage use of P-element or other transposon systems. P-elements were used for gene cloning [105], genetic rescue [106], and as potent mutagens by their insertion [107] or excision [108]. P-element insertions have enabled the creation of enhancer traps, thus allowing visualization of gene expression patterns using genetically encoded reporters [109]. The *Drosophila* Gene Disruption Project used P-elements to create single transposon insertions in over 30% of fly genes [110]. The remaining genes, due to target sequence bias of P-elements, are currently targeted using other transposons [111]. The catalog of transposons that can be used for fly transformation has been expanded over the years and includes mariner [112,113], Minos [114,115], and piggyBac [116,117]. Each of these transposable elements, except for Minos that seems to insert randomly, has its hot and cold spots, but failure to target a certain region can often be addressed by using a different transposon [111].

While random, transposon-mediated transgenesis is desirable for gene disruption or genomic targeting, but integration of reporter or rescue constructs calls for more control over the locus where these integrate, thus reducing the chance of position effects that can strongly influence gene expression [118,119]. Early attempts to repeatedly target a specific locus in the fruit fly genome were based on transposon homing [120]. Short regulatory sequences from *Polycomb* target genes or from the *linotte* locus included in the transposon were shown to increase the likelihood of such transposon landing in the vicinity of genomic regions bearing these sequences [120,121]. The low resolution (30 bp) and efficiency (20% of insertions) of this homing technique prompted further developments in the field. The introduction of an irreversible, site-specific recombinase from the phiC31 phage ushered in a new era in fly transgenesis. The phiC31 integrase catalyzes unidirectional recombination between two attachment sites, attP and attB, leading to the formation of attL and attR sites [122]. A circular construct harboring an attB site can be efficiently and specifically integrated into an attP site located on the genome [123]. The phiC31 integrase system has, for the first time, enabled transformation of flies with BAC-sized constructs [124]. The integrase can be expressed from mRNA co-injected with the construct [123] or from the genome under a germ-line specific promoter, the latter method being more efficient [125]. Several dozens of attP landing lines have been created and tested [123–126], creating unlimited possibilities to combine transgenes.

7.2. Express What You Want, Where You Want, and When You Want

P-element transgenesis has enabled the creation of a plethora of other *Drosophila* tools, of which the Gal4/UAS system is the most notable example. The system is based on yeast transcription factor gene *GAL4* fused to a minimal promoter. This construct is randomly inserted into the genome, for instance by means of P-element transgenesis, hitchhiking nearby enhancers and creating an enhancer trap. Alternatively, enhancer sequences can be cloned upstream of the minimal promoter and other arrangements, with enhancers cloned elsewhere, for example within introns, are also possible. The second component of the system is the Gal4 binding site, known as the upstream activation sequence (UAS), driving expression of the target gene. The combination of a *GAL4* enhancer trap with a UAS-driven target enables expression of the gene of interest in the desired tissues or cell types [127]. Several collections of *GAL4* enhancer traps have been created using P-element [128,129] and piggyBac [110,130] insertions. The enhancer trap resources have recently been supplemented by large collections of cloned enhancers driving expression of Gal4 [131,132]. The Gal4 expression pattern can be refined spatially [133] or temporarily [134] using the Gal80 repressor [135]. Further control over Gal4-driven expression can be obtained using variants requiring drugs for activation [136–139]. Today, the Gal4-UAS system is one of several binary expression systems available in *Drosophila*. Other examples include the LexA transactivator that binds LexOp sites [140] and the Q system with QF transactivator, QUAS binding sites and the QS repressor whose activity can additionally be drug controlled [141]. The existing binary systems can be combined to provide fine control over target expression pattern or for simultaneous targeting of different cellular populations [142].

7.3. Mutant Tissue on Demand

The Gal4-UAS system is an important component of yet another powerful fruit fly tool, the mosaic analysis with a repressible cell marker or MARCM. Induction of mosaicism in *Drosophila* is used either for studying an otherwise lethal phenotype within a tissue of interest [143] or for marking a clone of cells within a tissue of interest [144]. Mosaics can be created using flippase (Flp) mediated mitotic recombination between homologous chromosomes [145]. In this technique, homologous chromosomes carry an insertion (usually P-element mediated) of a flippase recognition target (FRT) site. One of the chromosomes carries a wild type and the other a mutant allele of the gene of interest. In the presence of flippase, recombination events between homologous chromosomes can occur during cell division, leading to the generation of homozygous mutant cells from heterozygous precursors. The MARCM technique (Figure 1) enhances Flp-mediated mitotic recombination by uniquely labeling mutant cells using a genetically encoded marker. The mutant clone is marked with a UAS-GFP (green fluorescent protein) construct, driven by ubiquitously expressed *GAL4*. These two transgenes are usually inserted together on any chromosome, except the wild type chromosome that carries the Gal4 repressor—*GAL80* under the control of a ubiquitous promoter. The presence of the Gal4 repressor on the wild-type chromosome prevents GFP expression in both heterozygous and homozygous wild type cells [135]. The MARCM technique was later extended to label wild type cells as well [146].

MARCM, among other mitotic-recombination-based approaches, has enabled the creation of tissue specific mutant cells for genes where a mutant exists. This, however, is not yet [111,147] the case for all fruit fly genes. Post-transcriptional gene silencing by double-stranded RNA (dsRNA) [148], commonly known as RNA interference (RNAi), allows the silencing of virtually any transcript encoded by the genome [149]. *Drosophila* is not only one of the first organisms where RNAi has been used to silence genes [150,151], but it has also played an important role in studying the mechanism of dsRNA dependent gene silencing [152]. Injections of dsRNA into the *Drosophila* embryos were used to pioneer RNAi in the fruit fly. However, this mode of delivery has limited use for studying gene function in the late stages of development or in a tissue specific manner. A combination of genetically encoded hairpin-loop RNAs with a Gal4/UAS system has been introduced to address these issues and place *Drosophila* RNAi under spatio-temporal control [153]. Efficient transformation techniques developed for *Drosophila* Schneider (S2) cells [154] combined with *Drosophila* dsRNA libraries allowed RNAi screens in cell culture on a genome-wide scale [155–158]. With genome-wide libraries of fly lines carrying UAS-driven hairpin RNAs, tissue-specific RNAi screens in the whole animals became possible [159,160]. While the first library used P-element for transgenesis, thus leading to variability of hairpin RNA expression levels in different lines, the next generation of libraries followed, using phiC31-mediated insertions into a defined locus [161,162]. RNAi in flies has proven very effective and allowed for a number of large scale screens to be performed, including ones targeting muscle development [162], heart function [163], obesity [164], pain [165], glial function [166], or piRNA pathways [167]. The off-target effect, a well-known pitfall of RNA interference, has been addressed in flies by specificity control using either cross-species rescue [168,169] or engineered RNAi-refractory transgenes [170].

7.4. Bright Rescue

Modern classical and reverse-genetic approaches often call for reliable sources of transgenes, both to induce new and rescue induced phenotypes. Classically, clones from cDNA libraries [69] combined with the Gal4/UAS system [127] have been used to specifically express a gene of interest in target tissue. These constructs could be used either to ectopically express a gene of interest [127], rescue a mutant phenotype [171], or by using a fusion of cDNA with a fluorescent protein coding sequence to visualize the localization of a protein of interest [172]. These approaches, however, do not allow simultaneous modification, such as introduction of point mutations, truncation, tagging, and expression of a protein of interest under native or nearly native control. This usually requires a larger genomic context.

Genome-wide libraries of fruit fly genomic DNA cloned in bacterial artificial chromosomes (BACs) or fosmids, spanning between 20 and over 100 kb, have been constructed for the purpose of genome sequencing [4]. The p[ACMAN] system (Figure 2B,C) has enabled turning them into reliable sources of modifiable genomic inserts, tailored for fly transgenesis. The centerpiece of the system is a single copy vector harboring a second, inducible medium copy origin of replication (oriV), a fly selectable marker (*white*), and attachment site (attB) for phiC31-mediated transgenesis [124]. Site-specific-recombinase-based transformation enables the insertion of constructs over 100 kb in size.

Genomic inserts are subcloned into the backbone using Red/ET homologous recombination (Figure 3), also known as recombineering [173–176]. The ability to arbitrarily modify and transform large genomic constructs has fostered the development of transformation ready genomic libraries of *Drosophila melanogaster* and other fly species. Two such resources have been created so far, the p[ACMAN] [177] and FlyFos [178]. The p[ACMAN] features BAC libraries with average insert sizes of 21 and 83 kb. The vector used is similar to the one in the p[ACMAN] subcloning kit. The FlyFos system (Figure 2A,C) features 36 kb fosmid libraries for *Drosophila melanogaster* and *pseudoobscura* [169,178]. The library vector also includes an inducible oriV, attB site, and a dominant fluorescent marker, selectable in diverse insect species [179]. The liquid culture recombineering pipeline [180] introduced in the system enables high-throughput gene tagging with a variety of tags in 96-well format [178].

Figure 1. Mosaic analysis with a repressible cell marker (MARCM). (**A**) Gal4 transcription factor (blue oval) drives expression of green fluorescent protein (GFP) gene (green box) by binding the upstream activation sequence (UAS) (white box). This expression is repressed when Gal80 (red oval) is present. As a consequence cells that do not carry a gene encoding Gal80 but carry genes encoding Gal4 and UAS-GFP are marked green. (**B**) In MARCM, the *GAL80* repressor gene (red box) is carried on a chromosome that bears the wild-type allele of a gene (yellow box) of interest and a flippase recognition target (FRT) site (grey triangle) placed pericentricaly. The homologous chromosome carries a FRT site in exactly the same position and a mutant allele (orange box), but does not carry the *GAL80* gene. Cells also carry the *GAL4* gene and UAS-GFP on the other chromosomes. During G2 phase (after DNA replication), flippase mediates recombination between two FRT sites of homologous chromosomes, thus generating sister chromatids; one of which carries the wild-type allele and *GAL80* repressor and the other the mutant allele. During mitosis, sister chromatids are distributed to daughter cells, generating cells that are homozygous wild-type or homozygous mutant. Cells that are homozygous mutant are the only cells lacking the *GAL80* gene and thus are labeled with GFP. Reproduced with permission from MacMillan: Nature Protocols ©2007 [181].

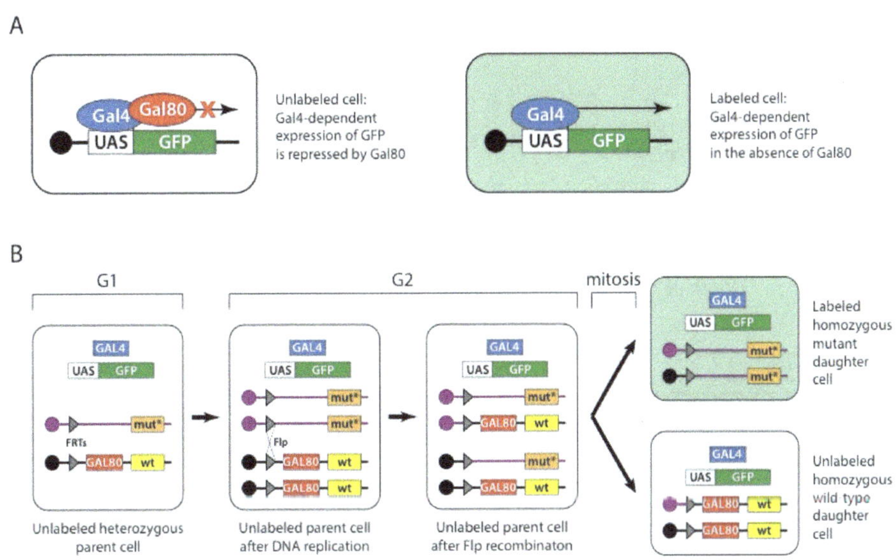

Figure 2. FlyFos and p[ACMAN] genomic libraries. (**A**) FlyFos library is cloned in a fosmid vector, pFlyFos. Genomic inserts were cloned into the *Pml*I site. pFlyFos features an inducible origin of replication (oriS for single copy and oriV for arabinose-inducible moderate copy maintenance), attB site for fly transgenesis, and 3xP3-dsRed as a fly-selectable marker. (**B**) p[ACMAN] libraries are cloned into the *Bam*HI site of a p[ACMAN] bacterial artificial chromosome (BAC) vector. This vector also features inducible oriS/oriV and attB site, but uses white as fly selectable marker. In addition to phiC31-mediated transgenesis, p[ACMAN] vector carrying small inserts can theoretically be used for P-element transformation. (**C**) Size distribution of FlyFos and p[ACMAN] *D. melanogaster* libraries. There are two p[ACMAN] libraries: CHORI-321 with average clone size of 83.3 kb and CHORI-322 with average clone size of 21 kb. The FlyFos library has an average clone size of 36 kb.

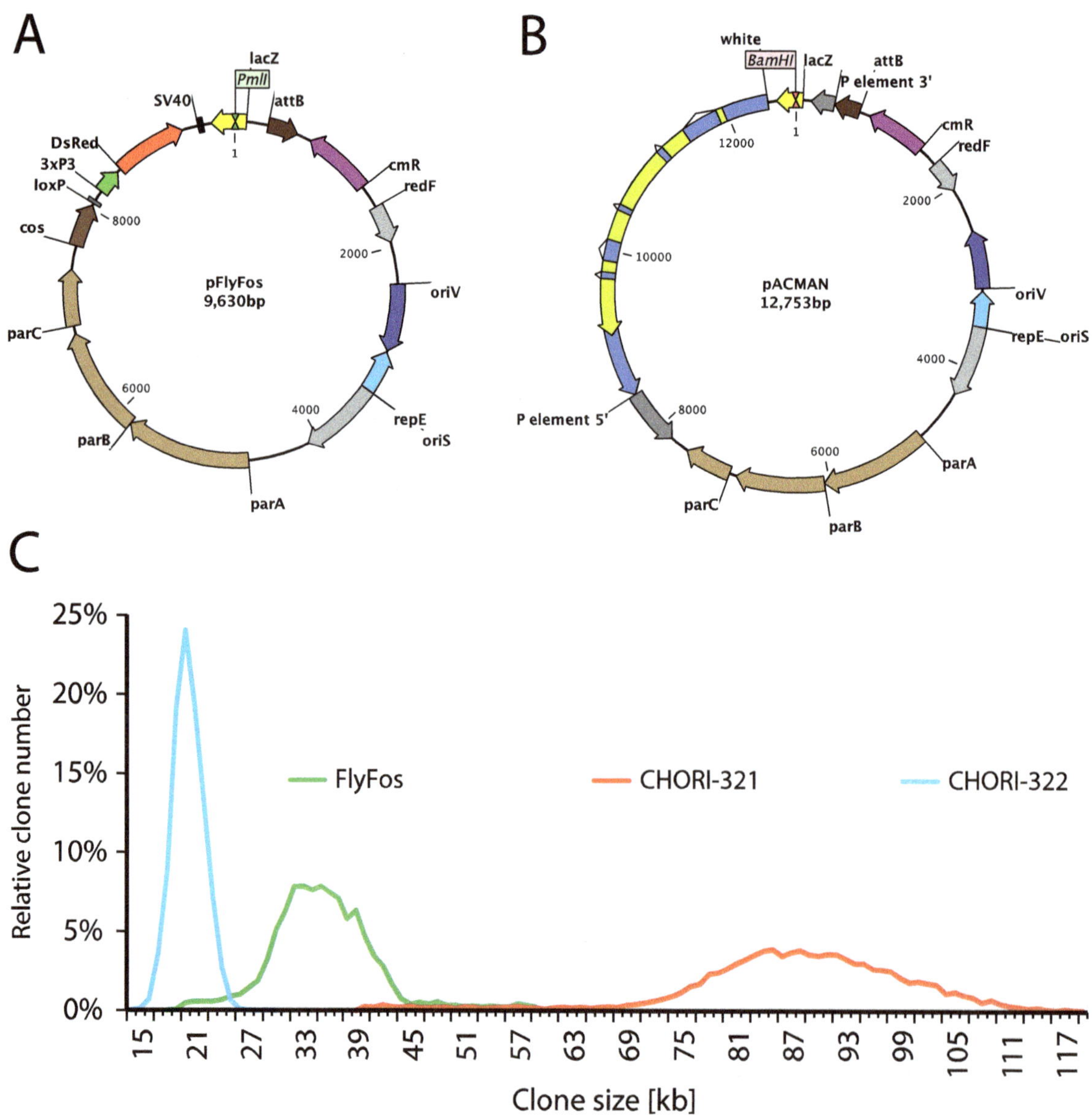

Figure 3. Recombineering principles. (**A**) The target sequence (orange box) is carried on a single copy fosmid or bacterial artificial chromosome (BAC) vector. The 50 bp fragments flanking the target sequence, called homology arms (HA), are depicted as yellow boxes. In this example, the target sequence will be replaced with the recombineering cassette. However, when homology arms are designed to directly follow each other, the cassette can be simply inserted into the target vector. (**B**) The PCR-amplified recombineering cassette harboring homology arms (introduced as primer overhangs) on its termini is electroporated into bacteria carrying the target vector. In the depicted example, the cassette contains a reporter (green arrow) and a flippase recognition target (FRT)-flanked (grey triangles) bacterial selectable marker (red arrow). Homologous recombinase, transiently expressed in bacteria mediates recombination between homology arms replacing the target sequence with the recombineering cassette. (**C**) Recombinant bacterial cells are selected using the selectable marker encoded in the recombineering cassette. If the selectable marker is flanked by FRT sites, it can now be removed (flipped-out) through transient expression of flippase. (**D**) The final recombineering product contains the desired sequence and a 34 bp FRT scar flanked by the homology arms.

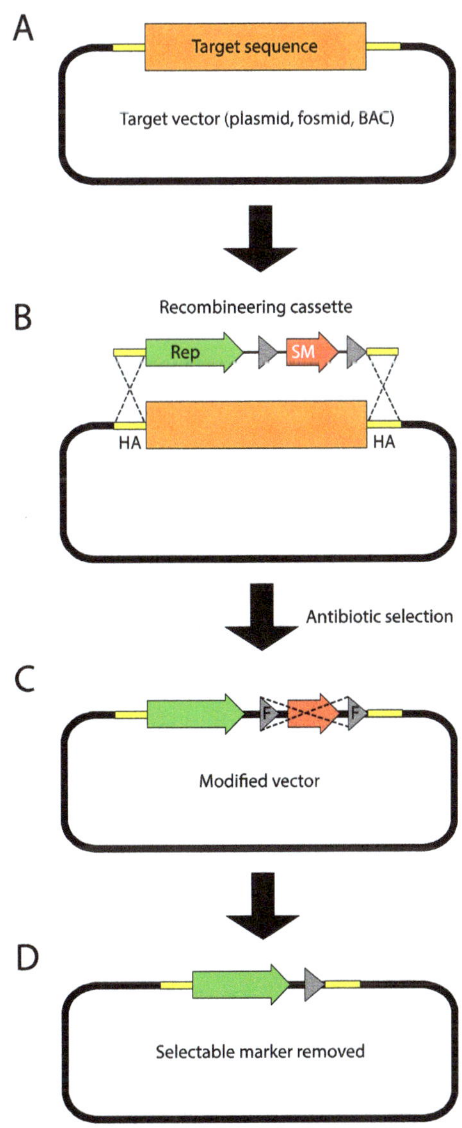

BAC and fosmid-based recombineering has enabled the introduction of modified "third alleles" of genes of interest. The powerful fruit fly genetics toolkit also allows for modifications of genes *in situ*, in their native loci. The first *in situ* genomic targeting in *Drosophila* was performed using the ends-in technique [182]. Ends-in genomic targeting relies on double strand break (DSB) repair through homologous recombination. The targeting construct contains homology arms, one of which is antiparallel to the genomic sequence, and leads to duplication of the targeted locus upon recombination. Initial targeting attempts involving linear DNA injection into the germline were unsuccessful. Inserting the targeting construct into a random locus first, via P-element transgenesis, has solved the issue. FRT sites present on the flanks of the construct were used to mobilize the targeting construct from the genome before generating DSB using I-*Sce*I nuclease [182]. Ends-out targeting uses very similar basic logic, but relies on homology arms that are both parallel to the genomic locus, therefore leading to a clean insertion or replacement [183]. Both ends-in and ends-out have provided reliable means to target genomic loci; however, at a cost of relatively low efficiency. This has made targeting the same locus with different cassettes a labor-intensive task. The integrase-mediated approach for gene knock-out (IMAGO) technique (Figure 4) combines ends-out targeting with phiC31-mediated recombinase-mediated cassette exchange (RMCE) [184]. IMAGO uses ends-out to replace the targeted locus with an attP-flanked selectable marker, which can subsequently be replaced with any desirable construct, thus enabling *in situ* gene tagging, conditional knock-outs, or functional analysis of orthologs. An alternative strategy uses a single attP site and a loxP-flanked selectable marker as the knock-out cassette [185]. Rescue constructs can then be integrated into the target locus using phiC31-mediated transgenesis, just like into any other landing site.

Genomic targeting techniques using DSBs induced in the targeting construct have proven to be robust tools. However, these approaches have a quite high price tag, because of their low efficiency. Homologous recombination with genomic loci is known to be much more effective if DSBs are introduced in the chromosome [186]. Induction of chromosomal DSBs in specific genomic loci requires designer nucleases that can target a sequence of choice. Currently three custom nuclease systems are in broad use: zinc finger nucleases (ZFNs) [187], transcription activator-like effector nucleases (TALENs), and the bacterial clustered regularly interspaced short palindromic repeat (CRISPR) system and its RNA-driven Cas9 nuclease. Double strand breaks are repaired using one of two cellular mechanisms: non-homologous end joining (NHEJ) and homologous recombination (HR). NHEJ involves processing and ligation of broken strands and usually leads to insertions and deletions [188]. However; it has also been shown to mediate efficient knock-ins in zebrafish [189]. HR requires a sequence homologous to the locus in which DSB has occurred, either from a sister chromatid, paralogous locus; or provided linear or plasmid DNA, and can, therefore, be exploited to insert or replace a genomic sequence with custom constructs [190].

Figure 4. Integrase-mediated approach for gene knock-out (IMAGO). (**A**) A targeting construct harboring *white* gene flanked by attP sites, 1 kb–5 kb homology arms, I-*Sce*I meganuclease site, and flippase recognition target (FRT) sites is inserted into the fly genome using transposition or site-specific integration. (**B**) The targeting cassette is mobilized into the circular episome by flippase and subsequently linearized by meganuclease. This linear fragment induces cellular double strand break repair mechanisms and (with certain frequency) replaces the genomic locus flanked by homology arms. (**C**) Recombinant progeny are selected for *white* dominant marker. The attP sites flanking the *white* gene can be used for recombinase-mediated cassette exchange. (**D**, **F**) A plasmid containing the attB-flanked construct (cKO or a mutant allele) is injected into phiC31-expressing fly embryos and exchanges the attP-flanked *white* gene. (**E**, **G**) Recombinant progeny carrying modified alleles are selected for loss of the *white* dominant marker.

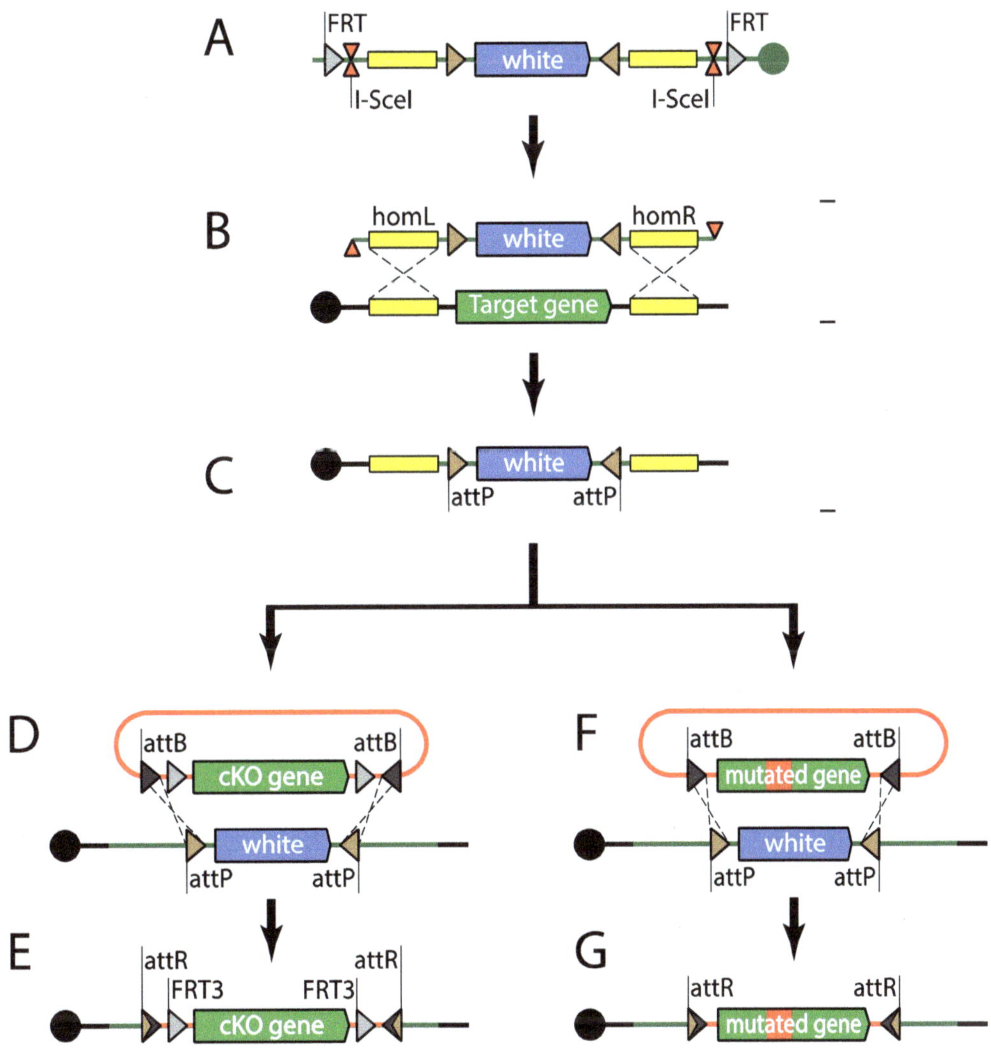

ZFNs were the first designer nuclease system (Figure 5A) to be introduced in flies [188]. They are comprised of three to four zinc-finger DNA binding modules, each recognizing three base pairs, and a FokI endonuclease. Since FokI needs to dimerize for activity, a pair of ZFNs is required for DNA cleavage [187]. The specificity and affinity of zinc-finger modules is context dependent, therefore, several strategies have been developed to achieve assembly of optimal DNA binding domains [191–194].

TALENs (Figure 5B), similar to ZFNs, are hybrids of DNA binding domains derived from transcription factors and FokI endonuclease and, as a consequence, two TALENs are required to form a functional nuclease [195–197]. The TALE (transcription activator-like effector) domains contain a tandem array of 15.5–19.5 repeats, each made of 34 residues, two of which provide DNA-binding specificity against a single nucleotide [198]. Due to the highly repetitive coding sequence of the TALE domain, special approaches have been developed for its efficient assembly using type IIs endonucleases [197]. CRISPR (Figure 5C), a defensive nuclease system from *Streptococcus pyogenes*, takes a completely different approach to DNA cleavage. The specificity is provided by a crRNA pairing with a 20 nt complimentary sequence within the DNA target. The cleavage is performed by Cas9 nuclease that requires trans-activating CRISPR RNA (tracrRNA) in addition to crRNA for activity. The complimentary region of the DNA target must be followed by a 3 bp PAM (protospacer adjacent motif) [199,200]. A pair of crRNA and tracrRNA can be replaced by a single hybrid guide RNA (sgRNA), thus reducing the system to two components [201]. To date, several implementations of the CRISPR system have been created in *Drosophila* [202–204], including transgenic flies with genomically encoded sources of Cas9 [205–207] and tracrRNA/sgRNA [208]. The CRISPR system has been combined with classical ends-out targeting and site-specific integrase approaches, resulting in a versatile toolkit for genome engineering [209]. It should be stated that at this stage the efficiency and specificity of all designer-nuclease-based approaches *in vivo* remains to be fully established, although CRISPR is showing great promise.

8. Conclusions

Drosophila occupies a paramount position among model organisms, largely due to the variety of genetic tools unique to the fruit fly, its short generation time, and ease of transformation. The *Drosophila* classic Gal4/UAS two-component expression system and its counterparts, LexA and Q, can be combined with one another and with site-specific recombination systems like Flp/FRT, Cre/LoxP or phiC31, yielding novel combinatorial systems for even tighter spatio-temporal gene expression control, clonal analysis, and lineage tracing [210]. The fruit fly genome is easily accessible using a broad range of genome engineering tools, including those based on classic transposition, site-specific recombinases and fosmid/BAC recombineering [211], as well as the emerging field of genome editing using designer nucleases [212]. Availability of an almost complete genomic sequence for over 12 species from genus *Drosophila* and dozens of various *D. melanogaster* strain genomes make fruit flies an excellent model for comparative genomics and population genetics. A large number of human disease-related genes that have homologs in the fruit fly [213,214] connected with powerful resources for QTL mapping and GWAS [89,90] make *Drosophila* an attractive model for studying the genetic basis of human disease.

Rapid development of genome engineering techniques, especially those introducing synthetic approaches using designer DNA binding domains of TALEs [215,216] and the CRISPR system [217], will undeniably affect the *Drosophila* field in the next years. Completing the genome and transcriptome sequencing effort for additional fly species [218] will aid in further functional annotation of the *Drosophila* genome and fuel the evolutionary developmental biology field.

Figure 5. Designer nucleases. (**A**) Zink finger nucleases (ZFNs) combine a zinc finger DNA binding domain with a FokI nickase. Each zinc finger recognizes a triplet of bases and usually three to six zinc fingers are present in the targeting domain. Cleavage occurs outside the target sequence and requires a pair of ZFNs, each binding one DNA strand. (**B**) Transcription activator-like effector nucleases (TALEN), similarly to ZFNs have two domains: a DNA binding domain and a FokI nickase domain. The targeting domain is composed of 33–35 amino acid repeats, each binding a single nucleotide. The cleavage mechanism of TALENs is identical to ZFNs. (**C**) Clustered regularly interspaced short palindromic repeats (CRISPR) is a RNA driven double-stranded DNA endonuclease system. Cleavage specificity is provided by crRNA (cyan) that hybridizes with the target sequence (green). Cleavage is performed by the Cas9 protein that, in addition to crRNA, requires tracrRNA for activity. The cleavage site (star) is located between the target sequence and NGG protospacer adjacent motif, complimentary to the sequence immediately downstream of the target. crRNA and tracrRNA can be fused to form guide RNA of similar activity.

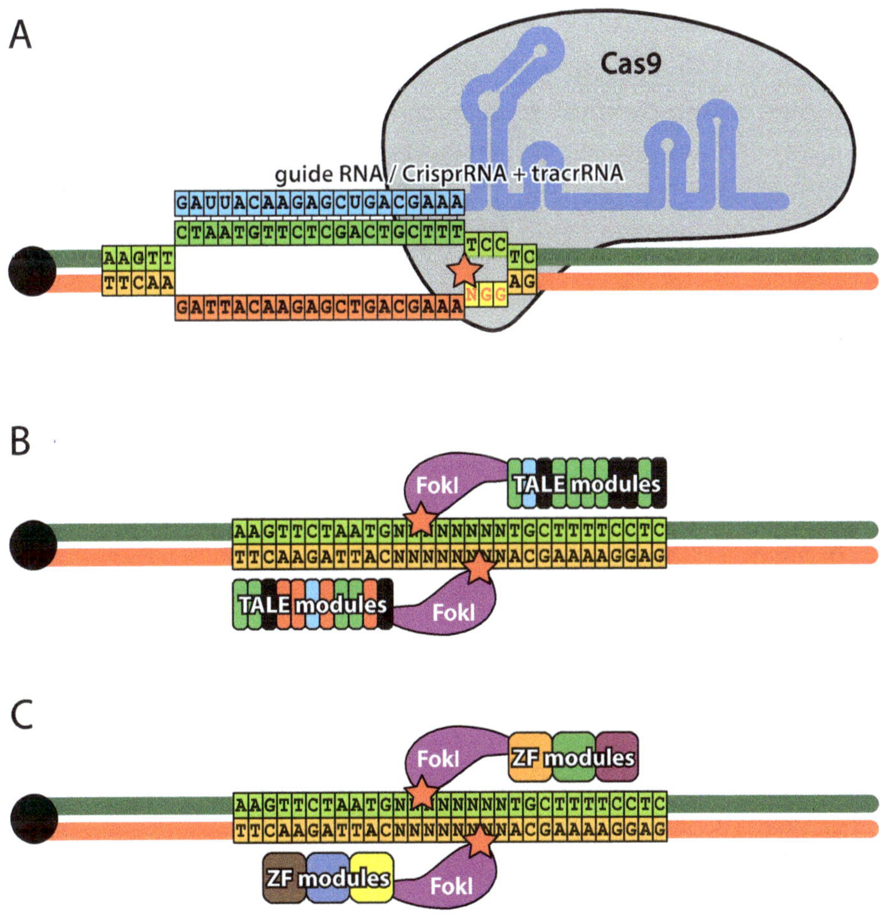

Acknowledgments

Work in the Hassan lab is supported by VIB and grants form FWO, BELSPO, and KU Leuven.

Radoslaw K. Ejsmont was supported by an EMBO ALTF 1056-2011 long-term fellowship and currently by a Marie Curie Actions co-funded Omics@VIB postdoctoral fellowship.

Author Contributions

Wrote the manuscript: Radoslaw K. Ejsmont and Bassem A. Hassan.

References

1. Fleischmann, R.D.; Adams, M.D.; White, O.; Clayton, R.A.; Kirkness, E.F.; Kerlavage, A.R.; Bult, C.J.; Tomb, J.F.; Dougherty, B.A.; Merrick, J.M.; *et al.* Whole-genome random sequencing and assembly of haemophilus influenzae Rd. *Science* **1995**, *269*, 496–512.
2. Clayton, R.A.; White, O.; Ketchum, K.A.; Venter, J.C. The first genome from the third domain of life. *Nature* **1997**, *387*, 459–462.
3. *C. elegans* Sequencing Consortium. Genome sequence of the nematode *C. elegans*: A platform for investigating biology. *Science* **1998**, *282*, 2012–2018.
4. Adams, M.D.; Celniker, S.E.; Holt, R.A.; Evans, C.A.; Gocayne, J.D.; Amanatides, P.G.; Scherer, S.E.; Li, P.W.; Hoskins, R.A.; Galle, R.F.; *et al.* The genome sequence of drosophila melanogaster. *Science* **2000**, *287*, 2185–2195.
5. Lewis, E.B. A gene complex controlling segmentation in drosophila. *Nature* **1978**, *276*, 565–570.
6. Nusslein-Volhard, C.; Wieschaus, E. Mutations affecting segment number and polarity in drosophila. *Nature* **1980**, *287*, 795–801.
7. Gehring, W.J.; Hiromi, Y. Homeotic genes and the homeobox. *Annu. Rev. Genet.* **1986**, *20*, 147–173.
8. Lawrence, P.A.; Morata, G. Homeobox genes: Their function in drosophila segmentation and pattern formation. *Cell* **1994**, *78*, 181–189.
9. Krumlauf, R. Hox genes in vertebrate development. *Cell* **1994**, *78*, 191–201.
10. Quiring, R.; Walldorf, U.; Kloter, U.; Gehring, W.J. Homology of the eyeless gene of drosophila to the small eye gene in mice and aniridia in humans. *Science* **1994**, *265*, 785–789.
11. Ton, C.C.; Hirvonen, H.; Miwa, H.; Weil, M.M.; Monaghan, P.; Jordan, T.; van Heyningen, V.; Hastie, N.D.; Meijers-Heijboer, H.; Drechsler, M.; *et al.* Positional cloning and characterization of a paired box- and homeobox-containing gene from the aniridia region. *Cell* **1991**, *67*, 1059–1074.
12. Hill, R.E.; Favor, J.; Hogan, B.L.; Ton, C.C.; Saunders, G.F.; Hanson, I.M.; Prosser, J.; Jordan, T.; Hastie, N.D.; van Heyningen, V. Mouse small eye results from mutations in a paired-like homeobox-containing gene. *Nature* **1991**, *354*, 522–525.
13. Halder, G.; Callaerts, P.; Flister, S.; Walldorf, U.; Kloter, U.; Gehring, W.J. Eyeless initiates the expression of both sine oculis and eyes absent during drosophila compound eye development. *Development* **1998**, *125*, 2181–2191.
14. Kumar, J.P. The sine oculis homeobox (six) family of transcription factors as regulators of development and disease. *Cell. Mol. Life Sci.* **2009**, *66*, 565–583.

15. Jarman, A.P.; Grell, E.H.; Ackerman, L.; Jan, L.Y.; Jan, Y.N. Atonal is the proneural gene for drosophila photoreceptors. *Nature* **1994**, *369*, 398–400.
16. Jarman, A.P.; Grau, Y.; Jan, L.Y.; Jan, Y.N. Atonal is a proneural gene that directs chordotonal organ formation in the drosophila peripheral nervous system. *Cell* **1993**, *73*, 1307–1321.
17. Brown, N.L.; Patel, S.; Brzezinski, J.; Glaser, T. Math5 is required for retinal ganglion cell and optic nerve formation. *Development* **2001**, *128*, 2497–2508.
18. Bermingham, N.A.; Hassan, B.A.; Price, S.D.; Vollrath, M.A.; Ben-Arie, N.; Eatock, R.A.; Bellen, H.J.; Lysakowski, A.; Zoghbi, H.Y. Math1: An essential gene for the generation of inner ear hair cells. *Science* **1999**, *284*, 1837–1841.
19. Tabata, T.; Eaton, S.; Kornberg, T.B. The drosophila hedgehog gene is expressed specifically in posterior compartment cells and is a target of engrailed regulation. *Genes Dev.* **1992**, *6*, 2635–2645.
20. Lee, J.J.; von Kessler, D.P.; Parks, S.; Beachy, P.A. Secretion and localized transcription suggest a role in positional signaling for products of the segmentation gene hedgehog. *Cell* **1992**, *71*, 33–50.
21. Hooper, J.E.; Scott, M.P. The drosophila patched gene encodes a putative membrane protein required for segmental patterning. *Cell* **1989**, *59*, 751–765.
22. Nakano, Y.; Guerrero, I.; Hidalgo, A.; Taylor, A.; Whittle, J.R.; Ingham, P.W. A protein with several possible membrane-spanning domains encoded by the drosophila segment polarity gene patched. *Nature* **1989**, *341*, 508–513.
23. Stone, D.M.; Hynes, M.; Armanini, M.; Swanson, T.A.; Gu, Q.; Johnson, R.L.; Scott, M.P.; Pennica, D.; Goddard, A.; Phillips, H.; *et al.* The tumour-suppressor gene patched encodes a candidate receptor for sonic hedgehog. *Nature* **1996**, *384*, 129–134.
24. Marigo, V.; Davey, R.A.; Zuo, Y.; Cunningham, J.M.; Tabin, C.J. Biochemical evidence that patched is the hedgehog receptor. *Nature* **1996**, *384*, 176–179.
25. Bhanot, P.; Brink, M.; Samos, C.H.; Hsieh, J.C.; Wang, Y.; Macke, J.P.; Andrew, D.; Nathans, J.; Nusse, R. A new member of the frizzled family from drosophila functions as a wingless receptor. *Nature* **1996**, *382*, 225–230.
26. Siegfried, E.; Chou, T.B.; Perrimon, N. Wingless signaling acts through zeste-white 3, the drosophila homolog of glycogen synthase kinase-3, to regulate engrailed and establish cell fate. *Cell* **1992**, *71*, 1167–1179.
27. Wehrli, M.; Dougan, S.T.; Caldwell, K.; O'Keefe, L.; Schwartz, S.; Vaizel-Ohayon, D.; Schejter, E.; Tomlinson, A.; DiNardo, S. Arrow encodes an ldl-receptor-related protein essential for wingless signalling. *Nature* **2000**, *407*, 527–530.
28. Brunner, E.; Peter, O.; Schweizer, L.; Basler, K. Pangolin encodes a lef-1 homologue that acts downstream of armadillo to transduce the wingless signal in drosophila. *Nature* **1997**, *385*, 829–833.
29. Adler, P.N. The genetic control of tissue polarity in drosophila. *BioEssays* **1992**, *14*, 735–741.
30. Eaton, S.; Julicher, F. Cell flow and tissue polarity patterns. *Curr. Opin. Genet. Dev.* **2011**, *21*, 747–752.

31. Wharton, K.A.; Johansen, K.M.; Xu, T.; Artavanis-Tsakonas, S. Nucleotide sequence from the neurogenic locus notch implies a gene product that shares homology with proteins containing egf-like repeats. *Cell* **1985**, *43*, 567–581.
32. Artavanis-Tsakonas, S.; Matsuno, K.; Fortini, M.E. Notch signaling. *Science* **1995**, *268*, 225–232.
33. Artavanis-Tsakonas, S.; Rand, M.D.; Lake, R.J. Notch signaling: Cell fate control and signal integration in development. *Science* **1999**, *284*, 770–776.
34. Wu, S.; Huang, J.; Dong, J.; Pan, D. Hippo encodes a STE-20 family protein kinase that restricts cell proliferation and promotes apoptosis in conjunction with salvador and warts. *Cell* **2003**, *114*, 445–456.
35. Harvey, K.F.; Pfleger, C.M.; Hariharan, I.K. The drosophila mst ortholog, hippo, restricts growth and cell proliferation and promotes apoptosis. *Cell* **2003**, *114*, 457–467.
36. Udan, R.S.; Kango-Singh, M.; Nolo, R.; Tao, C.; Halder, G. Hippo promotes proliferation arrest and apoptosis in the salvador/warts pathway. *Nat. Cell Biol.* **2003**, *5*, 914–920.
37. Fortini, M.E.; Bonini, N.M. Modeling human neurodegenerative diseases in drosophila: On a wing and a prayer. *Trends Genet.* **2000**, *16*, 161–167.
38. Lloyd, T.E.; Taylor, J.P. Flightless flies: Drosophila models of neuromuscular disease. *Ann. N. Y. Acad. Sci.* **2010**, *1184*, e1–e20.
39. Potter, C.J.; Turenchalk, G.S.; Xu, T. Drosophila in cancer research. An expanding role. *Trends Genet.* **2000**, *16*, 33–39.
40. Jackson, G.R.; Salecker, I.; Dong, X.; Yao, X.; Arnheim, N.; Faber, P.W.; MacDonald, M.E.; Zipursky, S.L. Polyglutamine-expanded human huntingtin transgenes induce degeneration of drosophila photoreceptor neurons. *Neuron* **1998**, *21*, 633–642.
41. Warrick, J.M.; Paulson, H.L.; Gray-Board, G.L.; Bui, Q.T.; Fischbeck, K.H.; Pittman, R.N.; Bonini, N.M. Expanded polyglutamine protein forms nuclear inclusions and causes neural degeneration in drosophila. *Cell* **1998**, *93*, 939–949.
42. Morales, J.; Hiesinger, P.R.; Schroeder, A.J.; Kume, K.; Verstreken, P.; Jackson, F.R.; Nelson, D.L.; Hassan, B.A. Drosophila fragile x protein, dfxr, regulates neuronal morphology and function in the brain. *Neuron* **2002**, *34*, 961–972.
43. Luo, L.Q.; Martin-Morris, L.E.; White, K. Identification, secretion, and neural expression of appl, a drosophila protein similar to human amyloid protein precursor. *J. Neurosci.* **1990**, *10*, 3849–3861.
44. Luo, L.; Tully, T.; White, K. Human amyloid precursor protein ameliorates behavioral deficit of flies deleted for appl gene. *Neuron* **1992**, *9*, 595–605.
45. Ye, Y.; Fortini, M.E. Apoptotic activities of wild-type and alzheimer's disease-related mutant presenilins in drosophila melanogaster. *J. Cell Biol.* **1999**, *146*, 1351–1364.
46. Wittmann, C.W.; Wszolek, M.F.; Shulman, J.M.; Salvaterra, P.M.; Lewis, J.; Hutton, M.; Feany, M.B. Tauopathy in drosophila: Neurodegeneration without neurofibrillary tangles. *Science* **2001**, *293*, 711–714.
47. Feany, M.B.; Bender, W.W. A drosophila model of parkinson's disease. *Nature* **2000**, *404*, 394–398.

48. Whitworth, A.J.; Theodore, D.A.; Greene, J.C.; Benes, H.; Wes, P.D.; Pallanck, L.J. Increased glutathione s-transferase activity rescues dopaminergic neuron loss in a drosophila model of parkinson's disease. *Proc. Natl. Acad. Sci. USA* **2005**, *102*, 8024–8029.
49. Parkes, T.L.; Elia, A.J.; Dickinson, D.; Hilliker, A.J.; Phillips, J.P.; Boulianne, G.L. Extension of drosophila lifespan by overexpression of human sod1 in motorneurons. *Nat. Genet.* **1998**, *19*, 171–174.
50. Shahidullah, M.; Le Marchand, S.J.; Fei, H.; Zhang, J.; Pandey, U.B.; Dalva, M.B.; Pasinelli, P.; Levitan, I.B. Defects in synapse structure and function precede motor neuron degeneration in drosophila models of fus-related als. *J. Neurosci.* **2013**, *33*, 19590–19598.
51. Van der Plas, M.C.; Pilgram, G.S.; Plomp, J.J.; de Jong, A.; Fradkin, L.G.; Noordermeer, J.N. Dystrophin is required for appropriate retrograde control of neurotransmitter release at the drosophila neuromuscular junction. *J. Neurosci.* **2006**, *26*, 333–344.
52. Moore, M.S.; DeZazzo, J.; Luk, A.Y.; Tully, T.; Singh, C.M.; Heberlein, U. Ethanol intoxication in drosophila: Genetic and pharmacological evidence for regulation by the camp signaling pathway. *Cell* **1998**, *93*, 997–1007.
53. Kaun, K.R.; Azanchi, R.; Maung, Z.; Hirsh, J.; Heberlein, U. A drosophila model for alcohol reward. *Nat. Neurosci.* **2011**, *14*, 612–619.
54. McClung, C.; Hirsh, J. Stereotypic behavioral responses to free-base cocaine and the development of behavioral sensitization in drosophila. *Curr. Biol.* **1998**, *8*, 109–112.
55. Al-Anzi, B.; Sapin, V.; Waters, C.; Zinn, K.; Wyman, R.J.; Benzer, S. Obesity-blocking neurons in drosophila. *Neuron* **2009**, *63*, 329–341.
56. Baker, K.D.; Thummel, C.S. Diabetic larvae and obese flies-emerging studies of metabolism in drosophila. *Cell Metab.* **2007**, *6*, 257–266.
57. Wolf, M.J.; Amrein, H.; Izatt, J.A.; Choma, M.A.; Reedy, M.C.; Rockman, H.A. Drosophila as a model for the identification of genes causing adult human heart disease. *Proc. Natl. Acad. Sci. USA* **2006**, *103*, 1394–1399.
58. Roeder, T.; Isermann, K.; Kallsen, K.; Uliczka, K.; Wagner, C. A drosophila asthma model—What the fly tells us about inflammatory diseases of the lung. *Adv. Exp. Med. Biol.* **2012**, *710*, 37–47.
59. Xu, T.; Wang, W.; Zhang, S.; Stewart, R.A.; Yu, W. Identifying tumor suppressors in genetic mosaics: The drosophila lats gene encodes a putative protein kinase. *Development* **1995**, *121*, 1053–1063.
60. Pagliarini, R.A.; Xu, T. A genetic screen in drosophila for metastatic behavior. *Science* **2003**, *302*, 1227–1231.
61. Januschke, J.; Gonzalez, C. Drosophila asymmetric division, polarity and cancer. *Oncogene* **2008**, *27*, 6994–7002.
62. Brumby, A.M.; Richardson, H.E. Using drosophila melanogaster to map human cancer pathways. *Nat. Rev. Cancer* **2005**, *5*, 626–639.
63. Vidal, M.; Cagan, R.L. Drosophila models for cancer research. *Curr. Opin. Genet. Dev.* **2006**, *16*, 10–16.
64. Miles, W.O.; Dyson, N.J.; Walker, J.A. Modeling tumor invasion and metastasis in drosophila. *Dis. Models Mech.* **2011**, *4*, 753–761.

65. Gonzalez, C. Drosophila melanogaster: A model and a tool to investigate malignancy and identify new therapeutics. *Nat. Rev. Cancer* **2013**, *13*, 172–183.
66. Bosco, G.; Campbell, P.; Leiva-Neto, J.T.; Markow, T.A. Analysis of drosophila species genome size and satellite DNA content reveals significant differences among strains as well as between species. *Genetics* **2007**, *177*, 1277–1290.
67. Hoskins, R.A.; Smith, C.D.; Carlson, J.W.; Carvalho, A.B.; Halpern, A.; Kaminker, J.S.; Kennedy, C.; Mungall, C.J.; Sullivan, B.A.; Sutton, G.G.; *et al.* Heterochromatic sequences in a drosophila whole-genome shotgun assembly. *Genome Biol.* **2002**, doi:10.1186/gb-2002-3-12-research0085.
68. Misra, S.; Crosby, M.A.; Mungall, C.J.; Matthews, B.B.; Campbell, K.S.; Hradecky, P.; Huang, Y.; Kaminker, J.S.; Millburn, G.H.; Prochnik, S.E.; *et al.* Annotation of the drosophila melanogaster euchromatic genome: A systematic review. *Genome Biol.* **2002**, doi:10.1186/gb-2002-3-12-research0083.
69. Stapleton, M.; Carlson, J.; Brokstein, P.; Yu, C.; Champe, M.; George, R.; Guarin, H.; Kronmiller, B.; Pacleb, J.; Park, S.; *et al.* A drosophila full-length cdna resource. *Genome Biol.* **2002**, doi:10.1186/gb-2002-3-12-research0080.
70. Tomancak, P.; Beaton, A.; Weiszmann, R.; Kwan, E.; Shu, S.; Lewis, S.E.; Richards, S.; Ashburner, M.; Hartenstein, V.; Celniker, S.E.; *et al.* Systematic determination of patterns of gene expression during drosophila embryogenesis. *Genome Biol.* **2002**, doi:10.1186/gb-2002-3-12-research0088.
71. Kaminker, J.S.; Bergman, C.M.; Kronmiller, B.; Carlson, J.; Svirskas, R.; Patel, S.; Frise, E.; Wheeler, D.A.; Lewis, S.E.; Rubin, G.M.; *et al.* The transposable elements of the drosophila melanogaster euchromatin: A genomics perspective. *Genome Biol.* **2002**, doi:10.1186/gb-2002-3-12-research0084.
72. Ohler, U.; Liao, G.C.; Niemann, H.; Rubin, G.M. Computational analysis of core promoters in the drosophila genome. *Genome Biol.* **2002**, doi:10.1186/gb-2002-3-12-research0087.
73. Marygold, S.J.; Leyland, P.C.; Seal, R.L.; Goodman, J.L.; Thurmond, J.; Strelets, V.B.; Wilson, R.J.; the FlyBase consortium. Flybase: Improvements to the bibliography. *Nucleic Acids Res.* **2013**, *41*, D751–D757.
74. Daines, B.; Wang, H.; Wang, L.; Li, Y.; Han, Y.; Emmert, D.; Gelbart, W.; Wang, X.; Li, W.; Gibbs, R.; *et al.* The drosophila melanogaster transcriptome by paired-end RNA sequencing. *Genome Res.* **2011**, *21*, 315–324.
75. Mungall, C.J.; Misra, S.; Berman, B.P.; Carlson, J.; Frise, E.; Harris, N.; Marshall, B.; Shu, S.; Kaminker, J.S.; Prochnik, S.E.; *et al.* An integrated computational pipeline and database to support whole-genome sequence annotation. *Genome Biol.* **2002**, doi:10.1186/gb-2002-3-12-research0081.
76. Lewis, S.E.; Searle, S.M.; Harris, N.; Gibson, M.; Lyer, V.; Richter, J.; Wiel, C.; Bayraktaroglir, L.; Birney, E.; Crosby, M.A.; *et al.* Apollo: A sequence annotation editor. *Genome Biol.* **2002**, doi:10.1186/gb-2002-3-12-research0082.
77. Reese, M.G.; Kulp, D.; Tammana, H.; Haussler, D. Genie—Gene finding in drosophila melanogaster. *Genome Res.* **2000**, *10*, 529–538.

78. Wood, V.; Gwilliam, R.; Rajandream, M.A.; Lyne, M.; Lyne, R.; Stewart, A.; Sgouros, J.; Peat, N.; Hayles, J.; Baker, S.; *et al.* The genome sequence of schizosaccharomyces pombe. *Nature* **2002**, *415*, 871–880.
79. Stein, L.D.; Bao, Z.; Blasiar, D.; Blumenthal, T.; Brent, M.R.; Chen, N.; Chinwalla, A.; Clarke, L.; Clee, C.; Coghlan, A.; *et al.* The genome sequence of caenorhabditis briggsae: A platform for comparative genomics. *PLoS Biol.* **2003**, *1*, e45.
80. Richards, S.; Liu, Y.; Bettencourt, B.R.; Hradecky, P.; Letovsky, S.; Nielsen, R.; Thornton, K.; Hubisz, M.J.; Chen, R.; Meisel, R.P.; *et al.* Comparative genome sequencing of drosophila pseudoobscura: Chromosomal, gene, and cis-element evolution. *Genome Res.* **2005**, *15*, 1–18.
81. Drosophila 12 Genomes Consortium; Clark, A.G.; Eisen, M.B.; Smith, D.R.; Bergman, C.M.; Oliver, B.; Markow, T.A.; Kaufman, T.C.; Kellis, M.; Gelbart, W.; *et al.* Evolution of genes and genomes on the drosophila phylogeny. *Nature* **2007**, *450*, 203–218.
82. Lin, M.F.; Carlson, J.W.; Crosby, M.A.; Matthews, B.B.; Yu, C.; Park, S.; Wan, K.H.; Schroeder, A.J.; Gramates, L.S.; St Pierre, S.E.; *et al.* Revisiting the protein-coding gene catalog of drosophila melanogaster using 12 fly genomes. *Genome Res.* **2007**, *17*, 1823–1836.
83. Stark, A.; Lin, M.F.; Kheradpour, P.; Pedersen, J.S.; Parts, L.; Carlson, J.W.; Crosby, M.A.; Rasmussen, M.D.; Roy, S.; Deoras, A.N.; *et al.* Discovery of functional elements in 12 drosophila genomes using evolutionary signatures. *Nature* **2007**, *450*, 219–232.
84. Majoros, W.H.; Ohler, U. Modeling the evolution of regulatory elements by simultaneous detection and alignment with phylogenetic pair hmms. *PLoS Comput. Biol.* **2010**, *6*, e1001037.
85. Arunachalam, M.; Jayasurya, K.; Tomancak, P.; Ohler, U. An alignment-free method to identify candidate orthologous enhancers in multiple drosophila genomes. *Bioinformatics* **2010**, *26*, 2109–2115.
86. Aerts, S.; Quan, X.J.; Claeys, A.; Naval Sanchez, M.; Tate, P.; Yan, J.; Hassan, B.A. Robust target gene discovery through transcriptome perturbations and genome-wide enhancer predictions in drosophila uncovers a regulatory basis for sensory specification. *PLoS Biol.* **2010**, *8*, e1000435.
87. Mackay, T.F. Quantitative trait loci in drosophila. *Nat. Rev. Genet.* **2001**, *2*, 11–20.
88. Hindorff, L.A.; Sethupathy, P.; Junkins, H.A.; Ramos, E.M.; Mehta, J.P.; Collins, F.S.; Manolio, T.A. Potential etiologic and functional implications of genome-wide association loci for human diseases and traits. *Proc. Natl. Acad. Sci. USA* **2009**, *106*, 9362–9367.
89. Mackay, T.F.; Richards, S.; Stone, E.A.; Barbadilla, A.; Ayroles, J.F.; Zhu, D.; Casillas, S.; Han, Y.; Magwire, M.M.; Cridland, J.M.; *et al.* The drosophila melanogaster genetic reference panel. *Nature* **2012**, *482*, 173–178.
90. King, E.G.; Merkes, C.M.; McNeil, C.L.; Hoofer, S.R.; Sen, S.; Broman, K.W.; Long, A.D.; Macdonald, S.J. Genetic dissection of a model complex trait using the drosophila synthetic population resource. *Genome Res.* **2012**, *22*, 1558–1566.
91. Ayroles, J.F.; Carbone, M.A.; Stone, E.A.; Jordan, K.W.; Lyman, R.F.; Magwire, M.M.; Rollmann, S.M.; Duncan, L.H.; Lawrence, F.; Anholt, R.R.; *et al.* Systems genetics of complex traits in drosophila melanogaster. *Nat. Genet.* **2009**, *41*, 299–307.

92. Massouras, A.; Waszak, S.M.; Albarca-Aguilera, M.; Hens, K.; Holcombe, W.; Ayroles, J.F.; Dermitzakis, E.T.; Stone, E.A.; Jensen, J.D.; Mackay, T.F.; *et al.* Genomic variation and its impact on gene expression in drosophila melanogaster. *PLoS Genet.* **2012**, *8*, e1003055.
93. Jordan, K.W.; Craver, K.L.; Magwire, M.M.; Cubilla, C.E.; Mackay, T.F.; Anholt, R.R. Genome-wide association for sensitivity to chronic oxidative stress in drosophila melanogaster. *PLoS One* **2012**, *7*, e38722.
94. Jumbo-Lucioni, P.; Bu, S.; Harbison, S.T.; Slaughter, J.C.; Mackay, T.F.; Moellering, D.R.; de Luca, M. Nuclear genomic control of naturally occurring variation in mitochondrial function in drosophila melanogaster. *BMC Genomics* **2012**, doi:10.1186/1471-2164-13-659.
95. Magwire, M.M.; Fabian, D.K.; Schweyen, H.; Cao, C.; Longdon, B.; Bayer, F.; Jiggins, F.M. Genome-wide association studies reveal a simple genetic basis of resistance to naturally coevolving viruses in drosophila melanogaster. *PLoS Genet.* **2012**, *8*, e1003057.
96. Harbison, S.T.; McCoy, L.J.; Mackay, T.F. Genome-wide association study of sleep in drosophila melanogaster. *BMC Genomics* **2013**, doi:10.1186/1471-2164-14-281.
97. King, E.G.; Macdonald, S.J.; Long, A.D. Properties and power of the drosophila synthetic population resource for the routine dissection of complex traits. *Genetics* **2012**, *191*, 935–949.
98. ENCODE Project Consortium. The encode (encyclopedia of DNA elements) project. *Science* **2004**, *306*, 636–640.
99. Gerstein, M.B.; Lu, Z.J.; van Nostrand, E.L.; Cheng, C.; Arshinoff, B.I.; Liu, T.; Yip, K.Y.; Robilotto, R.; Rechtsteiner, A.; Ikegami, K.; *et al.* Integrative analysis of the caenorhabditis elegans genome by the modencode project. *Science* **2010**, *330*, 1775–1787.
100. modENCODE Consortium; Roy, S.; Ernst, J.; Kharchenko, P.V.; Kheradpour, P.; Negre, N.; Eaton, M.L.; Landolin, J.M.; Bristow, C.A.; Ma, L.; *et al.* Identification of functional elements and regulatory circuits by drosophila modencode. *Science* **2010**, *330*, 1787–1797.
101. Riddle, N.C.; Minoda, A.; Kharchenko, P.V.; Alekseyenko, A.A.; Schwartz, Y.B.; Tolstorukov, M.Y.; Gorchakov, A.A.; Jaffe, J.D.; Kennedy, C.; Linder-Basso, D.; *et al.* Plasticity in patterns of histone modifications and chromosomal proteins in drosophila heterochromatin. *Genome Res.* **2011**, *21*, 147–163.
102. Kharchenko, P.V.; Alekseyenko, A.A.; Schwartz, Y.B.; Minoda, A.; Riddle, N.C.; Ernst, J.; Sabo, P.J.; Larschan, E.; Gorchakov, A.A.; Gu, T.; *et al.* Comprehensive analysis of the chromatin landscape in drosophila melanogaster. *Nature* **2011**, *471*, 480–485.
103. Nègre, N.; Brown, C.D.; Ma, L.; Bristow, C.A.; Miller, S.W.; Wagner, U.; Kheradpour, P.; Eaton, M.L.; Loriaux, P.; Sealfon, R.; *et al.* A cis-regulatory map of the drosophila genome. *Nature* **2011**, *471*, 527–531.
104. Spradling, A.C.; Rubin, G.M. Transposition of cloned p elements into drosophila germ line chromosomes. *Science* **1982**, *218*, 341–347.
105. Searles, L.L.; Jokerst, R.S.; Bingham, P.M.; Voelker, R.A.; Greenleaf, A.L. Molecular cloning of sequences from a drosophila RNA polymerase ii locus by p element transposon tagging. *Cell* **1982**, *31*, 585–592.
106. Rubin, G.M.; Spradling, A.C. Genetic transformation of drosophila with transposable element vectors. *Science* **1982**, *218*, 348–353.

107. Cooley, L.; Kelley, R.; Spradling, A. Insertional mutagenesis of the drosophila genome with single P elements. *Science* **1988**, *239*, 1121–1128.
108. Cooley, L.; Thompson, D.; Spradling, A.C. Constructing deletions with defined endpoints in drosophila. *Proc. Natl. Acad. Sci. USA* **1990**, *87*, 3170–3173.
109. O'Kane, C.J.; Gehring, W.J. Detection *in situ* of genomic regulatory elements in drosophila. *Proc. Natl. Acad. Sci. USA* **1987**, *84*, 9123–9127.
110. Bellen, H.J.; Levis, R.W.; Liao, G.; He, Y.; Carlson, J.W.; Tsang, G.; Evans-Holm, M.; Hiesinger, P.R.; Schulze, K.L.; Rubin, G.M.; *et al*. The bdgp gene disruption project: Single transposon insertions associated with 40% of drosophila genes. *Genetics* **2004**, *167*, 761–781.
111. Bellen, H.J.; Levis, R.W.; He, Y.; Carlson, J.W.; Evans-Holm, M.; Bae, E.; Kim, J.; Metaxakis, A.; Savakis, C.; Schulze, K.L.; *et al*. The drosophila gene disruption project: Progress using transposons with distinctive site specificities. *Genetics* **2011**, *188*, 731–743.
112. Jacobson, J.W.; Medhora, M.M.; Hartl, D.L. Molecular structure of a somatically unstable transposable element in drosophila. *Proc. Natl. Acad. Sci. USA* **1986**, *83*, 8684–8688.
113. Garza, D.; Medhora, M.; Koga, A.; Hartl, D.L. Introduction of the transposable element mariner into the germline of drosophila melanogaster. *Genetics* **1991**, *128*, 303–310.
114. Franz, G.; Savakis, C. Minos, a new transposable element from drosophila hydei, is a member of the tc1-like family of transposons. *Nucleic Acids Res.* **1991**, *19*, 6646.
115. Loukeris, T.G.; Arca, B.; Livadaras, I.; Dialektaki, G.; Savakis, C. Introduction of the transposable element minos into the germ line of drosophila melanogaster. *Proc. Natl. Acad. Sci. USA* **1995**, *92*, 9485–9489.
116. Handler, A.M.; McCombs, S.D.; Fraser, M.J.; Saul, S.H. The lepidopteran transposon vector, piggybac, mediates germ-line transformation in the mediterranean fruit fly. *Proc. Natl. Acad. Sci. USA* **1998**, *95*, 7520–7525.
117. Handler, A.M.; Harrell, R.A., 2nd. Germline transformation of drosophila melanogaster with the piggybac transposon vector. *Insect Mol. Biol.* **1999**, *8*, 449–457.
118. Spradling, A.C.; Rubin, G.M. The effect of chromosomal position on the expression of the drosophila xanthine dehydrogenase gene. *Cell* **1983**, *34*, 47–57.
119. Levis, R.; Hazelrigg, T.; Rubin, G.M. Effects of genomic position on the expression of transduced copies of the white gene of drosophila. *Science* **1985**, *229*, 558–561.
120. Taillebourg, E.; Dura, J.M. A novel mechanism for p element homing in drosophila. *Proc. Natl. Acad. Sci. USA* **1999**, *96*, 6856–6861.
121. Cheng, Y.; Kwon, D.Y.; Arai, A.L.; Mucci, D.; Kassis, J.A. P-element homing is facilitated by engrailed polycomb-group response elements in drosophila melanogaster. *PLoS One* **2012**, *7*, e30437.
122. Thorpe, H.M.; Smith, M.C. *In vitro* site-specific integration of bacteriophage DNA catalyzed by a recombinase of the resolvase/invertase family. *Proc. Natl. Acad. Sci. USA* **1998**, *95*, 5505–5510.
123. Groth, A.C.; Fish, M.; Nusse, R.; Calos, M.P. Construction of transgenic drosophila by using the site-specific integrase from phage phic31. *Genetics* **2004**, *166*, 1775–1782.
124. Venken, K.J.; He, Y.; Hoskins, R.A.; Bellen, H.J. P[acman]: A bac transgenic platform for targeted insertion of large DNA fragments in d. Melanogaster. *Science* **2006**, *314*, 1747–1751.

125. Bischof, J.; Maeda, R.K.; Hediger, M.; Karch, F.; Basler, K. An optimized transgenesis system for drosophila using germ-line-specific phic31 integrases. *Proc. Natl. Acad. Sci. USA* **2007**, *104*, 3312–3317.

126. Markstein, M.; Pitsouli, C.; Villalta, C.; Celniker, S.E.; Perrimon, N. Exploiting position effects and the gypsy retrovirus insulator to engineer precisely expressed transgenes. *Nat. Genet.* **2008**, *40*, 476–483.

127. Brand, A.H.; Perrimon, N. Targeted gene expression as a means of altering cell fates and generating dominant phenotypes. *Development* **1993**, *118*, 401–415.

128. Manseau, L.; Baradaran, A.; Brower, D.; Budhu, A.; Elefant, F.; Phan, H.; Philp, A.V.; Yang, M.; Glover, D.; Kaiser, K.; *et al.* Gal4 enhancer traps expressed in the embryo, larval brain, imaginal discs, and ovary of drosophila. *Dev. Dyn.* **1997**, *209*, 310–322.

129. Hayashi, S.; Ito, K.; Sado, Y.; Taniguchi, M.; Akimoto, A.; Takeuchi, H.; Aigaki, T.; Matsuzaki, F.; Nakagoshi, H.; Tanimura, T.; *et al.* Getdb, a database compiling expression patterns and molecular locations of a collection of gal4 enhancer traps. *Genesis* **2002**, *34*, 58–61.

130. Horn, C.; Offen, N.; Nystedt, S.; Hacker, U.; Wimmer, E.A. Piggybac-based insertional mutagenesis and enhancer detection as a tool for functional insect genomics. *Genetics* **2003**, *163*, 647–661.

131. Jenett, A.; Rubin, G.M.; Ngo, T.T.; Shepherd, D.; Murphy, C.; Dionne, H.; Pfeiffer, B.D.; Cavallaro, A.; Hall, D.; Jeter, J.; *et al.* A gal4-driver line resource for drosophila neurobiology. *Cell Rep.* **2012**, *2*, 991–1001.

132. Stark, A.; Dickson, B.J. Vt (vienna tile) gal4 Driver Lines. Available online: http://stockcenter.vdrc.at/control/vtlibrary/ (accessed on 10 January 2014).

133. Suster, M.L.; Seugnet, L.; Bate, M.; Sokolowski, M.B. Refining gal4-driven transgene expression in drosophila with a gal80 enhancer-trap. *Genesis* **2004**, *39*, 240–245.

134. McGuire, S.E.; Le, P.T.; Osborn, A.J.; Matsumoto, K.; Davis, R.L. Spatiotemporal rescue of memory dysfunction in drosophila. *Science* **2003**, *302*, 1765–1768.

135. Lee, T.; Luo, L. Mosaic analysis with a repressible cell marker for studies of gene function in neuronal morphogenesis. *Neuron* **1999**, *22*, 451–461.

136. Han, D.D.; Stein, D.; Stevens, L.M. Investigating the function of follicular subpopulations during drosophila oogenesis through hormone-dependent enhancer-targeted cell ablation. *Development* **2000**, *127*, 573–583.

137. Osterwalder, T.; Yoon, K.S.; White, B.H.; Keshishian, H. A conditional tissue-specific transgene expression system using inducible gal4. *Proc. Natl. Acad. Sci. USA* **2001**, *98*, 12596–12601.

138. Roman, G.; Endo, K.; Zong, L.; Davis, R.L. P[switch], a system for spatial and temporal control of gene expression in drosophila melanogaster. *Proc. Natl. Acad. Sci. USA* **2001**, *98*, 12602–12607.

139. Stebbins, M.J.; Urlinger, S.; Byrne, G.; Bello, B.; Hillen, W.; Yin, J.C. Tetracycline-inducible systems for drosophila. *Proc. Natl. Acad. Sci. USA* **2001**, *98*, 10775–10780.

140. Lai, S.L.; Lee, T. Genetic mosaic with dual binary transcriptional systems in drosophila. *Nat. Neurosci.* **2006**, *9*, 703–709.

141. Potter, C.J.; Tasic, B.; Russler, E.V.; Liang, L.; Luo, L. The q system: A repressible binary system for transgene expression, lineage tracing, and mosaic analysis. *Cell* **2010**, *141*, 536–548.

142. Yagi, R.; Mayer, F.; Basler, K. Refined lexa transactivators and their use in combination with the drosophila gal4 system. *Proc. Natl. Acad. Sci. USA* **2010**, *107*, 16166–16171.
143. Dang, D.T.; Perrimon, N. Use of a yeast site-specific recombinase to generate embryonic mosaics in drosophila. *Dev. Genet.* **1992**, *13*, 367–375.
144. Harrison, D.A.; Perrimon, N. Simple and efficient generation of marked clones in drosophila. *Curr. Biol.* **1993**, *3*, 424–433.
145. Golic, K.G. Site-specific recombination between homologous chromosomes in drosophila. *Science* **1991**, *252*, 958–961.
146. Yu, H.H.; Chen, C.H.; Shi, L.; Huang, Y.; Lee, T. Twin-spot marcm to reveal the developmental origin and identity of neurons. *Nat. Neurosci.* **2009**, *12*, 947–953.
147. Venken, K.J.; Schulze, K.L.; Haelterman, N.A.; Pan, H.; He, Y.; Evans-Holm, M.; Carlson, J.W.; Levis, R.W.; Spradling, A.C.; Hoskins, R.A.; *et al.* Mimic: A highly versatile transposon insertion resource for engineering drosophila melanogaster genes. *Nat. Methods* **2011**, *8*, 737–743.
148. Fire, A.; Xu, S.; Montgomery, M.K.; Kostas, S.A.; Driver, S.E.; Mello, C.C. Potent and specific genetic interference by double-stranded RNA in caenorhabditis elegans. *Nature* **1998**, *391*, 806–811.
149. Hannon, G.J. RNA interference. *Nature* **2002**, *418*, 244–251.
150. Kennerdell, J.R.; Carthew, R.W. Use of dsrna-mediated genetic interference to demonstrate that frizzled and frizzled 2 act in the wingless pathway. *Cell* **1998**, *95*, 1017–1026.
151. Misquitta, L.; Paterson, B.M. Targeted disruption of gene function in drosophila by RNA interference (RNA-i): A role for nautilus in embryonic somatic muscle formation. *Proc. Natl. Acad. Sci. USA* **1999**, *96*, 1451–1456.
152. Hammond, S.M.; Bernstein, E.; Beach, D.; Hannon, G.J. An RNA-directed nuclease mediates post-transcriptional gene silencing in drosophila cells. *Nature* **2000**, *404*, 293–296.
153. Kennerdell, J.R.; Carthew, R.W. Heritable gene silencing in drosophila using double-stranded RNA. *Nat. Biotechnol.* **2000**, *18*, 896–898.
154. Clemens, J.C.; Worby, C.A.; Simonson-Leff, N.; Muda, M.; Maehama, T.; Hemmings, B.A.; Dixon, J.E. Use of double-stranded RNA interference in drosophila cell lines to dissect signal transduction pathways. *Proc. Natl. Acad. Sci. USA* **2000**, *97*, 6499–6503.
155. Kiger, A.A.; Baum, B.; Jones, S.; Jones, M.R.; Coulson, A.; Echeverri, C.; Perrimon, N. A functional genomic analysis of cell morphology using RNA interference. *J. Biol.* **2003**, doi:10.1186/1475-4924-2-27.
156. Lum, L.; Yao, S.; Mozer, B.; Rovescalli, A.; von Kessler, D.; Nirenberg, M.; Beachy, P.A. Identification of hedgehog pathway components by RNAi in drosophila cultured cells. *Science* **2003**, *299*, 2039–2045.
157. Boutros, M.; Kiger, A.A.; Armknecht, S.; Kerr, K.; Hild, M.; Koch, B.; Haas, S.A.; Paro, R.; Perrimon, N.; Heidelberg Fly Array, C. Genome-wide RNAi analysis of growth and viability in drosophila cells. *Science* **2004**, *303*, 832–835.
158. Foley, E.; O'Farrell, P.H. Functional dissection of an innate immune response by a genome-wide RNAi screen. *PLoS Biol.* **2004**, *2*, E203.
159. Dietzl, G.; Chen, D.; Schnorrer, F.; Su, K.-C.; Barinova, Y.; Fellner, M.; Gasser, B.; Kinsey, K.; Oppel, S.; Scheiblauer, S.; *et al.* A genome-wide transgenic RNAi library for conditional gene inactivation in drosophila. *Nature* **2007**, *448*, 151–156.

160. Ueda, R. RNAi Fly—A Comprehensive RNAi-Mutant Fly Bank. Available online: http://www.shigen.nig.ac.jp/fly/nigfly/about/aboutrnai.jsp/ (accessed on 13 January 2014).
161. Ni, J.-Q.; Liu, L.-P.; Binari, R.; Hardy, R.; Shim, H.-S.; Cavallaro, A.; Booker, M.; Pfeiffer, B.D.; Markstein, M.; Wang, H.; *et al.* A drosophila resource of transgenic RNAi lines for neurogenetics. *Genetics* **2009**, *182*, 1089–1100.
162. Schnorrer, F.; Schonbauer, C.; Langer, C.C.; Dietzl, G.; Novatchkova, M.; Schernhuber, K.; Fellner, M.; Azaryan, A.; Radolf, M.; Stark, A.; *et al.* Systematic genetic analysis of muscle morphogenesis and function in drosophila. *Nature* **2010**, *464*, 287–291.
163. Neely, G.G.; Kuba, K.; Cammarato, A.; Isobe, K.; Amann, S.; Zhang, L.; Murata, M.; Elmen, L.; Gupta, V.; Arora, S.; *et al.* A global *in vivo* drosophila RNAi screen identifies not3 as a conserved regulator of heart function. *Cell* **2010**, *141*, 142–153.
164. Pospisilik, J.A.; Schramek, D.; Schnidar, H.; Cronin, S.J.; Nehme, N.T.; Zhang, X.; Knauf, C.; Cani, P.D.; Aumayr, K.; Todoric, J.; *et al.* Drosophila genome-wide obesity screen reveals hedgehog as a determinant of brown *versus* white adipose cell fate. *Cell* **2010**, *140*, 148–160.
165. Neely, G.G.; Hess, A.; Costigan, M.; Keene, A.C.; Goulas, S.; Langeslag, M.; Griffin, R.S.; Belfer, I.; Dai, F.; Smith, S.B.; *et al.* A genome-wide drosophila screen for heat nociception identifies alpha2delta3 as an evolutionarily conserved pain gene. *Cell* **2010**, *143*, 628–638.
166. Ghosh, A.; Kling, T.; Snaidero, N.; Sampaio, J.L.; Shevchenko, A.; Gras, H.; Geurten, B.; Gopfert, M.C.; Schulz, J.B.; Voigt, A.; *et al.* A global *in vivo* drosophila RNAi screen identifies a key role of ceramide phosphoethanolamine for glial ensheathment of axons. *PLoS Genet.* **2013**, *9*, e1003980.
167. Czech, B.; Preall, J.B.; McGinn, J.; Hannon, G.J. A transcriptome-wide RNAi screen in the drosophila ovary reveals factors of the germline piRNA pathway. *Mol. Cell* **2013**, *50*, 749–761.
168. Kondo, S.; Booker, M.; Perrimon, N. Cross-species RNAi rescue platform in drosophila melanogaster. *Genetics* **2009**, *183*, 1165–1173.
169. Langer, C.C.H.; Ejsmont, R.K.; Schönbauer, C.; Schnorrer, F.; Tomancak, P. *In vivo* RNAi rescue in drosophila melanogaster with genomic transgenes from drosophila pseudoobscura. *PLoS One* **2010**, *5*, e8928.
170. Schulz, J.G.; David, G.; Hassan, B.A. A novel method for tissue-specific RNAi rescue in drosophila. *Nucleic Acids Res.* **2009**, *37*, e93.
171. Lawrence, P.A.; Bodmer, R.; Vincent, J.P. Segmental patterning of heart precursors in drosophila. *Development* **1995**, *121*, 4303–4308.
172. Murray, M.J.; Merritt, D.J.; Brand, A.H.; Whitington, P.M. *In vivo* dynamics of axon pathfinding in the drosophilia CNS: A time-lapse study of an identified motorneuron. *J. Neurobiol.* **1998**, *37*, 607–621.
173. Murphy, K.C. Use of bacteriophage lambda recombination functions to promote gene replacement in escherichia coli. *J. Bacteriol.* **1998**, *180*, 2063–2071.
174. Zhang, Y.; Buchholz, F.; Muyrers, J.P.; Stewart, A.F. A new logic for DNA engineering using recombination in escherichia coli. *Nat. Genet.* **1998**, *20*, 123–128.
175. Muyrers, J.P.; Zhang, Y.; Testa, G.; Stewart, A.F. Rapid modification of bacterial artificial chromosomes by et-recombination. *Nucleic Acids Res.* **1999**, *27*, 1555–1557.

176. Yu, D.; Ellis, H.M.; Lee, E.C.; Jenkins, N.A.; Copeland, N.G.; Court, D.L. An efficient recombination system for chromosome engineering in *Escherichia coli*. *Proc. Natl. Acad. Sci. USA* **2000**, *97*, 5978–5983.
177. Venken, K.J.T.; Carlson, J.W.; Schulze, K.L.; Pan, H.; He, Y.; Spokony, R.; Wan, K.H.; Koriabine, M.; de Jong, P.J.; White, K.P.; *et al*. Versatile P[acman] bac libraries for transgenesis studies in drosophila melanogaster. *Nat. Methods* **2009**, *6*, 431–434.
178. Ejsmont, R.K.; Sarov, M.; Winkler, S.; Lipinski, K.A.; Tomancak, P. A toolkit for high-throughput, cross-species gene engineering in drosophila. *Nat. Methods* **2009**, *6*, 435–437.
179. Horn, C.; Jaunich, B.; Wimmer, E.A. Highly sensitive, fluorescent transformation marker for drosophila transgenesis. *Dev. Genes Evol.* **2000**, *210*, 623–629.
180. Poser, I.; Sarov, M.; Hutchins, J.R.; Heriche, J.K.; Toyoda, Y.; Pozniakovsky, A.; Weigl, D.; Nitzsche, A.; Hegemann, B.; Bird, A.W.; *et al*. Bac transgeneomics: A high-throughput method for exploration of protein function in mammals. *Nat. Methods* **2008**, *5*, 409–415.
181. Wu, J.S.; Luo, L. A protocol for mosaic analysis with a repressible cell marker (Marcm) in drosophila. *Nat. Protoc.* **2006**, *1*, 2583–2589.
182. Rong, Y.S.; Golic, K.G. Gene targeting by homologous recombination in drosophila. *Science* **2000**, *288*, 2013–2018.
183. Gong, W.J.; Golic, K.G. Ends-out, or replacement, gene targeting in drosophila. *Proc. Natl. Acad. Sci. USA* **2003**, *100*, 2556–2561.
184. Choi, C.M.; Vilain, S.; Langen, M.; van Kelst, S.; de Geest, N.; Yan, J.; Verstreken, P.; Hassan, B.A. Conditional mutagenesis in drosophila. *Science* **2009**, doi:10.1126/science.1168275.
185. Huang, J.; Zhou, W.; Dong, W.; Watson, A.M.; Hong, Y. From the cover: Directed, efficient, and versatile modifications of the drosophila genome by genomic engineering. *Proc. Natl. Acad. Sci. USA* **2009**, *106*, 8284–8289.
186. Gloor, G.B.; Nassif, N.A.; Johnson-Schlitz, D.M.; Preston, C.R.; Engels, W.R. Targeted gene replacement in drosophila via p element-induced gap repair. *Science* **1991**, *253*, 1110–1117.
187. Kim, Y.G.; Cha, J.; Chandrasegaran, S. Hybrid restriction enzymes: Zinc finger fusions to Fok I cleavage domain. *Proc. Natl. Acad. Sci. USA* **1996**, *93*, 1156–1160.
188. Bibikova, M.; Golic, M.; Golic, K.G.; Carroll, D. Targeted chromosomal cleavage and mutagenesis in drosophila using zinc-finger nucleases. *Genetics* **2002**, *161*, 1169–1175.
189. Auer, T.O.; Duroure, K.; de Cian, A.; Concordet, J.P.; Del Bene, F. Highly efficient crispr/cas9-mediated knock-in in zebrafish by homology-independent DNA repair. *Genome Res.* **2014**, *24*, 142–153.
190. Beumer, K.; Bhattacharyya, G.; Bibikova, M.; Trautman, J.K.; Carroll, D. Efficient gene targeting in drosophila with zinc-finger nucleases. *Genetics* **2006**, *172*, 2391–2403.
191. Maeder, M.L.; Thibodeau-Beganny, S.; Osiak, A.; Wright, D.A.; Anthony, R.M.; Eichtinger, M.; Jiang, T.; Foley, J.E.; Winfrey, R.J.; Townsend, J.A.; *et al*. Rapid "open-source" engineering of customized zinc-finger nucleases for highly efficient gene modification. *Mol. Cell* **2008**, *31*, 294–301.
192. Sander, J.D.; Dahlborg, E.J.; Goodwin, M.J.; Cade, L.; Zhang, F.; Cifuentes, D.; Curtin, S.J.; Blackburn, J.S.; Thibodeau-Beganny, S.; Qi, Y.; *et al*. Selection-free zinc-finger-nuclease engineering by context-dependent assembly (coda). *Nat. Methods* **2011**, *8*, 67–69.

193. Kim, S.; Lee, M.J.; Kim, H.; Kang, M.; Kim, J.S. Preassembled zinc-finger arrays for rapid construction of zfns. *Nat. Methods* **2011**, doi:10.1038/nmeth0111-7a.
194. Doyon, Y.; Vo, T.D.; Mendel, M.C.; Greenberg, S.G.; Wang, J.; Xia, D.F.; Miller, J.C.; Urnov, F.D.; Gregory, P.D.; Holmes, M.C. Enhancing zinc-finger-nuclease activity with improved obligate heterodimeric architectures. *Nat. Methods* **2011**, *8*, 74–79.
195. Mahfouz, M.M.; Li, L.; Shamimuzzaman, M.; Wibowo, A.; Fang, X.; Zhu, J.K. *De novo*-engineered transcription activator-like effector (tale) hybrid nuclease with novel DNA binding specificity creates double-strand breaks. *Proc. Natl. Acad. Sci. USA* **2011**, *108*, 2623–2628.
196. Miller, J.C.; Tan, S.; Qiao, G.; Barlow, K.A.; Wang, J.; Xia, D.F.; Meng, X.; Paschon, D.E.; Leung, E.; Hinkley, S.J.; *et al.* A tale nuclease architecture for efficient genome editing. *Nat. Biotechnol.* **2011**, *29*, 143–148.
197. Morbitzer, R.; Elsaesser, J.; Hausner, J.; Lahaye, T. Assembly of custom tale-type DNA binding domains by modular cloning. *Nucleic Acids Res.* **2011**, *39*, 5790–5799.
198. Moscou, M.J.; Bogdanove, A.J. A simple cipher governs DNA recognition by tal effectors. *Science* **2009**, doi:10.1126/science.1178817.
199. Brouns, S.J.; Jore, M.M.; Lundgren, M.; Westra, E.R.; Slijkhuis, R.J.; Snijders, A.P.; Dickman, M.J.; Makarova, K.S.; Koonin, E.V.; van der Oost, J. Small crispr rnas guide antiviral defense in prokaryotes. *Science* **2008**, *321*, 960–964.
200. Gasiunas, G.; Barrangou, R.; Horvath, P.; Siksnys, V. Cas9-crrna ribonucleoprotein complex mediates specific DNA cleavage for adaptive immunity in bacteria. *Proc. Natl. Acad. Sci. USA* **2012**, *109*, E2579–E2586.
201. Jinek, M.; Chylinski, K.; Fonfara, I.; Hauer, M.; Doudna, J.A.; Charpentier, E. A programmable dual-RNA-guided DNA endonuclease in adaptive bacterial immunity. *Science* **2012**, *337*, 816–821.
202. Bassett, A.R.; Tibbit, C.; Ponting, C.P.; Liu, J.L. Highly efficient targeted mutagenesis of drosophila with the crispr/cas9 system. *Cell Rep.* **2013**, *4*, 220–228.
203. Gratz, S.J.; Cummings, A.M.; Nguyen, J.N.; Hamm, D.C.; Donohue, L.K.; Harrison, M.M.; Wildonger, J.; O'Connor-Giles, K.M. Genome engineering of drosophila with the crispr RNA-guided cas9 nuclease. *Genetics* **2013**, *194*, 1029–1035.
204. Yu, Z.; Ren, M.; Wang, Z.; Zhang, B.; Rong, Y.S.; Jiao, R.; Gao, G. Highly efficient genome modifications mediated by crispr/cas9 in drosophila. *Genetics* **2013**, *195*, 289–291.
205. Sebo, Z.L.; Lee, H.B.; Peng, Y.; Guo, Y. A simplified and efficient germline-specific crispr/cas9 system for drosophila genomic engineering. *Fly* **2014**, *8*, 52–57.
206. Ren, X.; Sun, J.; Housden, B.E.; Hu, Y.; Roesel, C.; Lin, S.; Liu, L.-P.; Yang, Z.; Mao, D.; Sun, L.; *et al.* Optimized gene editing technology for drosophila melanogaster using germ line-specific cas9. *Proc. Natl. Acad. Sci. USA* **2013**, *110*, 19012–19017.
207. Gratz, S.J.; Ukken, F.P.; Rubinstein, C.D.; Thiede, G.; Donohue, L.K.; Cummings, A.M.; O'Connor-Giles, K.M. Highly specific and efficient crispr/cas9-catalyzed homology-directed repair in drosophila. *Genetics* **2014**, doi:10.1534/genetics.113.160713.
208. Kondo, S.; Ueda, R. Highly improved gene targeting by germline-specific cas9 expression in drosophila. *Genetics* **2013**, *195*, 715–721.

209. Baena-Lopez, L.A.; Alexandre, C.; Mitchell, A.; Pasakarnis, L.; Vincent, J.P. Accelerated homologous recombination and subsequent genome modification in drosophila. *Development* **2013**, *140*, 4818–4825.
210. Del Valle Rodriguez, A.; Didiano, D.; Desplan, C. Power tools for gene expression and clonal analysis in drosophila. *Nat. Methods* **2012**, *9*, 47–55.
211. Venken, K.J.; Bellen, H.J. Genome-wide manipulations of drosophila melanogaster with transposons, flp recombinase, and phic31 integrase. *Methods Mol. Biol.* **2012**, *859*, 203–228.
212. Gratz, S.J.; Wildonger, J.; Harrison, M.M.; O'Connor-Giles, K.M. Crispr/cas9-mediated genome engineering and the promise of designer flies on demand. *Fly* **2013**, doi:10.4161/fly.26566.
213. Reiter, L.T.; Potocki, L.; Chien, S.; Gribskov, M.; Bier, E. A systematic analysis of human disease-associated gene sequences in drosophila melanogaster. *Genome Res.* **2001**, *11*, 1114–1125.
214. Chien, S.; Reiter, L.T.; Bier, E.; Gribskov, M. Homophila: Human disease gene cognates in drosophila. *Nucleic Acids Res.* **2002**, *30*, 149–151.
215. Richter, A.; Boch, J. Designer tales team up for highly efficient gene induction. *Nat. Methods* **2013**, *10*, 207–208.
216. Crocker, J.; Stern, D.L. Tale-mediated modulation of transcriptional enhancers *in vivo*. *Nat. Methods* **2013**, *10*, 762–767.
217. Mali, P.; Esvelt, K.M.; Church, G.M. Cas9 as a versatile tool for engineering biology. *Nat. Methods* **2013**, *10*, 957–963.
218. modENCODE Consortium Modencode Comparative Genomics Whitepaper. Available online: http://www.genome.gov/Pages/Research/Sequencing/SeqProposals/modENCODE_Comparative Genomics_WhitePaper.pdf (accessed on 16 January 2014).

10

Somatic Mosaicism in the Human Genome

Donald Freed [1,2], Eric L. Stevens [2,3,‡] and Jonathan Pevsner [1,2,3,*]

[1] Program in Biochemistry, Cellular and Molecular Biology, Johns Hopkins School of Medicine, Baltimore, MD 21205, USA; E-Mail: freedd@kennedykrieger.org

[2] Department of Neurology, Kennedy Krieger Institute, 707 N. Broadway, Baltimore, MD 21205, USA

[3] Department of Psychiatry and Behavioral Sciences, Johns Hopkins School of Medicine, Baltimore, MD 21205, USA

‡ Present address: CFSAN Division of Microbiology, Food and Drug Administration, College Park, MD 20740, USA; E-Mail: eric.stevens@fda.hhs.gov

* Author to whom correspondence should be addressed; E-Mail: pevsner@kennedykrieger.org;

Abstract: Somatic mosaicism refers to the occurrence of two genetically distinct populations of cells within an individual, derived from a postzygotic mutation. In contrast to inherited mutations, somatic mosaic mutations may affect only a portion of the body and are not transmitted to progeny. These mutations affect varying genomic sizes ranging from single nucleotides to entire chromosomes and have been implicated in disease, most prominently cancer. The phenotypic consequences of somatic mosaicism are dependent upon many factors including the developmental time at which the mutation occurs, the areas of the body that are affected, and the pathophysiological effect(s) of the mutation. The advent of second-generation sequencing technologies has augmented existing array-based and cytogenetic approaches for the identification of somatic mutations. We outline the strengths and weaknesses of these techniques and highlight recent insights into the role of somatic mosaicism in causing cancer, neurodegenerative, monogenic, and complex disease.

Keywords: mutation; somatic; germline; mosaicism; complex disease; retrotransposition; neurodegeneration; aging

---

1. Introduction to Somatic Mosaicism

1.1. Early Studies of Mosaicism

Somatic mosaic mutations are defined as mutations that occur in some cells of the soma of a single individual (Figure 1) [1,2]. The mixture of mutation-positive cells with non-mutated cells results in an individual who is a mosaic, or contains different DNA within different cells of his or her body. Mosaic mutations may be present in the germline or soma; however, typically only mutations in the soma have phenotypic consequences or are detectable by current genotyping methods. Mosaic mutations in germ cells are usually only discovered when they lead to inherited conditions in multiple progeny. *De novo* mutations are operationally defined as mutations found in all cells of an individual but not detected in that individual's parents (Figure 1d,e) [3]. *De novo* mutations only present in the offspring may occur very early in development; however, this is rare and increasingly sensitive genetic assays are discovering low-level parental mosaicism in supposedly *de novo* cases (Figure 1b) [4,5].

The role of somatic genetic changes in human health has been considered at least since 1914 when Theodor Boveri recognized that cancers frequently have abnormal karyotypes [6]. Alfred Knudson built upon the work of Boveri and others and in 1971 described a two-hit model of cancer resulting from both an inherited germline mutation and a later somatic mutation [7]. The model of metastatic cancer occurring as a result of multiple mutations in a single cell lineage has remained largely unchanged for over 40 years [8,9].

The scientific community was slower to realize the importance of postzygotic mutational events outside of cancer. In the early 1950s, Barbara McClintock demonstrated the phenotypic importance of somatic transposition in *Zea mays*, and in 1959 Sir Macfarlane Burnet proposed a role for somatic mutation in disease [10,11]. Nonetheless, few studies indicated a role for somatic mosaicism in human health. This changed in the 1970s with the discovery that somatic gene rearrangement creates functional diversity of immunoglobulin and T-cell receptor genes [12–14]. Today, it is known that somatic mutations are ubiquitous [15] and have important roles in cancer [9], aging [16,17], neurodegeneration [18], monogenic disease [19–21], reversion of inherited disease [22–25], and numerous neurocutaneous disorders [26].

1.2. Categories of Somatic Variation

Somatic variation has been observed at all genomic scales from point mutations to aneuploidies. At the level of whole chromosomes and large chromosomal segments, complex genomic rearrangements occur somatically (as well as in the germline). The loss or gain of entire chromosomes is thought to be caused by errors in chromosomal segregation during anaphase, while non-allelic homologous recombination may cause the loss, gain, or rearrangement of large genomic regions [27,28]. The phenotypic consequences of these events vary considerably based on the size of the event and the genomic region involved.

Figure 1. Overview of categories of variation including inherited (panels A–C), *de novo* (panels D,E), and somatic variation (panels F,G). Inherited mutations are always transmitted through the germline (**A**); although a parent may also have a mosaic mutation (this combination of somatic and germline mosaicism is occasionally termed gonadal mosaicism) (**B**); In such cases, a child may inherit the variant as a heterozygous mutation with a more severe clinical phenotype. A parent may also have germline mosaicism that may be inherited by progeny (**C**); *De novo* mutations are operationally defined as genotypes observed in a child but not in either parent. They may originate in a parental germ cell (as may be inferred in a pedigree having multiple affected offspring) (**D**) or postzygotically (**E**); Somatic mutation may occur relatively early in development (**F**) or at any later time throughout the lifespan (**G**), generally affecting fewer cells.

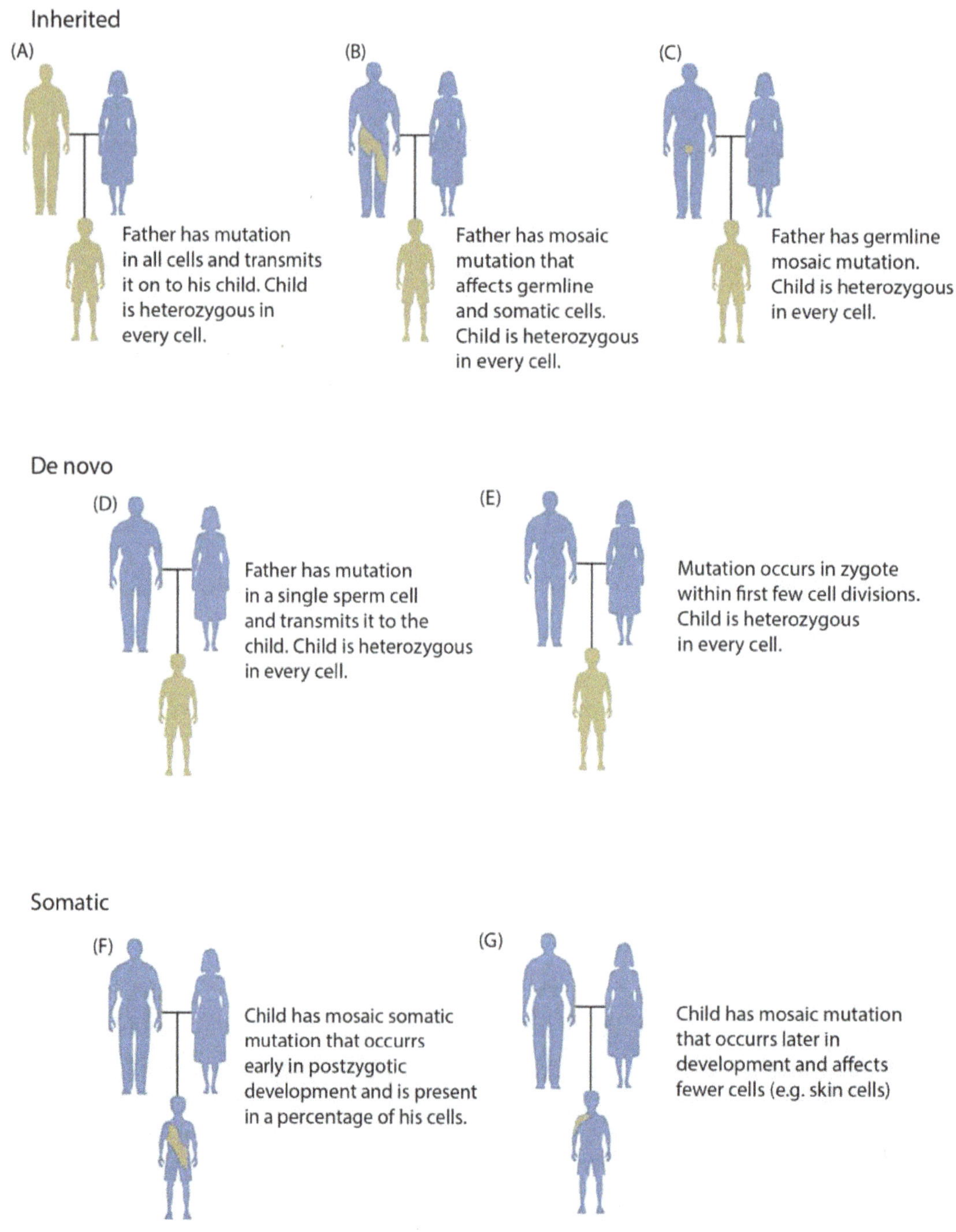

In many instances, both copies of a chromosome pair (or of a chromosomal segment) are inherited from one parent, a phenomenon termed uniparental disomy (UPD) [29,30]. UPD may involve two copies from a parent that are identical (uniparental isodisomy) or different (uniparental heterodisomy). Either form may disrupt epigenetically imprinted regions (defined as undergoing differential expression depending on the parent of origin), while uniparental isodisomy may also expose two copies of a recessive mutation. One mechanism for the occurrence of UPD involves trisomic rescue in which an extra (third) copy of a chromosome is rejected, producing a diploid cell line in which one parent's monoploid copy is lost [31]. Frequently, the trisomic rescue is restricted to a fraction of cells in an individual resulting in mosaic trisomy/UPD [32]. UPD may also result from somatic recombination occurring from a reciprocal exchange during mitosis, leading to loss of heterozygosity.

RNA-templated DNA polymerases are another cause of genomic instability. While numerous types of repetitive elements are present in human genomes, only non-long terminal repeat retrotransposons are currently competent for retrotransposition [33]. Successful retrotransposition of these elements is dependent upon functional protein products from long interspersed elements (LINEs). In most somatic tissues, LINEs are epigenetically suppressed; however, these elements escape epigenetic repression during early embryonic development, and their integration into other functional genomic elements occasionally results in disease such as choroideremia (Online Mendelian Inheritance in Man [OMIM] #303100) [34]. Retrotransposition may also occur in somatic tissues with unusual epigenetic states [35].

Low complexity regions, including trinucleotide repeats, are scattered throughout the mammalian genome. Trinucleotide repeats can be hypervariable and expansions of some trinucleotide repeats are the causes of nearly 30 disorders [36,37]. The molecular mechanisms underlying expansion or contraction of these regions are complex and cause these regions to have variable length throughout the body of those afflicted with disease [38–44].

Small genetic aberrations may be caused by a number of mechanisms. Polymerase errors may result in nucleotide misincorporation or small insertions or deletions in the germline or soma. Over time, DNA will accumulate numerous lesions and DNA polymerization across these lesions is especially error-prone. DNA lesions may be detected and repaired prior to DNA polymerization, but lesion repair may also create single nucleotide variants, or small insertions or deletions [45,46].

In linear mammalian genomes, DNA replication starts at multiple origins with DNA polymerases ε and δ [47,48]. Polymerase ε moves with high processivity 5'–3' along the genome on the leading strand, moving in the same direction as the replication fork. On the lagging strand replication by polymerase δ also proceeds 5'–3' but in the opposite direction as the replication fork, causing replication of that strand to be iterative. This process works well for the majority of the genome, but replication of the lagging strand leads to loss of genetic information at the ends of the chromosome during every replication [49]. This end replication problem is solved in the germline because the ends of chromosomes, telomeres, are protected by repetitive DNA which is synthesized by a dedicated RNA-templated DNA polymerase called telomerase [49]. However, telomerase is not usually expressed in somatic tissues, likely as a method of protection against malignant transformation, and decreased telomere length is a form of somatic variation.

1.3. Mosaicism during Development

A defining characteristic of mosaic mutations is that they occur postzygotically and are inherited by all subsequent cells in their lineage (Figure 1). Somatic errors in chromosomal segregation in early development induce an extraordinarily high rate of aneuploidy. Fifteen to 20% of clinically recognized pregnancies result in spontaneous abortion, and half of these are attributed to aneuploidy [29]. A review of 36 published studies showed that of 815 human preimplantation embryos, only 177 (22%) were diploid while 73% were mosaic [50]. In most cases, these were diploid-aneuploid mosaic embryos, having one or more diploid cells as well as other cells that were haploid or polyploid for a particular chromosome. Mitotic errors could account for the high rate of chromosomal mosaicism.

Due to the exponential rate of growth during development, somatic mutations must occur early in development to have phenotypic effects over large portions of the body. Severe somatic mutations, which would be embryonic lethal if inherited, have a short window during development in which they must occur to be observed in adults [19]. If these severe mutations occur early in development, they will be embryonically or prenatally lethal; occurring later in development they may have little or no obvious phenotypic effect.

Mutations that alter cellular growth do not necessarily have to occur within such a short developmental window. Inactivating mutations in genes encoding tumor suppressors or activating mutations in oncogenes may have functional consequences regardless of when they occur, as evident from their known roles in cancer. On the other hand, somatic growth-retarding mutations, such as inactivating mutations in oncogenes or certain cyclins, are unlikely to have phenotypic effect in adults regardless of when they occur in development as the total number of cells containing the mutation is likely to be small.

Somatic mutations are thought to occur in all cells during replication. On average, 50 mutations occur in microsatellite regions during every mitotic division of a given cell [15]. Mutations in microsatellites and other regions of the genome, assessed by either single-cell or deep sequencing, can then be used to infer cell lineage trees [51]. To date, the most successful lineage tracing experiments have made use of increasingly sophisticated microscopy techniques [52]. However, microscopy-based approaches have practical and technological barriers such as the requirement that non-transgenic cells must be monitored over time. Recent advances in whole genome amplification (WGA) and second-generation sequencing offer genetic-based approaches that do not have the same limitations. Already, these techniques have been used to provide a detailed view of the genetics of cancer metastasis [53,54].

1.4. Mosaicism across the Body

By definition, somatic mosaic mutations affect only a subset of cells within an individual (Figure 1). This is most easily visible in monogenic mutations affecting pigmentation patterns. While such patterns may be mistaken for stochastic X chromosome inactivation or autoimmune response, somatic mutation is generally localized over a small portion of the body and in many cases occurs along lines of Blaschko [55]. To date, almost all non-cancerous somatic mutations characterized at the molecular level result in visible abnormalities, usually involving hypertrophy (cellular overgrowth) or abnormal pigmentation [26,55]. Some of our inability to identify mutations that do not result in visible phenotypes is practical; during dissection it is difficult to distinguish affected from unaffected tissue. However, due to the current emphasis on visible phenotypes, few data are available on the extent to which non-visible somatic mutations influence important biological processes.

An important consideration is that somatic mutations occur in varying cell types and tissues as well as different developmental stages. This raises the possibility that a specific mutation may vary in its clinical importance depending on where the mutation occurs across the body. Mutations in *GNAQ* provide an example. We identified p.Arg183Gln mutations in *GNAQ*, encoding the G protein alpha subunit Gαq, as the cause of both Sturge-Weber syndrome (OMIM #185300) and port-wine stain birthmarks (OMIM #163000) [56]. Port-wine stains are non-syndromic vascular abnormalities, while the Sturge-Weber syndrome is a severe neurocutaneous disorder, although both conditions likely affect some of the same cell types (e.g., endothelial cells). The milder phenotype of the birthmarks could result from a later developmental origin of the mutation during fetal development. The identical p.Arg183Gln mutation in *GNAQ*, when occurring in melanocytes later in life, is a frequent driver mutation in uveal melanoma (OMIM #155720), highlighting the importance of both the location and timing of the mutation. p.Arg183Gln mutations in different cell types and developmental stages could have different phenotypic consequences, if any [57].

Other mosaic mutations also differ in their clinical importance based on cell or tissue-specific involvement. McCune-Albright syndrome (OMIM #174800) is characterized by increased function of endocrine glands, sexual precocity, café-au-lait macules, and fibrous dysplasia. These symptoms can vary considerably based, in part, on the bodily extent of the mutation [58]. Like Sturge-Weber syndrome, this disorder is caused by somatic activating mutations in a gene encoding a G protein alpha subunit (*GNAS* encoding Gαs). Expression of this gene highlights another dimension of mosaicism. *GNAS* is expressed biallelically through most of the body, but the maternal allele is imprinted in particular tissues such as the pituitary. The disorders progressive osseous heteroplasia (OMIM #166350) and pseudopseudohypoparathyroidism (OMIM #612463) result from loss of function mutations in the paternal allele of *GNAS* [59].

Somatic mutations in three *AKT* genes also have cell-specific effects [60–62]. Somatic *AKT1* mutations are associated with somatic breast cancer, colorectal cancer, and ovarian cancer as well as the Proteus syndrome. The *AKT2* gene is expressed selectively in insulin-responsive tissues and mutations are associated with diabetes. Somatic mutations in *AKT3* cause Megalencephaly-polymicrogyria-polydactyly-hydrocephalus syndrome 2 (OMIM *615937). Given the localized nature of somatic mutations in *AKT* discovered to date, it is likely that mutations in these genes occurring outside of vulnerable cell types have few effects. These examples highlight the complex interaction of localized somatic mutation with tissue or cell-specific gene expression and signaling pathways (Figure 2).

Numerous studies have aimed to assess the prevalence of mosaic alterations in tissues of apparently normal individuals. Reanalysis of data from multiple large genome-wide association studies have determined that the number of detectable mosaic events rises sharply after age 50. Furthermore, individuals with increased numbers of mosaic events have higher risk for developing cancer [63,64]. While this measured increase of mosaicism may be due to increased rates mutation rates in elderly individuals, it is much more likely that these events are the result of clonal expansion and positive selection within the stem cell niche or decline in the total number of hematopoietic stem cell progenitors later in life. Notably, increased rates of mosaicism in apparently normal tissues have been linked to poorer prognosis in individuals with ovarian cancer [65].

Figure 2. Tissue-specific effects of mutations in *GNAQ* (**A**); *GNAS* (**B**); and *AKT1*, *AKT2*, and *AKT3* (**C**). Constitutively activating mutations in *GNAQ* may lead to either Sturge-Weber syndrome, nonsyndromic port-wine stains, or uveal melanoma (**A**). Somatic activating mutations in *GNAS* lead to McCune-Albright syndrome, which may involve variable hyperthyroidism, *café au lait* macules and sexual precocity (**B**). Activating mutations in all three of the *AKT* genes cause cellular overgrowth phenotypes with mutations in *AKT2* also implicated in abnormal insulin signaling (**C**).

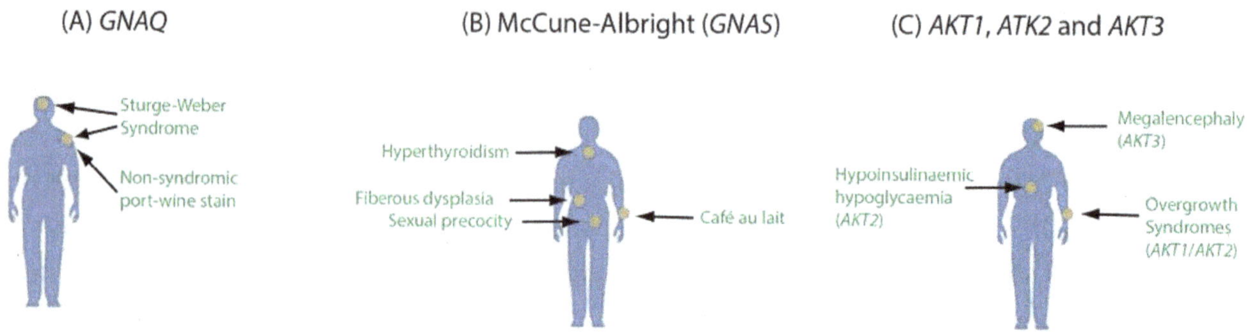

Studies of twins have demonstrated that post-zygotic mutations may be phenotypically important. Notable examples are monozygotic twins who are discordant for phenotypic sex due to mosaic loss of chromosome Y [66,67]. Numerous examples of monozygotic twins exist where either the presence [68,69] or severity [70] of disease is discordant between twin pairs due to variable proportions of mosaic cells.

Studies of multiple tissues of apparently normal individuals have also found evidence for mosaic events. Analysis of CNVs using hybridization of DNA from multiple tissues of three apparently normal individuals to bacterial artifical chromosome arrays found evidence for six somatic CNVs [71]. Higher resolution examination of a total 33 tissues from six individuals using array comparative genomic hybridization found evidence for 73 high-confidence mosaic CNVs, although a majority of high-confidence events (54/73) were found in one of two particular tissues [72]. It has been noted that induced pluripotent stem cells (iPSC) frequently contain CNVs which may cause genomic instability inheritent to the process of immortalization. Abyzov *et al.* performed a detailed study of this phenomenon and concluded that almost half of CNVs present in iPSC lines can be found in the parental fibroblasts. Furthermore, they conclude that approximately 30% of all fibroblasts in their sample contain some mosaic CNVs [73].

While experimentation with bulk tissues has shown that somatic mosaicism occurs frequently in normal populations, the combination of DNA from many cells limits the ability of an assay to detect mosaic events unique to single or few cells. As a result, sequencing of single-cells has been recently used to assay mosaicism in normal tissues. These methods have been used to sensitively reexamine conclusions regarding the extent of mosaicism in the brain. Previous reports had indicated that up to 33% of neuroblasts were aneuploid while up 80 retrotransposon insertions occur per neuron [74–77]. Single-cell experiments of the same phenomena have shown that large copy-number variants occur in over 14% of neurons but whole chromosome aneuplodies and retrotransposition events are relatively rare [78–80].

Single-cell studies have also been used to investigate the extent to which mosaicism occurs in early development. It has been known since 1983 that chorionic villus sampling may indicate the presence of a trisomy, while the fetus is diploid without the presence of mosaicism, a condition termed confined placental mosaicism [81–83]. Single-cell studies of young embryos cultured *in vitro* also demonstrate that chromosomal aneuploidies are common and were found in 83% of tested embryos [84]. While it is likely that many aneuploid embryos are unlikley to result in viable pregnancies, recent advances in prenatal testing allow for the sensitive and specific detection of numerous trisomies by sequencing of circulating fetal DNA from maternal plasma [85].

2. Detection of Somatic Mosaicism

2.1. Technical Considerations

Almost every type of genetic variation has been implicated as a source of somatic variation including expansion of trinucleotide repeats, point mutation, copy-number loss/gain, uniparental disomy, mitotic recombination, aneuploidy, translocation, and retrotransposition [39–41,43,44,69,77,79,86–94]. The techniques summarized below vary widely in their ability to detect specific types of somatic variation and more specialized techniques exist for the sensitive detection of some types of variation.

A primary consideration during the analysis of mosaic samples is the purity of the dissection from tissue samples. The presence of normal cells in affected tissue significantly decreases the ability of downstream analyses to detect mosaic alterations. This problem can be compounded by the prevalence of cellular migration during development in some tissues. Thus, in a tissue affected by a somatic mutation, two neighboring cells may both be affected if they share a common lineage from the mutated cell. Alternatively, cellular migration could cause neighboring cells to originate from distinct precursors with only one cell affected. Cellular migration can place an important biological constraint on the visible frequency of driver somatic mutations in affected tissues (e.g., in the brain) [3,95,96].

While contamination of normal cells is known to decrease the observed frequency of mosaic mutations, other mechanisms may decrease the detectable fraction of mosaic cells within a sample. Two possibilities are cell-type specific lethality and mosaic absence of essential juxtacrine or paracrine signaling factors. Cellular signaling pathways are known to have cell-type specific effects raising the possibility that a mosaic mutation may be lethal in only one type of cell within a tissue (Figure 3a). Furthermore, some paracrine or juxtacrine signaling factors are essential for cell viability [97–99]. Mosaic loss of these factors could result in affected tissue that is dependent upon surrounding normal tissue for survival, reducing the total number of mutant cells (Figure 3b).

In Sturge-Weber affected tissues, we detected *GNAQ* mutant allele frequencies between 1% and 18% [56]. Other studies using similar techniques have detected mutant allele frequencies of 1%–47% [61], 3%–30% [100], and 3%–35% [60] for causative mutations in individuals with Proteus syndrome (OMIM #176920), CLOVE (Congenital Lipomatous Overgrowth, Vascular anomalies, and Epidermal nevi) syndrome (OMIM #612918), and hemimegalencephaly (e.g., OMIM #615937), respectively. Such relatively low allele frequencies are likely explained by the presence of low proportions of affected cells in a given tissue due to cellular migration or impure dissection. However, the occurrence of mosaic cell death due to either cell-specific lethality or loss of essential signaling factors should be considered.

Figure 3. Cell death may reduce the total number of cells harboring somatic mutation. Mosaic mutations may cause cell-type-specific lethality (**A**); Mosaic loss of an essential juxtacrine signaling factor may cause localized death of cells that are not adjacent to unaffected tissues (**B**).

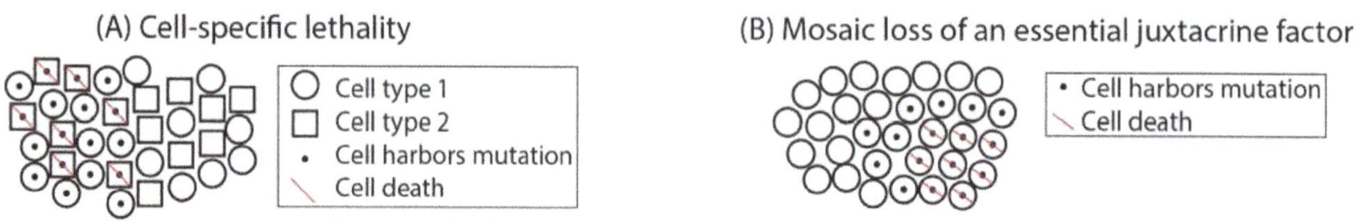

In second-generation sequencing experiments, sequencing and mapping errors are a major concern, as some portions of the genome are known to be prone to false-positive variant calls [101]. Recent improvements in sequencing chemistry have lowered the frequency of sequencing errors. However, biased errors in sequencing are still problematic for the detection of somatic variation, especially when the mutant allele frequency may be close to the technology's inherent error rate. Generally, ultra-high depth sequencing (>500 reads) of normal and affected tissues will permit detection of these errors. However, exploratory studies generally do not reach this level of depth. It is likely that without validation, these errors are a source of false positives in somatic variation databases. Comparing suspected somatic mutations across multiple tissue types from multiple individuals to estimate local error profiles may be a possible solution to this problem [102].

2.2. Cytogenetics

Microscopy-based methods allow for the detection of large mosaic events in single cells. Early cytogenetic methods for identifying extra or fewer chromosomes involved counting condensed metaphase chromosomes under a microscope [103]. Later methods using Giemsa staining and other dyes produced unique chromosomal bands allowing for the identification of intra- and interchromosomal translocations, duplications, deletions, and large structural rearrangements. However, banding techniques can only resolve aberrations larger than 3–10 Mb [104]. Other methods, such as fluorescent *in situ* hybridization (FISH), label a specific region of the genome by hybridization of a fluorescent probe allowing for the detection of deletions and some duplications [105]. Variations in this methodology exist using multiple probes of different color to detect several unique fragments at a time (e.g., multicolor FISH). These methods are able to achieve resolutions below 100 kilobases or, in some cases, as few as several kilobases [106]. Potential probe binding to off-target regions is a major consideration in most FISH experiments and adequate controls are required to confirm locus specificity [106]. Variants on classical FISH methods continue to be developed which promise to increase the ability of fluorescent probes to detect small chromosomal abnormalities across increasingly large portions of the genome [106,107]. In combination with high-throughput techniques, these approaches may be used to screen large numbers of cells from a single individual allowing for the detection of low levels of mosaicism.

2.3. Genome-Wide Arrays

Comparative genomic hybridization (CGH) is a technique in which fluorophore-labeled DNA from a control and test individual are hybridized to a metaphase reference chromosome [108]. The ratio of fluorescence emission is then measured to allow for the detection of duplication or deletions. A ratio of 1:1 indicates that both samples of DNA carry the same copy number while deviations from this ratio indicate a copy number variant [109].

Two principal array-based techniques that have emerged as alternatives to CGH are array CGH (aCGH) and single nucleotide polymorphism microarrays (SNP microarrays) [110–112]. Similar to CGH, both aCGH and SNP microarrays have the ability to detect changes in copy number over large regions of the genome. SNP microarrays further have the ability to genotype individuals at the probed sites, which may be useful in the detection of low-level somatic events [113]. Array-based approaches offer increased sensitivity over the entire genome for small CNVs relative to genome-wide microscopy-based approaches. aCGH and SNP microarray analysis can resolve regions less than 100 kb in size. However, the sensitivity of array-based approaches for somatic CNVs is dependent on having at least 5%–10% of the cells assayed containing the genetic variant. For larger CNVs affecting a smaller fraction of cells, microscopy-based approaches are more sensitive.

In both aCGH and SNP microarrays, deviations in relative probe intensities indicate deletion or insertion events. Normalized probe intensities are commonly reported as log-R ratios, with higher intensities indicating insertions while lower intensities indicate deletions. For SNP microarrays, the relative intensities of the two probes (one specific to each allele) at a locus is informative, and normalization of these intensities is measured as a B-allele frequency. For normal diploid tissues, B-allele frequencies approximate 0.0, 0.5, and 1.0 for AA, AB, and BB genotypes, respectively, while log-R ratios approximate 0 indicating no copy number change.

The hybridization of genomic DNA to microarrays is inherently noisy and can be subject to large batch effects [114]. Furthermore, individual probes or even whole arrays may have errors caused by faulty manufacture. Together these artifacts make the detection of statistically significant mosaic CNVs difficult, but many software packages detect these events. Numerous tools use hidden Markov Models (HMMs) to integrate B-allele frequency and log-R ratio information for the detection of mosaic events, including PennCNV-2, GPHMM and MixHMM [115–117]. gBPCR uses an approach similar to the Bayesian Piecewise Constant Regression for the detection of mosaic abnormalities but has a long runtime per sample [118]. We developed triPOD which uses multiple algorithms for the detection of mosaic events and is unique in that it utilizes parental genotypes allowing for more sensitive detection of haplotype-specific mosaic abnormalities [113].

2.4. Second-Generation Sequencing

Second-generation sequencing techniques have revolutionized human genetics in the last decade. Sequencing is performed either on single cells, a discrete number of cells, or bulk tissue. In the typical sequencing experiment, DNA is extracted from the input material and is fragmented, size-selected, and sequenced to produce strings of inferred nucleotides and their respective quality scores [119]. This information is used to align the sequencing reads to a reference genome. Differences between the aligned reads and the reference can be used to infer genetic variants including single-nucleotide variants or polymorphisms (SNVs or SNPs), insertions, deletions, translocations, and retrotransposition events. Furthermore, the total number of reads aligned to certain regions of the genome can be used to infer copy-number changes [120,121]. Numerous variations on this basic approach exist and here we will discuss the methods most applicable for the detection of mosaic events.

Somatic genetic variants have been discovered via whole-exome or whole-genome sequencing of bulk tissue from paired affected and unaffected portions of the body [56,60,61,100]. Whole-exome sequencing relies upon an oligonucleotide bead or array-based enrichment of DNA fragments corresponding to exonic regions to reduce the representation of sequence from noncoding regions of the genome [122–124]. At similar depth, exome and whole-genome sequencing are considered to have similar sensitivity for most pathogenic SNVs and small insertions or deletions. Whole-exome sequencing is considered less sensitive for the identification of medium to large insertions or deletions or the detection of copy-number changes by analysis of read depth due to introduced biases. However, exome sequencing experiments are typically performed at higher depth due to the lower cost of the method.

Numerous software packages allow the identification of somatic variants from these data. Somatic variant callers typically evaluate second-generation sequence data from paired tumor/normal (or other affected/unaffected) samples. Examples include VarScan2 [121], SomaticSniper [125], JointSNVMix [126], Strelka [127], and MuTect [128]. After removal of low-quality reads, sequences are aligned to a reference genome to generate aligned binary sequence alignment/map (BAM) files [129]. At least three approaches have been employed for the detection of somatic SNVs and small insertions or deletions. (1) Allele frequencies can be compared. For example, VarScan2 performs pairwise comparisons of base calls and normalized sequence depth at each position, accounting for factors such base quality scores, coverage and variant allele frequencies; (2) Bayesian comparison of joint diploid genotype likelihood can be estimated for both samples. The SomaticSniper algorithm calculates the statistical significance of all somatic variants at positions above a minimum threshold of coverage using this method; (3) Other Bayesian approaches have been applied. For example, Strelka models the normal sample as germline variation plus noise, while the affected sample includes noise along with germline and somatic variation. Other types of somatic variation may be detected from bulk sequencing. Tools such as VarScan2, ADTeX, Control-FREEC, SomatiCA, and LUMPY may be used for the detection of somatic CNVs or structural variants [121,130–133].

Besides variant identification, quantification of the fraction of cells affected by particular somatic changes provides a better understanding of the extent of the mosaic mutation and the period during development at which it occurred. Several tools have been developed to deconvolute somatic mutations into distinct populations as reviewed by Yadav and De and Ding *et al.* [134,135].

An alternative approach to sequencing bulk tissue is sequencing single cells or small numbers of cells. As single or hundreds of cells contain very little DNA, most experiments utilize multiple displacement amplification (MDA) or PCR based methods to amplify genomic DNA. Amplification can greatly increase the total amount of available DNA for sequencing at the expense of introduced biases such as allele dropout and chimeric amplification of genomic fragments [79,136–138]. Despite these introduced biases, amplification and subsequent second-generation sequencing or array-based analysis of single cells has been used to reliably find somatic copy number variation and retrotransposition events within the human brain as well as to map cell lineage within a bulk tumor dissection [53,54,79,139]. Numerous groups have also used single-cell techniques to discover SNVs or indels in single cells, however, allelic dropout and chimeric amplification are more problematic for these analyses as biases can be reduced for analysis of CNVs by increasing bin sizes but are more difficult to account for in analysis of SNVs [140–142].

3. Somatic Mosaicism in Disease

3.1. Cancer and Aging

The relationship between somatic mutation and cancer has been extensively reviewed elsewhere [9,17,87,143–145] and comprehensive lists of known oncogenes or tumor suppressors or genes significantly and recurrently mutated in cancer have been previously described [9,87]. Cancer has been described as having six hallmarks: proliferative signaling, evading growth suppressors, resisting cell death, enabling replicative immortality, induction of angiogenesis, and inactivating invasion and metastasis [146]. Driver gene mutations are defined as conferring a selective growth advantage in tumor cells [9]. This may be achieved by elevating the activity of growth factors and/or their receptors, but more commonly driver mutations constitutively activate intracellular signal transduction cascades. Three of these are depicted in Figure 4 (in simplified form): Ras/Raf/MEK/ERK, Ras/PI3K/PTEN/Akt/mTOR [147], and *GNAQ*. These pathways contain both oncogenes (*RAS*, *RAF*, *MEK*, *PIK3CA*, *AKT*, *GNAQ*) and tumor suppressor genes (*NF1*, *PTEN*, *TSC1*, *TSC2*). For example the *RAS* family of oncogenes were the first oncogenes to be identified in cancer. Comprised of *HRAS*, *KRAS* and *NRAS*, activating mutations in these genes occur in approximately 20% of all cancers [148]. Germline variants are also well known to contribute to cancer morbidity [149–151]. Frequently, these variants affect proteins involved in DNA repair, highlighting the role of somatic mutations in tumorigenesis [152–155].

In common solid tumors, ~95% of protein altering mutations consist of single base substitutions, >90% of which are missense mutations, <8% are nonsense mutations, and <2% affect splice sites or untranslated regions [9]. Relatively large numbers of somatic mutations occur in tumors that are associated with mutagens such as ultraviolet light and cigarette smoke. For example, in non-small cell lung carcinomas the average mutation frequency is greater than ten-fold higher in smokers compared to those who never smoke [156].

Large-scale projects and databases have been developed to provide comprehensive catalogues of somatic mutations found in cancer [157,158]. COSMIC (Catalogue Of Somatic Mutations In Cancer) includes information on more than 1.6 million mutations from nearly 1 million cancer samples and includes various types of mutations (fusions, genomic rearrangements, whole genomes, and copy number variants) [157].

The combination of well-characterized somatic mutation databases and low-cost sequencing technologies may lead to improved patient outcomes in the near future. Biopsied tumors may be screened rapidly for putative driver mutations based on cancer type, informing treatment. Furthermore, once a cancer is in remission, tumor-specific DNA may be assayed at low cost with ultra-sensitive second-generation sequencing-based techniques [159]. These advances will likely improve prognosis for millions of cancer patients within the next decade.

The primary risk factor for cancer is age, and cancers offer insight into age or mutagen-associated mutational processes [160]. Somatic mutations have long been suspected to be an important part of the molecular mechanism of aging, and accumulation of DNA lesions and mutations occurs in both the germline and soma over time [63,64,161,162]. By chance, these mutations may result in malignant transformation, apoptosis, or otherwise hampered cellular function. As visible in cancers, the characteristics of acquired mutations differ by tissue type and are dependent upon environmental exposure [9]. Furthermore, frequently dividing stem cells and frequently transcribed genomic regions have different patterns of mutation that are cell-type specific.

In both mouse and human, increased rates of somatic mutation and numbers of DNA lesions due to either error-prone DNA polymerases or faulty DNA repair mechanisms cause cancer predisposition, early aging, and neurodegenerative phenotypes [17]. Increased rates of somatic mutation in the nuclear genome cause cancer predisposition, likely due to increased rates of mutation in somatic stem cell populations. This has been demonstrated in transgenic mice whose processive DNA polymerases lack proofreading. Notably, mice with mutated polymerases δ and ε develop distinct cancers but do not demonstrate premature aging phenotypes [163–165]. While these mice may not live long enough to demonstrate early aging phenotypes, their predisposition towards the development of cancer demonstrates a strong link between cancer and somatic mutation.

Mutations in genes affecting other pathways demonstrate a strong relationship between somatic mutations and aging. Mice with error-prone mitochondrial polymerases demonstrate a premature aging phenotype without cancer predisposition, although subsequent data by some of the same authors demonstrate that mitochondrial point mutations are unlikely the primary cause of aging in normal mice [166,167]. Individuals with defects in DNA repair also demonstrate symptoms of progeria. Cockayne syndrome (OMIM #216400) is caused by defects in transcription-coupled exonucleotide repair leading to an early aging phenotype combined with intellectual disability and neurodegeneration without noted predisposition to development of cancer [168]. Mutations in the genes encoding RecQ helicases cause Werner syndrome (OMIM #277700) and Rothmund-Thomson syndrome (OMIM #268400) [169]. The most prominent phenotype of individuals affected by these diseases is premature aging, although these individuals are also predisposed to developing cancer [169]. Bloom Syndrome (OMIM #210900) is notable in that it is also caused by mutations in a RecQ helicase-like protein and also increases cancer incidence, but does not appear to result in progeria. Mutations in numerous other genes are known to cause cancer predisposition. One such example is *BUB1B*. Loss of BUB1B protein function leads to premature chromatid separation and mosaic variegated aneuploidy syndrome 1 (OMIM #257300) typically resulting in cancer predisposition and intellectual disability [170].

Cancer is associated with many genomic changes. Large chromosomal changes occur in a variety of noncancerous conditions. An example is Pallister-Killian syndrome (OMIM #601803) is a dysmorphic condition caused by mosaicism for tetrasomy 12p. Affected individuals display tissue mosaicism, typically with apparently normal karyotypes from lymphocytes but 47 chromosomes in skin fibroblasts and chorionic villus and amniotic fluid cells. The extra chromosome is an isochromosome for a portion of chromosome 12p. In several cases hexasomy of chromosome 12p has been observed.

3.2. Neurodegenerative Disease

Somatic mutation is suspected to have a role in neurodegenerative disease [17,18]. As in cancer, mutations in genes directly involved in DNA repair are implicated in neurodegenerative diseases such as ataxia-telangiectasia (OMIM #208900) and ataxia-ocular apraxia 1 (OMIM #208920) [16,169,171–174]. These neurodegenerative phenotypes are likely caused by an increase of somatic mutation in the nervous system leading to cellular dysfunction, indicating a possible role for somatic changes and DNA lesions in age-related related neurodegenerative disorders.

There is evidence that mosaic mutations or accumulated damage to other macromolecules play a role in Alzheimer's disease (OMIM #104300) and Creutzfeldt-Jakob disease (CJD) (OMIM #123400). Alzheimer's disease is characterized by the accumulation of β-amyloid (Aβ) plaques while CJD is caused by misfolded protein PRNP [175,176]. Significant incidence of both diseases is attributed to familial risk and causal mosaic mutations have been found in sporadic cases [177,178]. Aβ plaques have long been implicated in the formation of prions and introduction of Aβ plaques into the brains of mice overexpressing Aβ leads to disease progression [179–181]. Consistent with the link to prions, the pathology of inoculated mice displays phenotypes dependent upon the infecting host [180]. This has been corroborated by more recent experiments, which demonstrate that Aβ aggregates from distinct sources have unique biophysical characteristics depending on the seeding protein [182–184]. While it is possible that sporadic misfolded or damaged proteins act as seeds in Alzheimer's, this is unlikely given the steep increase in disease incidence later in life and the constant turnover of cellular proteins [185]. This steep rise in incidence mirrors the rise in incidence of CJD in individuals who have predisposing mutations [186]. It is possible that in both diseases misfolded proteins arising as a result of age-related somatic mutation or damage to other macromolecules in single cells act as seeds for the initial protein aggregates.

3.3. Monogenic Disease

A list of diseases suspected to be caused by obligatory somatic mutations has been previously described [21] and subsequently updated [19,20]. We note that somatic mutation likely contributes significantly to nearly all Mendelian diseases.

We have described a series of oncogenes and tumor suppressor genes that undergo somatic mutation in cancer. These same genes can also acquire somatic mutations that result in neurocutaneous disorders or overgrowth syndromes, depending the particular cell type and developmental stage at which the mutation occurs. Mutations in *GNAQ* cause Sturge-Weber syndrome and port-wine stain birthmarks as well as uveal melanoma, as discussed above. Similarly, somatic mutations in *GNAS* can cause McCune-Albright syndrome or benign tumors such as adenomas. We next highlight several specific examples of such disorders affecting genes encoding intracellular signaling pathways (Figure 4).

Figure 4. Three intracellular signaling pathways are shown schematically. (**At left**), receptor tyrosine kinase activity leads to activation of PIK3CA, AKT, and mechanistic target of rapamycin (mTOR) [187,188]. mTOR participates in complexes (TORC1, activated by RHEB; TORC2, inhibited by RHEB) that regulate cell growth, proliferation, survival, and cell cycle progression. This pathway includes genes that are frequently mutated in tumors such as *PIK3CA* and *PTEN* (not shown); (**At center**), secreted growth factors bind to receptor tyrosine kinase receptors on the cell surface leading to activation of the low molecular weight G protein Ras and subsequent activation of Raf, MEK 1/2, and ERK 1/2 (official gene symbols *MAPK3*, *MAPK1*); (**At right**), a G-protein coupled receptor (GPCR) pathway is shown [189,190]. Ligands such as vasopressin, endothelin, glutamate, or norepinephrine bind to a GPCR. When bound by ligand, the receptor activates a G protein alpha subunit such as Gαq that binds and hydrolyzes GTP. This leads to activation of phospholipase Cβ producing inositol 1,4,5-triphosphate (IP_3) and membrane-associated diacylglycerol (DAG). DAG, through activation of protein kinase C, may activate the Raf/MEK/ERK pathway. IP_3 may bind to an IP_3 receptor activating calcium signaling pathways (not shown). Other G protein α subunits (such as Gαs encoded by *GNAS*) activate membrane-bound adenylate cyclase, producing cyclic AMP (cAMP) that activates protein kinase A (not shown).

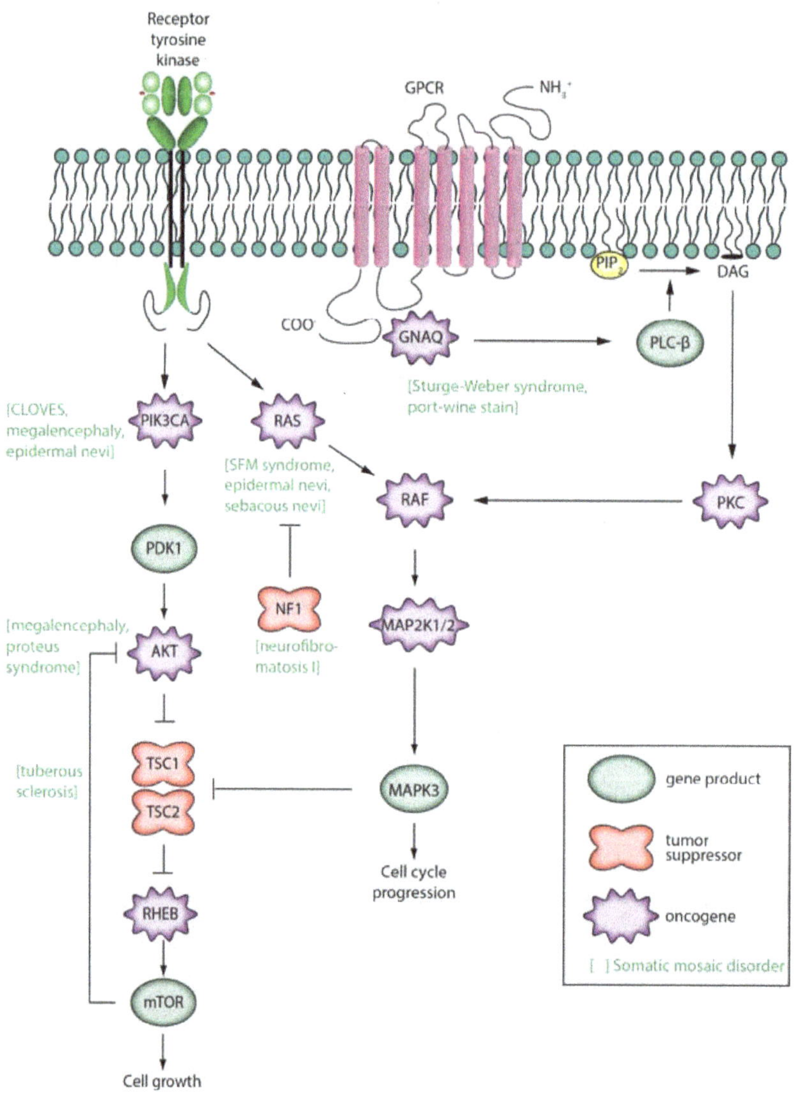

Phosphatidylinositol 3-kinases (PIK3s) are lipid kinases that phosphorylate phosphatidylinositol and other phosphoinositides, catalyzing intracellular signaling pathways involving a PI3K/AKT/mTOR network (Figure 4). Somatic, mosaic, gain-of-function mutations in *PIK3CA* (OMIM *171834) are associated with several syndromes involving overgrowth of the brain or lipomatous body overgrowth [191]. These include CLOVE syndrome, megalencephaly-capillary malformation syndrome, fibroadipose hyperplasia, and hemimegalencephaly. These conditions are often characterized by early segmental overgrowth, abnormal vasculogenesis, digital anomalies, cortical brain malformations, and connective tissue dysplasia. Somatic gain-of-function mutations in *PIK3CA* are also found in a broad range of cancers (ovarian, breast, lung, stomach, colorectal, and brain). While over 100 activating mutations in *PIK3CA* are known, mutations in two domains of the protein account for 80% of cancer-associated somatic mutations, and these same sites can be mutated in overgrowth disorders [192].

Clinical presentation of Proteus syndrome (OMIM #176920) includes distorting, progressive overgrowth of various tissues including skin, skeleton, adipose, and central nervous system. In most patients it is caused by somatic mosaic mutation of *AKT1* involving c.49G > A (p.Glu17Lys) [61]. This identical mutation is associated with breast, colorectal and ovarian cancers [193]. Mutations in the homologs of *AKT1*, *AKT2* and *AKT3* are also known to cause somatic disorder. p.Glu17Lys mutations in *AKT3* cause hemimegalencephaly and other brain malformations, while the identical mutation in *AKT2* is causative for hypoglycemia [60,62,194,195].

Germline inactivating mutations in the *TSC1* gene encoding hamartin cause tuberous sclerosis 1 (OMIM #191100), while mutations in *TSC2* encoding tuberin cause tuberous sclerosis 2 (OMIM #613254). Hamartin and tuberin act as tumor suppressors by activating the GTPase function of RHEB [196]. Inactivating mutations in a single allele are sufficient to cause tuberous sclerosis. Rare somatic inactivating mutations, lack of expression of the second allele or mosaic UDP events give rise to the multiple benign tumors, tubers and macules characteristic of the disease [197,198].

Neurofibromatosis 1 (OMIM #162200) (NF1) is characterized by the occurrence of at least two (of a list of seven) features such as *café au lait* spots, cutaneous neurofibromas, Lisch nodules (hamartomas) of the iris, and inguinal freckles [199]. Clinical diagnosis requires a first-degree relative with the condition. It is inherited in an autosomal dominant manner (and is among the most common such disorders with a prevalence of 1:3000). Most cases of NF1 are caused by heterozygous loss-of-function mutations of the tumor suppressor gene encoding neurofibromin 1. Only 50% of NF1 individuals have an affected parent, with another 50% having a *de novo* mutation. Neurofibromin 1 is a negative regulator of the RAS signal transduction pathway, with loss of function mutations in neurofibromin 1 leading to RAS activation.

It is possible that mosaic variation occurring during development may result in disease across numerous tissues. One such example is somatic mutation of *IDH1* and *IDH2* that has been shown to cause Ollier disease and Maffucci syndrome. These syndromes are characterized by multiple enchondromas (benign bone tumors originating from cartilage). The causative variants for disease are typically not detectable outside of the tumors indicating that relatively few cells harbor the mutation [200].

The application of sensitive approaches for the detection of mosaicism to a smaller subset of genes based on a patient's phenotype may increase the likelihood of finding causative variants. Jamuar *et al.* applied this approach examining two sets of previously implicated genes in 158 individuals with cerebral cortical defects. Causal mutations were found in 27 individuals, eight of who harbored the causative variant in a mosaic fashion. Notably, causal mutations were only validated at extremely high read depth

(>500×) highlighting both the importance of sequence coverage for the detection of mosaic variation and the utility of targeted approaches [201].

Somatic mutations are also known to cause reversion to normal mutations in individuals with monogenic disease [22,23,25,202,203]. Revertant mosaicism occurs when cells harboring a disease-causing mutation revert *in vivo* to a wild-type allele. The disease-causing mutation could be inherited from the germline or somatic. This has been observed for heritable skin diseases such as ichthyosis with confetti (OMIM #609165) and epidermolysis bullosa (OMIM #226650) [202,204] as well as rare blood disorders such as Fanconi anemia (OMIM #227646) and severe combined immunodeficiency resulting from adenosine deaminase deficiency (OMIM #102700) [205,206]. These somatic reversions to normal events may significantly ameliorate disease symptoms if the reversion occurs early enough in development.

For many other overgrowth syndromes somatic mutations have yet to be identified. Examples include Klippel-Trenaunay-Weber syndrome (OMIM %149,000), which involves cutaneous hemangiomata and clinically resembles Sturge-Weber syndrome; and Cobb syndrome (cutaneomeningospinal angiomatosis), which involves vascular cutaneous, muscular, osseous, or other lesions of spinal segments.

3.4. Complex Disease

Multiple recent papers have proposed that somatic mutation may play a role in the etiology of complex disease [3,207,208]. Studies of simplex autism probands have determined that *de novo* mutations account for 2%–15% of disease incidence and that at least 30% of *de novo* mutations can be causally implicated in simplex cases [209–212]. With *de novo* mutations playing such a large role, it is likely that post-zygotic somatic variation also contributes to disease in some individuals. To date, most genetic analysis has found few genetic variants to explain complex disease incidence, suggesting the occurrence of "missing heritability" [213]. A possible model is that somatic variation occurs in conjunction with common and rare inherited variation to cause disease. While this model is not directly supported by current evidence, recent experiments indicate that it warrants investigation. One surprising result from *in situ* hybridization experiments on postmortem brain tissue is the increased presence of patches of cortical disorganization in individuals with autism relative to controls [214]. The authors note that they examined only a small subsection of the brain and therefore cortical disorganization is likely widespread in individuals with autism. Furthermore, an interesting conclusion of recent large-scale examination of exonic *de novo* mutations in simplex autism is that most *de novo* variation implicated as causal occurs opposite wild type alleles [212]. Given that large CNVs are common in neurons of the cortex [78,80], we propose a model of brain-specific somatic mutation occurring opposite inherited *de novo* or rare mutation resulting in sporadic brain-specific loss of gene function and patches of cortical disorganization.

4. Conclusions

While the role of somatic mosaicism in disease is currently under active investigation, it is clear that functional somatic mosaicism has a significant role in human disease. In the last decade, major advances in both cytogenetic and second-generation sequencing techniques have enabled researchers to discover causative somatic mutations for an increasing number of diseases, and driver mutations in an increasing number of cancers. Furthermore, this increased understanding of the genetic underpinnings of disease is likely to lead to improved patient outcomes in the near future.

Acknowledgments

Donald Freed is supported in part by NIH grant 5T32GM007445-37. Jonathan Pevsner is supported by NIH grants R44 MH086192-02A1, P50 MH094268-01, and a grant from the Sturge-Weber Foundation.

Author Contributions

Donald Freed, Eric L. Stevens and Jonathan Pevsner wrote the manuscript and created the figures.

References

1. Edwards, J.H. Familiarity, recessivity and germline mosaicism. *Ann. Hum. Genet.* **1989**, *53*, 33–47.
2. Hartl, D.L. Recurrence risks for germinal mosaics. *Am. J. Hum. Genet.* **1971**, *23*, 124–134.
3. Poduri, A.; Evrony, G.D.; Xuyu, C.; Walsh, C.A. Somatic mutation, genomic variation, and neurological disease. *Science* **2013**, *341*, doi:10.1126/science.1237758.
4. Campbell, I.M.; Yuan, B.; Robberecht, C.; Pfundt, R.; Szafranski, P.; McEntagart, M.E.; Nagamani, S.C.; Erez, A.; Bartnik, M.; Wisniowiecka-Kowalnik, B.; *et al.* Parental somatic mosaicism is underrecognized and influences recurrence risk of genomic disorders. *Am. J. Hum. Genet.* **2014**, *95*, 173–182.
5. Van der Maarel, S.M.; Deidda, G.; Lemmers, R.J.L.F.; van Overveld, P.G.M.; van der Wielen, M.; Hewitt, J.E.; Sandkuijl, L.; Bakker, B.; van Ommen, G.-J.B.; Padberg, G.W.; *et al. De novo* facioscapulohumeral muscular dystrophy: Frequent somatic mosaicism, sex-dependent phenotype, and the role of mitotic transchromosomal repeat interaction between chromosomes 4 and 10. *Am. J. Hum. Genet.* **2000**, *66*, 26–35.
6. Boveri, T. *The Origin of Malignant Tumors*; The Williams and Wilkins Company: Baltimore, MD, USA, 1929.
7. Knudson, A.G. Mutation and cancer: Statistical study of retinoblastoma. *Proc. Natl. Acad. Sci. USA* **1971**, *68*, 820–823.
8. Nowell, P.C. The clonal evolution of tumor cell populations. *Science* **1976**, *194*, 23–28.
9. Vogelstein, B.; Papadopoulos, N.; Velculescu, V.E.; Zhou, S.; Diaz, L.A., Jr.; Kinzler, K.W. Cancer genome landscapes. *Science* **2013**, *339*, 1546–1558.
10. McClintock, B. Chromosome organization and genic expression. *Cold Spring Harb. Symp. Quant. Biol.* **1951**, *16*, 13–47.
11. Burnet, M. *The Clonal Selection Theory of Acquired Immunity*; Vanderbilt University Press: Nashville, TN, USA, 1959.
12. Brack, C.; Hirama, M.; Lenhard-Schuller, R.; Tonegawa, S. A complete immunoglobulin gene is created by somatic recombination. *Cell* **1978**, *15*, 1–14.
13. Tonegawa, S. Somatic generation of antibody diversity. *Nature* **1983**, *302*, 575–581.

14. Krangel, M.S. Mechanics of t cell receptor gene rearrangement. *Curr. Opin. Immunol.* **2009**, *21*, 133–139.
15. Frumkin, D.; Wasserstrom, A.; Kaplan, S.; Feige, U.; Shapiro, E. Genomic variability within an organism exposes its cell lineage tree. *PLOS Comput. Biol.* **2005**, *1*, e50.
16. Hoeijmakers, J.H.J. DNA damage, aging, and cancer. *N. Engl. J. Med.* **2009**, *361*, 1475–1485.
17. Kennedy, S.R.; Loeb, L.A.; Herr, A.J. Somatic mutations in aging, cancer and neurodegeneration. *Mech. Ageing Dev.* **2012**, *133*, 118–126.
18. Jeppesen, D.K.; Bohr, V.A.; Stevnsner, T. DNA repair deficiency in neurodegeneration. *Prog. Neurobiol.* **2011**, *94*, 166–200.
19. Erickson, R.P. Somatic gene mutation and human disease other than cancer: An update. *Mutat. Res.* **2010**, *705*, 96–106.
20. Erickson, R.P. Recent advances in the study of somatic mosaicism and diseases other than cancer. *Curr. Opin. Genet. Dev.* **2014**, *26*, 73–78.
21. Erickson, R.P. Somatic gene mutation and human disese other than cancer. *Mutat. Res.* **2003**, *543*, 125–136.
22. Hirschhorn, R. *In vivo* reversion to normal of inherited mutations in humans. *J. Med. Genet.* **2003**, *40*, 721–728.
23. Jonkman, M.F.; Castellanos Nuijts, M.; van Essen, A.J. Natural repair mechanisms in correcting pathogenic mutations in inherited skin disorders. *Clin. Exp. Dermatol.* **2003**, *28*, 625–631.
24. Lai-Cheong, J.E.; McGrath, J.A.; Uitto, J. Revertant mosaicism in skin: Natural gene therapy. *Trends Mol. Med.* **2011**, *17*, 140–148.
25. Jonkman, M.F. Revertant mosaicism in human genetic disorders. *Am. J. Med. Genet.* **1999**, *85*, 361–364.
26. Happle, R. Lethal genes surviving by mosaicism: A possible explanation for sporadic birth defects involving the skin. *J. Am. Acad. Dermatol.* **1987**, *16*, 899–906.
27. Liu, P.; Carvalho, C.M.; Hastings, P.J.; Lupski, J.R. Mechanisms for recurrent and complex human genomic rearrangements. *Curr. Opin. Genet. Dev.* **2012**, *22*, 211–220.
28. Cimini, D.; Howell, B.; Maddox, P.; Khodjakov, A.; Degrassi, F.; Salmon, E.D. Merotelic kinetochore orientation is a major mechanism of aneuploidy in mitotic mammalian tissue cells. *J. Cell Biol.* **2001**, *153*, 517–528.
29. Robinson, W.P. Mechanisms leading to uniparental disomy and their clinical consequences. *Bioessays* **2000**, *22*, 452–459.
30. Kotzot, D. Complex and segmental uniparental disomy updated. *J. Med. Genet.* **2008**, *45*, 545–556.
31. Conlin, L.K.; Thiel, B.D.; Bonnemann, C.G.; Medne, L.; Ernst, L.M.; Zackai, E.H.; Deardorff, M.A.; Krantz, I.D.; Hakonarson, H.; Spinner, N.B. Mechanisms of mosaicism, chimerism and uniparental disomy identified by single nucleotide polymorphism array analysis. *Hum. Mol. Genet.* **2010**, *19*, 1263–1275.
32. Liehr, T. Cytogenetic contribution to uniparental disomy (UPD). *Mol. Cytogenet.* **2010**, *3*, 8.
33. Hancks, D.C.; Kazazian, H.H., Jr. Active human retrotransposons: Variation and disease. *Curr. Opin. Genet. Dev.* **2012**, *22*, 191–203.

34. van den Hurk, J.A.; Meij, I.C.; Seleme, M.C.; Kano, H.; Nikopoulos, K.; Hoefsloot, L.H.; Sistermans, E.A.; de Wijs, I.J.; Mukhopadhyay, A.; Plomp, A.S.; *et al.* L1 retrotransposition can occur early in human embryonic development. *Hum. Mol. Genet.* **2007**, *16*, 1587–1592.
35. Lee, E.; Iskow, R.; Yang, L.; Gokcumen, O.; Haseley, P.; Luquette, L.J.; Lohr, J.G.; Harris, C.C.; Ding, L.; Wilson, R.K.; *et al.* Landscape of somatic retrotransposition in human cancers. *Science* **2012**, *337*, 967–971.
36. Mirkin, S.M. Expandable DNA repeats and human disease. *Nature* **2007**, *447*, 932–940.
37. Kim, J.C.; Mirkin, S.M. The balancing act of DNA repeat expansions. *Curr. Opin. Genet. Dev.* **2013**, *23*, 280–288.
38. McMurray, C.T. Mechanisms of trinucleotide repeat instability during human development. *Nat. Rev. Genet.* **2010**, *11*, 786–799.
39. Hellenbroich, Y.; Schwinger, E.; Zühlke, C.H. Limited somatic mosaicism for friedreich's ataxia GAA triplet repeat expansions identified by small pool PCR in blood leukocytes. *Acta Neurol. Scand.* **2001**, *103*, 188–192.
40. Hashida, H.; Goto, J.; Suzuki, T.; Jeong, S.-Y.; Masuda, N.; Ooie, T.; Tachiiri, Y.; Tsuchiya, H.; Kanazawa, I. Single cell analysis of cag repeat in brains of dentatorubral-pallidoluysian atrophy (DRPLA). *J. Neurol. Sci.* **2001**, *190*, 87–93.
41. Kahlem, P.; Djian, P. The expanded CAG repeat associated with juvenile Huntington disease shows a common origin of most or all neurons and glia in human cerebrum. *Neurosci. Lett.* **2000**, *286*, 203–207.
42. Møllersen, L.; Rowe, A.D.; Larsen, E.; Rognes, T.; Klungland, A. Continuous and periodic expansion of CAG repeats in Huntington's disease R6/1 mice. *PLOS Genet.* **2010**, *6*, e1001242.
43. Ueno, S.-i.; Kondoh, K.; Komure, Y.; Komure, O.; Kuno, S.; Kawai, J.; Hazama, F.; Sano, A. Somatic mosaicism of CAG repeat in dentatorubral-pallidoluysian atrophy (drpla). *Hum. Mol. Genet.* **1995**, *4*, 663–666.
44. Montermini, L.; Kish, S.J.; Jiralerspong, S.; Lamarche, J.B.; Pandolfo, M. Somatic mosaicism for friedreich's ataxia GAA triplet repeat expansions in the central nervous system. *Neurology* **1997**, *49*, 606–610.
45. Lindahl, T.; Wood, R.D. Quality control by DNA repair. *Science* **1999**, *286*, 1897–1905.
46. Hoeijmakers, J.H.J. Genome maintenance mechanisms for preventing cancer. *Nature* **2001**, *411*, 366–374.
47. Gilbert, D.M. Making sense of eukaryotic DNA replication origins. *Science* **2001**, *294*, 96–100.
48. Gilbert, D.M. Evaluating genome-scale approaches to eukaryotic DNA replication. *Nat. Rev. Genet.* **2010**, *11*, 673–684.
49. Levy, M.Z.; Allsopp, R.C.; Futcher, A.B.; Greider, C.W.; Harley, C.B. Telomere end-replication problem and cell aging. *J. Mol. Biol.* **1992**, *225*, 951–960.
50. Van Echten-Arends, J.; Mastenbroek, S.; Sikkema-Raddatz, B.; Korevaar, J.C.; Heineman, M.J.; van der Veen, F.; Repping, S. Chromosomal mosaicism in human preimplantation embryos: A systematic review. *Hum. Reprod. Update* **2011**, *17*, 620–627.
51. Shapiro, E.; Biezuner, T.; Linnarsson, S. Single-cell sequencing-based technologies will revolutionize whole-organism science. *Nat. Rev. Genet.* **2013**, *14*, 618–630.

52. Amat, F.; Lemon, W.; Mossing, D.P.; McDole, K.; Wan, Y.; Branson, K.; Myers, E.W.; Keller, P.J. Fast, accurate reconstruction of cell lineages from large-scale fluorescence microscopy data. *Nat. Methods* **2014**, doi:10.1038/nmeth.3036.
53. Navin, N.; Kendall, J.; Troge, J.; Andrews, P.; Rodgers, L.; McIndoo, J.; Cook, K.; Stepansky, A.; Levy, D.; Esposito, D.; *et al.* Tumour evolution inferred by single-cell sequencing. *Nature* **2011**, *472*, 90–94.
54. Wang, Y.; Waters, J.; Leung, M.L.; Unruh, A.; Roh, W.; Shi, X.; Chen, K.; Scheet, P.; Vattathil, S.; Liang, H.; *et al.* Clonal evolution in breast cancer revealed by single nucleus genome sequencing. *Nature* **2014**, *512*, 155–160.
55. Bolognia, J.L.; Orlow, S.J.; Glick, S.A. Lines of blaschko. *J. Am. Acad. Dermatol.* **1994**, *31*, 157–190.
56. Shirley, M.D.; Tang, H.T.; Gallione, B.A.; Baugher, J.D.; Frelin, L.P.; Cohen, B.; North, P.E.; Marchuk, D.A.; Comi, A.M.; Pevsner, J. Sturge-weber syndrome and port-wine stains caused by somatic mutation in *GNAQ*. *N. Engl. J. Med.* **2013**, *368*, 1971–1979.
57. Van Raamsdonk, C.D.; Griewank, K.G.; Crosby, M.B.; Garrido, M.C.; Vemula, S.; Wiesner, T.; Obenauf, A.C.; Wackernagel, W.; Green, G.; Bouvier, N.; *et al.* Mutations in GNA11 in uveal melanoma. *N. Engl. J. Med.* **2010**, *363*, 2191–2199.
58. Collins, M.T.; Singer, F.R.; Eugster, E. Mccune-albright syndrome and the extraskeletal manifestations of fibrous dysplasia. *Orphanet J. Rare Dis.* **2012**, *7* (*Suppl 1*), S4.
59. Bastepe, M.; Juppner, H. Gnas locus and pseudohypoparathyroidism. *Horm. Res.* **2005**, *63*, 65–74.
60. Poduri, A.; Evrony, G.D.; Cai, X.; Elhosary, P.C.; Beroukhim, R.; Lehtinen, M.K.; Hills, L.B.; Heinzen, E.L.; Hill, A.; Hill, R.S.; *et al.* Somatic activation of AKT3 causes hemispheric developmental brain malformations. *Neuron* **2012**, *74*, 41–48.
61. Lindhurst, M.J.; Sapp, J.C.; Teer, J.K.; Johnston, J.J.; Finn, E.M.; Peters, K.; Turner, J.; Cannons, J.L.; Bick, D.; Blakemore, L.; *et al.* A mosaic activating mutation in AKT1 associated with the proteus syndrome. *N. Engl. J. Med.* **2011**, *365*, 611–619.
62. Hussain, K.; Challis, B.; Rocha, N.; Payne, F.; Minic, M.; Thompson, A.; Daly, A.; Scott, C.; Harris, J.; Smillie, B.J.; *et al.* An activating mutation of AKT2 and human hypoglycemia. *Science* **2011**, *334*, 474.
63. Jacobs, K.B.; Yeager, M.; Zhou, W.; Wacholder, S.; Wang, Z.; Rodriguez-Santiago, B.; Hutchinson, A.; Deng, X.; Liu, C.; Horner, M.-J.; *et al.* Detectable clonal mosaicism and its relationship to aging and cancer. *Nat. Genet.* **2012**, *44*, 651–658.
64. Laurie, C.C.; Laurie, C.A.; Rice, K.; Doheny, K.F.; Zelnick, L.R.; McHugh, C.P.; Ling, H.; Hetrick, K.N.; Pugh, E.W.; Amos, C.; *et al.* Detectable clonal mosaicism from birth to old age and its relationship to cancer. *Nat. Genet.* **2012**, *44*, 642–650.
65. Aghili, L.; Foo, J.; DeGregori, J.; De, S. Patterns of somatically acquired amplifications and deletions in apparently normal tissues of ovarian cancer patients. *Cell Rep.* **2014**, *7*, 1310–1319.
66. Costa, T.; Lambert, M.; Teshima, I.; Ray, P.N.; Richer, C.-L.; Dallaire, L. Monozygotic twins with 45, X/46, XY mosaicism discordant for phenotypic sex. *Am. J. Med. Genet.* **1998**, *75*, 40–44.
67. Fujimoto, A.; Boelter, W.D.; Sparkes, R.S.; Lin, M.S.; Battersby, K. Monozygotic twins of discordant sex both with 45,X/46,X,idic(Y) mosaicism. *Am. J. Med. Genet.* **1991**, *41*, 239–245.

68. Kaplan, L.; Foster, R.; Shen, Y.; Parry, D.M.; McMaster, M.L.; O'Leary, M.C.; Gusella, J.F. Monozygotic twins discordant for neurofibromatosis 1. *Am. J. Med. Genet. A* **2010**, *152A*, 601–606.
69. Zeng, S.; Patil, S.R.; Yankowitz, J. Prenatal detection of mosaic trisomy 1q due to an unbalanced translocation in one fetus of a twin pregnancy following in vitro fertilization: A postzygotic error. *Am. J. Med. Genet. A* **2003**, *120A*, 464–469.
70. Helderman-van den Enden, A.T.J.M.; Maaswinkel-Mooij, P.D.; Hoogendoorn, E.; Willemsen, R.; Maat-Kievit, J.A.; Losekoot, M.; Oostra, B.A. Monozygotic twin brothers with the fragile X syndrome: Different CGG repeats and different mental capacities. *J. Med. Genet.* **1999**, *36*, 253–257.
71. Piotrowski, A.; Bruder, C.E.; Andersson, R.; Diaz de Stahl, T.; Menzel, U.; Sandgren, J.; Poplawski, A.; von Tell, D.; Crasto, C.; Bogdan, A.; *et al.* Somatic mosaicism for copy number variation in differentiated human tissues. *Hum. Mutat.* **2008**, *29*, 1118–1124.
72. O'Huallachain, M.; Karczewski, K.J.; Weissman, S.M.; Urban, A.E.; Snyder, M.P. Extensive genetic variation in somatic human tissues. *Proc. Natl. Acad. Sci. USA* **2012**, *109*, 18018–18023.
73. Abyzov, A.; Mariani, J.; Palejev, D.; Zhang, Y.; Haney, M.S.; Tomasini, L.; Ferrandino, A.F.; Rosenberg Belmaker, L.A.; Szekely, A.; Wilson, M.; *et al.* Somatic copy number mosaicism in human skin revealed by induced pluripotent stem cells. *Nature* **2012**, *492*, 438–442.
74. Rehen, S.K.; McConnell, M.J.; Kaushal, D.; Kingsbury, M.A.; Yang, A.H.; Chun, J. Chromosomal variation in neurons of the developing and adult mammalian nervous system. *Proc. Natl. Acad. Sci. USA* **2001**, *98*, 13361–13366.
75. Baillie, J.K.; Barnett, M.W.; Upton, K.R.; Gerhardt, D.J.; Richmond, T.A.; de Sapio, F.; Brennan, P.M.; Rizzu, P.; Smith, S.; Fell, M.; *et al.* Somatic retrotransposition alters the genetic landscape of the human brain. *Nature* **2011**, *479*, 534–537.
76. Coufal, N.G.; Garcia-Perez, J.L.; Peng, G.E.; Yeo, G.W.; Mu, Y.; Lovci, M.T.; Morell, M.; O'Shea, K.S.; Moran, J.V.; Gage, F.H. L1 retrotransposition in human neural progenitor cells. *Nature* **2009**, *460*, 1127–1131.
77. Muotri, A.R.; Chu, V.T.; Marchetto, M.C.; Deng, W.; Moran, J.V.; Gage, F.H. Somatic mosaicism in neuronal precursor cells mediated by l1 retrotransposition. *Nature* **2005**, *435*, 903–910.
78. Cai, X.; Evrony, G.D.; Lehmann, H.S.; Elhosary, P.C.; Mehta, B.K.; Poduri, A.; Walsh, C.A. Single-cell, genome-wide sequencing identifies clonal somatic copy-number variation in the human brain. *Cell Rep.* **2014**, *8*, 1280–1289.
79. Evrony, G.D.; Cai, X.; Lee, E.; Hills, L.B.; Elhosary, P.C.; Lehmann, H.S.; Parker, J.J.; Atabay, K.D.; Gilmore, E.C.; Poduri, A.; *et al.* Single-neuron sequencing analysis of L1 retrotransposition and somatic mutation in the human brain. *Cell* **2012**, *151*, 483–496.
80. McConnell, M.J.; Lindberg, M.R.; Brennand, K.J.; Piper, J.C.; Voet, T.; Cowing-Zitron, C.; Shumilina, S.; Lasken, R.S.; Vermeesch, J.R.; Hall, I.M.; *et al.* Mosaic copy number variation in human neurons. *Science* **2013**, *342*, 632–637.
81. Kalousek, D.K.; Vekemans, M. Confined placental mosaicism. *J. Med. Genet.* **1996**, *33*, 529–533.
82. Kalousek, D.K.; Dill, F.J. Chromosomal mosaicism confined to the placenta in human conceptions. *Science* **1983**, *221*, 665–667.
83. Taylor, T.H.; Gitlin, S.A.; Patrick, J.L.; Crain, J.L.; Wilson, J.M.; Griffin, D.K. The origin, mechanisms, incidence and clinical consequences of chromosomal mosaicism in humans. *Hum. Reprod. Update* **2014**, *20*, 571–581.

84. Vanneste, E.; Voet, T.; Le Caignec, C.; Ampe, M.; Konings, P.; Melotte, C.; Debrock, S.; Amyere, M.; Vikkula, M.; Schuit, F.; *et al.* Chromosome instability is common in human cleavage-stage embryos. *Nat. Med.* **2009**, *15*, 577–583.
85. Chen, E.Z.; Chiu, R.W.K.; Sun, H.; Akolekar, R.; Chan, K.C.A.; Leung, T.Y.; Jiang, P.; Zheng, Y.W.L.; Lun, F.M.F.; Chan, L.Y.S.; *et al.* Noninvasive prenatal diagnosis of fetal trisomy 18 and trisomy 13 by maternal plasma DNA sequencing. *PLOS ONE* **2011**, *6*, e21791.
86. Youssoufian, H.; Pyeritz, R.E. Mechanisms and consequences of somatic mosaicism in humans. *Nat. Rev. Genet.* **2002**, *3*, 748–758.
87. Watson, I.R.; Takahashi, K.; Futreal, P.A.; Chin, L. Emerging patterns of somatic mutations in cancer. *Nat. Rev. Genet.* **2013**, *14*, 703–718.
88. Ito, Y.; Tanaka, F.; Yamamoto, M.; Doyu, M.; Nagamatsu, M.; Riku, S.; Mitsuma, T.; Sobue, G. Somatic mosaicism of the expanded cag trinucleotide repeat in mrnas for the responsible gene of machado-joseph disease (MJD), dentatorubral-pallidoluysian atrophy (DRPLA), and spinal and bulbar muscular atrophy (SBMA). *Neurochem. Res.* **1998**, *23*, 25–32.
89. James, C.D.; Carlbom, E.; Nordenskjold, M.; Collins, V.P.; Cavenee, W.K. Mitotic recombination of chromosome 17 in astrocytomas. *Proc. Natl. Acad. Sci. USA* **1989**, *86*, 2858–2862.
90. Kleczkowska, A.; Fryns, J.P.; Van den Berghe, H. On the variable effect of mosaic normal/balanced chromosomal rearrangements in man. *J. Med. Genet.* **1990**, *27*, 505–507.
91. Kotzot, D.; Schmitt, S.; Bernasconi, F.; Robinson, W.P.; Lurie, I.W.; Ilyina, H.; Méhes, K.; Hamel, B.C.J.; Otten, B.J.; Hergersberg, M.; *et al.* Uniparental disomy 7 in silver-russell syndrome and primordial growth retardation. *Hum. Mol. Genet.* **1995**, *4*, 583–587.
92. Rodríguez-Santiago, B.; Malats, N.; Rothman, N.; Armengol, L.; Garcia-Closas, M.; Kogevinas, M.; Villa, O.; Hutchinson, A.; Earl, J.; Marenne, G.; *et al.* Mosaic uniparental disomies and aneuploidies as large structural variants of the human genome. *Am. J. Hum. Genet.* **2010**, *87*, 129–138.
93. Slatter, R.E.; Elliott, M.; Welham, K.; Carrera, M.; Schofield, P.N.; Barton, D.E.; Maher, E.R. Mosaic uniparental disomy in beckwith-wiedemann syndrome. *J. Med. Genet.* **1994**, *31*, 749–753.
94. Zori, R.T.; Gray, B.A.; Bent-Williams, A.; Driscoll, D.J.; Williams, C.A.; Zackowski, J.L. Preaxial acrofacial dysostosis (nager syndrome) associated with an inherited and apparently balanced X;9 translocation: Prenatal and postnatal late replication studies. *Am. J. Med. Genet.* **1993**, *46*, 379–383.
95. Walsh, C.; Cepko, C.L. Widespread dispersion of neuronal clones across functional regions of the cerebral cortex. *Science* **1992**, *255*, 434–440.
96. Pleasure, S.J.; Anderson, S.; Hevner, R.; Bagri, A.; Marin, O.; Lowenstein, D.H.; Rubenstein, J.L. Cell migration from the ganglionic eminences is required for the development of hippocampal gabaergic interneurons. *Neuron* **2000**, *28*, 727–740.
97. Hohn, A.; Leibrock, J.; Bailey, K.; Barde, Y.-A. Identification and characterization of a novel member of the nerve growth factor/brain-derived neurotrophic factor family. *Nature* **1990**, *344*, 339–341.
98. Leibrock, J.; Lottspeich, F.; Hohn, A.; Hofer, M.; Hengerer, B.; Masiakowski, P.; Thoenen, H.; Barde, Y.-A. Molecular cloning and expression of brain-derived neurotrophic factor. *Nature* **1989**, *341*, 149–152.

99. Levi-Montalcini, R. Growth control of nerve cells by a protein factor and its antiserum: Discovery of this factor may provide new leads to understanding of some neurogenetic processes. *Science* **1964**, *143*, 105–110.
100. Kurek, K.C.; Luks, V.L.; Ayturk, U.M.; Alomari, A.I.; Fishman, S.J.; Spencer, S.A.; Mulliken, J.B.; Bowen, M.E.; Yamamoto, G.L.; Kozakewich, H.P.; *et al.* Somatic mosaic activating mutations in PIK3CA cause cloves syndrome. *Am. J. Hum. Genet.* **2012**, *90*, 1108–1115.
101. Li, H. Toward better understanding of artifacts in variant calling from high-coverage samples. *Bioinformatics* **2014**, *30*, 2843–2851.
102. Gerstung, M.; Papaemmanuil, E.; Campbell, P.J. Subclonal variant calling with multiple samples and prior knowledge. *Bioinformatics* **2014**, *30*, 1198–1204.
103. Crotwell, P.L.; Hoyme, H.E. Advances in whole-genome genetic testing: From chromosomes to microarrays. *Curr. Probl. Pediatr. Adolesc. Health Care* **2012**, *42*, 47–73.
104. Bushman, D.M.; Chun, J. The genomically mosaic brain: Aneuploidy and more in neural diversity and disease. *Semin. Cell Dev. Biol.* **2013**, *24*, 357–369.
105. Notini, A.J.; Craig, J.M.; White, S.J. Copy number variation and mosaicism. *Cytogenet. Genome Res.* **2008**, *123*, 270–277.
106. Vorsanova, S.G.; Yurov, Y.B.; Iourov, I.Y. Human interphase chromosomes: A review of available molecular cytogenetic technologies. *Mol. Cytogenet.* **2010**, *3*, 1.
107. Imataka, G.; Arisaka, O. Chromosome analysis using spectral karyotyping (sky). *Cell Biochem. Biophys.* **2012**, *62*, 13–17.
108. Kallioniemi, A.; Kallioniemi, O.P.; Sudar, D.; Rutovitz, D.; Gray, J.W.; Waldman, F.; Pinkel, D. Comparative genomic hybridization for molecular cytogenetic analysis of solid tumors. *Science* **1992**, *258*, 818–821.
109. Pinkel, D.; Albertson, D.G. Array comparative genomic hybridization and its applications in cancer. *Nat. Genet.* **2005**, *37*, S11–S17.
110. Alkan, C.; Coe, B.P.; Eichler, E.E. Genome structural variation discovery and genotyping. *Nat. Rev. Genet.* **2011**, *12*, 363–376.
111. Bignell, G.R.; Huang, J.; Greshock, J.; Watt, S.; Butler, A.; West, S.; Grigorova, M.; Jones, K.W.; Wei, W.; Stratton, M.R.; *et al.* High-resolution analysis of DNA copy number using oligonucleotide microarrays. *Genome Res.* **2004**, *14*, 287–295.
112. Mohr, S.; Leikauf, G.D.; Keith, G.; Rihn, B.H. Microarrays as cancer keys: An array of possibilities. *J. Clin. Oncol.* **2002**, *20*, 3165–3175.
113. Baugher, J.D.; Baugher, B.D.; Shirley, M.D.; Pevsner, J. Sensitive and specific detection of mosaic chromosomal abnormalities using the parent-of-origin-based detection (POD) method. *BMC Genom.* **2013**, *14*, 367.
114. Leek, J.T.; Scharpf, R.B.; Bravo, H.C.; Simcha, D.; Langmead, B.; Johnson, W.E.; Geman, D.; Baggerly, K.; Irizarry, R.A. Tackling the widespread and critical impact of batch effects in high-throughput data. *Nat. Rev. Genet.* **2010**, *11*, 733–739.
115. Chen, G.K.; Chang, X.; Curtis, C.; Wang, K. Precise inference of copy number alterations in tumor samples from SNP arrays. *Bioinformatics* **2013**, *29*, 2964–2970.

116. Li, A.; Liu, Z.; Lezon-Geyda, K.; Sarkar, S.; Lannin, D.; Schulz, V.; Krop, I.; Winer, E.; Harris, L.; Tuck, D. GPHMM: An integrated hidden Markov model for identification of copy number alteration and loss of heterozygosity in complex tumor samples using whole genome SNP arrays. *Nucleic Acids Res.* **2011**, *39*, 4928–4941.
117. Liu, Z.; Li, A.; Schulz, V.; Chen, M.; Tuck, D. Mixhmm: Inferring copy number variation and allelic imbalance using SNP arrays and tumor samples mixed with stromal cells. *PLOS ONE* **2010**, *5*, e10909.
118. Rancoita, P.; Hutter, M.; Bertoni, F.; Kwee, I. An integrated bayesian analysis of LOH and copy number data. *BMC Bioinform.* **2010**, *11*, 321.
119. Cock, P.J.; Fields, C.J.; Goto, N.; Heuer, M.L.; Rice, P.M. The sanger FASTQ file format for sequences with quality scores, and the solexa/illumina FASTQ variants. *Nucleic Acids Res.* **2010**, *38*, 1767–1771.
120. Yoon, S.; Xuan, Z.; Makarov, V.; Ye, K.; Sebat, J. Sensitive and accurate detection of copy number variants using read depth of coverage. *Genome Res.* **2009**, *19*, 1586–1592.
121. Koboldt, D.C.; Zhang, Q.; Larson, D.E.; Shen, D.; McLellan, M.D.; Lin, L.; Miller, C.A.; Mardis, E.R.; Ding, L.; Wilson, R.K. Varscan 2: Somatic mutation and copy number alteration discovery in cancer by exome sequencing. *Genome Res.* **2012**, *22*, 568–576.
122. Ng, S.B.; Turner, E.H.; Robertson, P.D.; Flygare, S.D.; Bigham, A.W.; Lee, C.; Shaffer, T.; Wong, M.; Bhattacharjee, A.; Eichler, E.E.; *et al.* Targeted capture and massively parallel sequencing of 12 human exomes. *Nature* **2009**, *461*, 272–276.
123. Choi, M.; Scholl, U.I.; Ji, W.; Liu, T.; Tikhonova, I.R.; Zumbo, P.; Nayir, A.; Bakkaloğlu, A.; Özen, S.; Sanjad, S.; *et al.* Genetic diagnosis by whole exome capture and massively parallel DNA sequencing. *Proc. Natl. Acad. Sci.* **2009**, *106*, 19096–19101.
124. Hodges, E.; Xuan, Z.; Balija, V.; Kramer, M.; Molla, M.N.; Smith, S.W.; Middle, C.M.; Rodesch, M.J.; Albert, T.J.; Hannon, G.J.; *et al.* Genome-wide in situ exon capture for selective resequencing. *Nat. Genet.* **2007**, *39*, 1522–1527.
125. Larson, D.E.; Harris, C.C.; Chen, K.; Koboldt, D.C.; Abbott, T.E.; Dooling, D.J.; Ley, T.J.; Mardis, E.R.; Wilson, R.K.; Ding, L. Somaticsniper: Identification of somatic point mutations in whole genome sequencing data. *Bioinformatics* **2012**, *28*, 311–317.
126. Roth, A.; Ding, J.; Morin, R.; Crisan, A.; Ha, G.; Giuliany, R.; Bashashati, A.; Hirst, M.; Turashvili, G.; Oloumi, A.; *et al.* Jointsnvmix: A probabilistic model for accurate detection of somatic mutations in normal/tumour paired next-generation sequencing data. *Bioinformatics* **2012**, *28*, 907–913.
127. Saunders, C.T.; Wong, W.S.; Swamy, S.; Becq, J.; Murray, L.J.; Cheetham, R.K. Strelka: Accurate somatic small-variant calling from sequenced tumor-normal sample pairs. *Bioinformatics* **2012**, *28*, 1811–1817.
128. Cibulskis, K.; Lawrence, M.S.; Carter, S.L.; Sivachenko, A.; Jaffe, D.; Sougnez, C.; Gabriel, S.; Meyerson, M.; Lander, E.S.; Getz, G. Sensitive detection of somatic point mutations in impure and heterogeneous cancer samples. *Nature Biotechnol.* **2013**, *31*, 213–219.
129. Li, H.; Handsaker, B.; Wysoker, A.; Fennell, T.; Ruan, J.; Homer, N.; Marth, G.; Abecasis, G.; Durbin, R. The sequence alignment/map format and samtools. *Bioinformatics* **2009**, *25*, 2078–2079.

130. Amarasinghe, K.C.; Li, J.; Halgamuge, S.K. Convex: Copy number variation estimation in exome sequencing data using HMM. *BMC Bioinform.* **2013**, *14*, S2.
131. Boeva, V.; Popova, T.; Bleakley, K.; Chiche, P.; Cappo, J.; Schleiermacher, G.; Janoueix-Lerosey, I.; Delattre, O.; Barillot, E. Control-freec: A tool for assessing copy number and allelic content using next-generation sequencing data. *Bioinformatics* **2012**, *28*, 423–425.
132. Layer, R.M.; Chiang, C.; Quinlan, A.R.; Hall, I.M. Lumpy: A probabilistic framework for structural variant discovery. *Genome Biol.* **2014**, *15*, R84.
133. Chen, M.; Gunel, M.; Zhao, H. Somatica: Identifying, characterizing and quantifying somatic copy number aberrations from cancer genome sequencing data. *PLOS ONE* **2013**, *8*, e78143.
134. Ding, L.; Wendl, M.C.; McMichael, J.F.; Raphael, B.J. Expanding the computational toolbox for mining cancer genomes. *Nat. Rev. Genet.* **2014**, *15*, 556–570.
135. Yadav, V.K.; De, S. An assessment of computational methods for estimating purity and clonality using genomic data derived from heterogeneous tumor tissue samples. *Brief. Bioinform.* **2014**, doi:10.1093/bib/bbu002.
136. Dean, F.B.; Hosono, S.; Fang, L.; Wu, X.; Faruqi, A.F.; Bray-Ward, P.; Sun, Z.; Zong, Q.; Du, Y.; Du, J.; *et al.* Comprehensive human genome amplification using multiple displacement amplification. *Proc. Natl. Acad. Sci. USA* **2002**, *99*, 5261–5266.
137. Hosono, S.; Faruqi, A.F.; Dean, F.B.; Du, Y.; Sun, Z.; Wu, X.; Du, J.; Kingsmore, S.F.; Egholm, M.; Lasken, R.S. Unbiased whole-genome amplification directly from clinical samples. *Genome Res.* **2003**, *13*, 954–964.
138. Pugh, T.J.; Delaney, A.D.; Farnoud, N.; Flibotte, S.; Griffith, M.; Li, H.I.; Qian, H.; Farinha, P.; Gascoyne, R.D.; Marra, M.A. Impact of whole genome amplification on analysis of copy number variants. *Nucleic Acids Res.* **2008**, *36*, e80.
139. Baslan, T.; Kendall, J.; Rodgers, L.; Cox, H.; Riggs, M.; Stepansky, A.; Troge, J.; Ravi, K.; Esposito, D.; Lakshmi, B.; *et al.* Genome-wide copy number analysis of single cells. *Nat. Protoc.* **2012**, *7*, 1024–1041.
140. Gundry, M.; Li, W.; Maqbool, S.B.; Vijg, J. Direct, genome-wide assessment of DNA mutations in single cells. *Nucleic Acids Res.* **2012**, *40*, 2032–2040.
141. Hou, Y.; Song, L.; Zhu, P.; Zhang, B.; Tao, Y.; Xu, X.; Li, F.; Wu, K.; Liang, J.; Shao, D.; *et al.* Single-cell exome sequencing and monoclonal evolution of a JAK2-negative myeloproliferative neoplasm. *Cell* **2012**, *148*, 873–885.
142. Xu, X.; Hou, Y.; Yin, X.; Bao, L.; Tang, A.; Song, L.; Li, F.; Tsang, S.; Wu, K.; Wu, H.; *et al.* Single-cell exome sequencing reveals single-nucleotide mutation characteristics of a kidney tumor. *Cell* **2012**, *148*, 886–895.
143. Sjoblom, T.; Jones, S.; Wood, L.D.; Parsons, D.W.; Lin, J.; Barber, T.D.; Mandelker, D.; Leary, R.J.; Ptak, J.; Silliman, N.; *et al.* The consensus coding sequences of human breast and colorectal cancers. *Science* **2006**, *314*, 268–274.
144. Stephens, P.J.; Tarpey, P.S.; Davies, H.; van Loo, P.; Greenman, C.; Wedge, D.C.; Nik-Zainal, S.; Martin, S.; Varela, I.; Bignell, G.R.; *et al.* The landscape of cancer genes and mutational processes in breast cancer. *Nature* **2012**, *486*, 400–404.
145. Stratton, M.R.; Campbell, P.J.; Futreal, P.A. The cancer genome. *Nature* **2009**, *458*, 719–724.
146. Hanahan, D.; Weinberg, R.A. Hallmarks of cancer: The next generation. *Cell* **2011**, *144*, 646–674.

147. McCubrey, J.A.; Steelman, L.S.; Chappell, W.H.; Abrams, S.L.; Montalto, G.; Cervello, M.; Nicoletti, F.; Fagone, P.; Malaponte, G.; Mazzarino, M.C.; *et al.* Mutations and deregulation of Ras/Raf/MEK/ERK and PI3K/PTEN/Akt/mTOR cascades which alter therapy response. *Oncotarget* **2012**, *3*, 954–987.
148. Downward, J. Targeting RAS signalling pathways in cancer therapy. *Nat. Rev. Cancer* **2003**, *3*, 11–22.
149. Liaw, D.; Marsh, D.J.; Li, J.; Dahia, P.L.M.; Wang, S.I.; Zheng, Z.; Bose, S.; Call, K.M.; Tsou, H.C.; Peacoke, M.; *et al.* Germline mutations of the pten gene in cowden disease, an inherited breast and thyroid cancer syndrome. *Nat. Genet.* **1997**, *16*, 64–67.
150. Malkin, D.; Li, F.; Strong, L.; Fraumeni, J.; Nelson, C.; Kim, D.; Kassel, J.; Gryka, M.; Bischoff, F.; Tainsky, M.; *et al.* Germ line p53 mutations in a familial syndrome of breast cancer, sarcomas, and other neoplasms. *Science* **1990**, *250*, 1233–1238.
151. Morin, P.J.; Sparks, A.B.; Korinek, V.; Barker, N.; Clevers, H.; Vogelstein, B.; Kinzler, K.W. Activation of β-catenin-TCF signaling in colon cancer by mutations in β-catenin or APC. *Science* **1997**, *275*, 1787–1790.
152. Wooster, R.; Bignell, G.; Lancaster, J.; Swift, S.; Seal, S.; Mangion, J.; Collins, N.; Gregory, S.; Gumbs, C.; Micklem, G.; *et al.* Identification of the breast cancer susceptibility gene BRCA2. *Nature* **1995**, *378*, 789–792.
153. Moynahan, M.E.; Chiu, J.W.; Koller, B.H.; Jasin, M. Brca1 controls homology-directed DNA repair. *Mol. Cell* **1999**, *4*, 511–518.
154. Miyaki, M.; Konishi, M.; Tanaka, K.; Kikuchi-Yanoshita, R.; Muraoka, M.; Yasuno, M.; Igari, T.; Koike, M.; Chiba, M.; Mori, T. Germline mutation of MSH6 as the cause of hereditary nonpolyposis colorectal cancer. *Nat. Genet.* **1997**, *17*, 271–272.
155. Cleaver, J.E. Defective repair replication of DNA in xeroderma pigmentosum. *Nature* **1968**, *218*, 652–656.
156. Govindan, R.; Ding, L.; Griffith, M.; Subramanian, J.; Dees, N.D.; Kanchi, K.L.; Maher, C.A.; Fulton, R.; Fulton, L.; Wallis, J.; *et al.* Genomic landscape of non-small cell lung cancer in smokers and never-smokers. *Cell* **2012**, *150*, 1121–1134.
157. Forbes, S.A.; Bindal, N.; Bamford, S.; Cole, C.; Kok, C.Y.; Beare, D.; Jia, M.; Shepherd, R.; Leung, K.; Menzies, A.; *et al.* Cosmic: Mining complete cancer genomes in the catalogue of somatic mutations in cancer. *Nucleic Acids Res.* **2011**, *39*, D945–D950.
158. Collins, F.S.; Barker, A.D. Mapping the cancer genome. Pinpointing the genes involved in cancer will help chart a new course across the complex landscape of human malignancies. *Sci. Am.* **2007**, *296*, 50–57.
159. Debeljak, M.; Freed, D.N.; Welch, J.A.; Haley, L.; Beierl, K.; Iglehart, B.S.; Pallavajjala, A.; Gocke, C.D.; Leffell, M.S.; Lin, M.-T.; *et al.* Haplotype counting by next-generation sequencing for ultrasensitive human DNA detection. *J. Mol. Diagn.* **2014**, *16*, 495–503.
160. Armitage, P.; Doll, R. The age distribution of cancer and a multi-stage theory of carcinogenesis. *Br. J. Cancer* **1954**, *8*, 1–12.
161. Szilard, L. On the nature of the aging process. *Proc. Natl. Acad. Sci. USA* **1959**, *45*, 30–45.
162. Curtis, H.J. Biological mechanisms underlying the aging process. *Science* **1963**, *141*, 686–694.

163. Goldsby, R.E.; Hays, L.E.; Chen, X.; Olmsted, E.A.; Slayton, W.B.; Spangrude, G.J.; Preston, B.D. High incidence of epithelial cancers in mice deficient for DNA polymerase δ proofreading. *Proc. Natl. Acad. Sci. USA* **2002**, *99*, 15560–15565.

164. Goldsby, R.E.; Lawrence, N.A.; Hays, L.E.; Olmsted, E.A.; Chen, X.; Singh, M.; Preston, B.D. Defective DNA polymerase-[delta] proofreading causes cancer susceptibility in mice. *Nat. Med.* **2001**, *7*, 638–639.

165. Albertson, T.M.; Ogawa, M.; Bugni, J.M.; Hays, L.E.; Chen, Y.; Wang, Y.; Treuting, P.M.; Heddle, J.A.; Goldsby, R.E.; Preston, B.D. DNA polymerase ε and δ proofreading suppress discrete mutator and cancer phenotypes in mice. *Proc. Natl. Acad. Sci. USA* **2009**, *106*, 17101–17104.

166. Vermulst, M.; Bielas, J.H.; Kujoth, G.C.; Ladiges, W.C.; Rabinovitch, P.S.; Prolla, T.A.; Loeb, L.A. Mitochondrial point mutations do not limit the natural lifespan of mice. *Nat. Genet.* **2007**, *39*, 540–543.

167. Trifunovic, A.; Wredenberg, A.; Falkenberg, M.; Spelbrink, J.N.; Rovio, A.T.; Bruder, C.E.; Bohlooly-Y, M.; Gidlof, S.; Oldfors, A.; Wibom, R.; *et al.* Premature ageing in mice expressing defective mitochondrial DNA polymerase. *Nature* **2004**, *429*, 417–423.

168. Marteijn, J.A.; Lans, H.; Vermeulen, W.; Hoeijmakers, J.H.J. Understanding nucleotide excision repair and its roles in cancer and ageing. *Nat. Rev. Mol. Cell. Biol.* **2014**, *15*, 465–481.

169. Mohaghegh, P.; Hickson, I.D. DNA helicase deficiencies associated with cancer predisposition and premature ageing disorders. *Hum. Mol. Genet.* **2001**, *10*, 741–746.

170. Hanks, S.; Coleman, K.; Reid, S.; Plaja, A.; Firth, H.; Fitzpatrick, D.; Kidd, A.; Mehes, K.; Nash, R.; Robin, N.; *et al.* Constitutional aneuploidy and cancer predisposition caused by biallelic mutations in bub1b. *Nat. Genet.* **2004**, *36*, 1159–1161.

171. Date, H.; Onodera, O.; Tanaka, H.; Iwabuchi, K.; Uekawa, K.; Igarashi, S.; Koike, R.; Hiroi, T.; Yuasa, T.; Awaya, Y.; *et al.* Early-onset ataxia with ocular motor apraxia and hypoalbuminemia is caused by mutations in a new HIT superfamily gene. *Nat. Genet.* **2001**, *29*, 184–188.

172. Moreira, M.-C.; Barbot, C.; Tachi, N.; Kozuka, N.; Uchida, E.; Gibson, T.; Mendonca, P.; Costa, M.; Barros, J.; Yanagisawa, T.; *et al.* The gene mutated in ataxia-ocular apraxia 1 encodes the new HIT/Zn-finger protein aprataxin. *Nat. Genet.* **2001**, *29*, 189–193.

173. Niedernhofer, L.J. Tissue-specific accelerated aging in nucleotide excision repair deficiency. *Mech. Ageing Dev.* **2008**, *129*, 408–415.

174. Monnat Jr, R.J. Human RECQ helicases: Roles in DNA metabolism, mutagenesis and cancer biology. *Semin. Cancer Biol.* **2010**, *20*, 329–339.

175. Burdick, D.; Soreghan, B.; Kwon, M.; Kosmoski, J.; Knauer, M.; Henschen, A.; Yates, J.; Cotman, C.; Glabe, C. Assembly and aggregation properties of synthetic Alzheimer's A4/beta amyloid peptide analogs. *J. Biol. Chem.* **1992**, *267*, 546–554.

176. Goldfarb, L.G.; Brown, P.; McCombie, W.R.; Goldgaber, D.; Swergold, G.D.; Wills, P.R.; Cervenakova, L.; Baron, H.; Gibbs, C.J.; Gajdusek, D.C. Transmissible familial Creutzfeldt-Jakob disease associated with five, seven, and eight extra octapeptide coding repeats in the PRNP gene. *Proc. Natl. Acad. Sci.* USA **1991**, *88*, 10926–10930.

177. Beck, J.A.; Poulter, M.; Campbell, T.A.; Uphill, J.B.; Adamson, G.; Geddes, J.F.; Revesz, T.; Davis, M.B.; Wood, N.W.; Collinge, J.; *et al.* Somatic and germline mosaicism in sporadic early-onset alzheimer's disease. *Hum. Mol. Genet.* **2004**, *13*, 1219–1224.

178. Alzualde, A.; Moreno, F.; Martínez-Lage, P.; Ferrer, I.; Gorostidi, A.; Otaegui, D.; Blázquez, L.; Atares, B.; Cardoso, S.; Martínez de Pancorbo, M.; *et al.* Somatic mosaicism in a case of apparently sporadic Creutzfeldt-Jakob disease carrying a *de novo* D178n mutation in the PRNP gene. *Am. J. Med. Genet. B Neuropsychiatr. Genet.* **2010**, *153B*, 1283–1291.

179. Eikelenboom, P.; Bate, C.; Van Gool, W.A.; Hoozemans, J.J.M.; Rozemuller, J.M.; Veerhuis, R.; Williams, A. Neuroinflammation in alzheimer's disease and prion disease. *Glia* **2002**, *40*, 232–239.

180. Meyer-Luehmann, M.; Coomaraswamy, J.; Bolmont, T.; Kaeser, S.; Schaefer, C.; Kilger, E.; Neuenschwander, A.; Abramowski, D.; Frey, P.; Jaton, A.L.; *et al.* Exogenous induction of cerebral β-amyloidogenesis is governed by agent and host. *Science* **2006**, *313*, 1781–1784.

181. Kane, M.D.; Lipinski, W.J.; Callahan, M.J.; Bian, F.; Durham, R.A.; Schwarz, R.D.; Roher, A.E.; Walker, L.C. Evidence for seeding of beta-amyloid by intracerebral infusion of alzheimer brain extracts in beta-amyloid precursor protein-transgenic mice. *J. Neurosci.* **2000**, *20*, 3606–3611.

182. Lu, J.-X.; Qiang, W.; Yau, W.-M.; Schwieters, C.D.; Meredith, S.C.; Tycko, R. Molecular structure of β-amyloid fibrils in alzheimer's disease brain tissue. *Cell* **2013**, *154*, 1257–1268.

183. Stöhr, J.; Condello, C.; Watts, J.C.; Bloch, L.; Oehler, A.; Nick, M.; DeArmond, S.J.; Giles, K.; DeGrado, W.F.; Prusiner, S.B. Distinct synthetic aβ prion strains producing different amyloid deposits in bigenic mice. *Proc. Natl. Acad. Sci.* **2014**, *111*, 10329–10334.

184. Watts, J.C.; Condello, C.; Stöhr, J.; Oehler, A.; Lee, J.; DeArmond, S.J.; Lannfelt, L.; Ingelsson, M.; Giles, K.; Prusiner, S.B. Serial propagation of distinct strains of aβ prions from alzheimer's disease patients. *Proc. Natl. Acad. Sci. USA* **2014**, *111*, 10323–10328.

185. Ott, A.; Breteler, M.M.B.; van Harskamp, F.; Claus, J.J.; van der Cammen, T.J.M.; Grobbee, D.E.; Hofman, A. Prevalence of alzheimer's disease and vascular dementia: Association with education. The Rotterdam study. *BMJ* **1995**, *310*, 970.

186. Chapman, J.; Ben-Israel, J.; Goldhammer, Y.; Korczyn, A.D. The risk of developing creutzfeldt-jakob disease in subjects with the PRNP gene codon 200 point mutation. *Neurology* **1994**, *44*, 1683–1686.

187. Fruman, D.A.; Rommel, C. PI3K and cancer: Lessons, challenges and opportunities. *Nat. Rev. Drug Discov.* **2014**, *13*, 140–156.

188. Hay, N. The AKT-mtor tango and its relevance to cancer. *Cancer Cell* **2005**, *8*, 179–183.

189. Rozengurt, E. Mitogenic signaling pathways induced by G protein-coupled receptors. *J. Cell. Physiol.* **2007**, *213*, 589–602.

190. Dorsam, R.T.; Gutkind, J.S. G-protein-coupled receptors and cancer. *Nat. Rev. Cancer* **2007**, *7*, 79–94.

191. Mirzaa, G.; Conaway, R.; Graham, J.M., Jr.; Dobyns, W.B. PIK3CA-related segmental overgrowth. In *GeneReviews*; Pagon, R.A.; Adam, M.P.; Ardinger, H.H.; *et al.*, Eds. University of Washington Seattle: Seattle, WA, USA, 2013.

192. Samuels, Y.; Wang, Z.; Bardelli, A.; Silliman, N.; Ptak, J.; Szabo, S.; Yan, H.; Gazdar, A.; Powell, S.M.; Riggins, G.J.; *et al.* High frequency of mutations of the PIK3CA gene in human cancers. *Science* **2004**, *304*, 554.

193. Carpten, J.D.; Faber, A.L.; Horn, C.; Donoho, G.P.; Briggs, S.L.; Robbins, C.M.; Hostetter, G.; Boguslawski, S.; Moses, T.Y.; Savage, S.; *et al.* A transforming mutation in the pleckstrin homology domain of AKT1 in cancer. *Nature* **2007**, *448*, 439–444.

194. Riviere, J.B.; Mirzaa, G.M.; O'Roak, B.J.; Beddaoui, M.; Alcantara, D.; Conway, R.L.; St-Onge, J.; Schwartzentruber, J.A.; Gripp, K.W.; Nikkel, S.M.; *et al. De novo* germline and postzygotic mutations in AKT3, PIK3R2 and PIK3CA cause a spectrum of related megalencephaly syndromes. *Nat. Genet.* **2012**, *44*, 934–940.
195. Lee, J.H.; Huynh, M.; Silhavy, J.L.; Kim, S.; Dixon-Salazar, T.; Heiberg, A.; Scott, E.; Bafna, V.; Hill, K.J.; Collazo, A.; *et al. De novo* somatic mutations in components of the PI3K-AKT3-mtor pathway cause hemimegalencephaly. *Nat. Genet.* **2012**, *44*, 941–945.
196. Zhang, Y.; Gao, X.; Saucedo, L.J.; Ru, B.; Edgar, B.A.; Pan, D. RHEB is a direct target of the tuberous sclerosis tumour suppressor proteins. *Nat. Cell. Biol.* **2003**, *5*, 578–581.
197. Curatolo, P.; Bombardieri, R.; Jozwiak, S. Tuberous sclerosis. *Lancet* **2008**, *372*, 657–668.
198. Henske, E.P.; Wessner, L.L.; Golden, J.; Scheithauer, B.W.; Vortmeyer, A.O.; Zhuang, Z.; Klein-Szanto, A.J.; Kwiatkowski, D.J.; Yeung, R.S. Loss of tuberin in both subependymal giant cell astrocytomas and angiomyolipomas supports a two-hit model for the pathogenesis of tuberous sclerosis tumors. *Am. J. Pathol.* **1997**, *151*, 1639–1647.
199. Tsang, E.; Birch, P.; Friedman, J.M. Valuing gene testing in children with possible neurofibromatosis 1. *Clin. Genet.* **2012**, *82*, 591–593.
200. Pansuriya, T.C.; van Eijk, R.; d'Adamo, P.; van Ruler, M.A.; Kuijjer, M.L.; Oosting, J.; Cleton-Jansen, A.M.; van Oosterwijk, J.G.; Verbeke, S.L.; Meijer, D.; *et al.* Somatic mosaic IDH1 and IDH2 mutations are associated with enchondroma and spindle cell hemangioma in ollier disease and maffucci syndrome. *Nat. Genet.* **2011**, *43*, 1256–1261.
201. Jamuar, S.S.; Lam, A.T.; Kircher, M.; D'Gama, A.M.; Wang, J.; Barry, B.J.; Zhang, X.; Hill, R.S.; Partlow, J.N.; Rozzo, A.; *et al.* Somatic mutations in cerebral cortical malformations. *N. Engl. J. Med.* **2014**, *371*, 733–743.
202. Choate, K.A.; Lu, Y.; Zhou, J.; Choi, M.; Elias, P.M.; Farhi, A.; Nelson-Williams, C.; Crumrine, D.; Williams, M.L.; Nopper, A.J.; *et al.* Mitotic recombination in patients with ichthyosis causes reversion of dominant mutations in KRT10. *Science* **2010**, *330*, 94–97.
203. Pasmooij, A.M.; Jonkman, M.F.; Uitto, J. Revertant mosaicism in heritable skin diseases: Mechanisms of natural gene therapy. *Discov. Med.* **2012**, *14*, 167–179.
204. Pasmooij, A.M.; Pas, H.H.; Bolling, M.C.; Jonkman, M.F. Revertant mosaicism in junctional epidermolysis bullosa due to multiple correcting second-site mutations in LAMB3. *J. Clin. Invest.* **2007**, *117*, 1240–1248.
205. Hirschhorn, R.; Yang, D.R.; Puck, J.M.; Huie, M.L.; Jiang, C.K.; Kurlandsky, L.E. Spontaneous *in vivo* reversion to normal of an inherited mutation in a patient with adenosine deaminase deficiency. *Nat. Genet.* **1996**, *13*, 290–295.
206. Soulier, J.; Leblanc, T.; Larghero, J.; Dastot, H.; Shimamura, A.; Guardiola, P.; Esperou, H.; Ferry, C.; Jubert, C.; Feugeas, J.-P.; *et al.* Detection of somatic mosaicism and classification of fanconi anemia patients by analysis of the FA/BRCA pathway. *Blood* **2005**, *105*, 1329–1336.
207. De, S. Somatic mosaicism in healthy human tissues. *Trends Genet.* **2011**, *27*, 217–223.
208. Insel, T.R. Brain somatic mutations: The dark matter of psychiatric genetics [quest]. *Mol. Psychiatry* **2014**, *19*, 156–158.
209. Krumm, N.; O'Roak, B.J.; Shendure, J.; Eichler, E.E. A *de novo* convergence of autism genetics and molecular neuroscience. *Trends Neurosci.* **2014**, *37*, 95–105.

210. Gaugler, T.; Klei, L.; Sanders, S.J.; Bodea, C.A.; Goldberg, A.P.; Lee, A.B.; Mahajan, M.; Manaa, D.; Pawitan, Y.; Reichert, J.; *et al.* Most genetic risk for autism resides with common variation. *Nat. Genet.* **2014**, *46*, 881–885.
211. De Rubeis, S.; He, X.; Goldberg, A.P.; Poultney, C.S.; Samocha, K.; Ercument Cicek, A.; Kou, Y.; Liu, L.; Fromer, M.; Walker, S.; *et al.* Synaptic, transcriptional and chromatin genes disrupted in autism. *Nature* **2014**, *515*, 209–215.
212. Iossifov, I.; O'Roak, B.J.; Sanders, S.J.; Ronemus, M.; Krumm, N.; Levy, D.; Stessman, H.A.; Witherspoon, K.T.; Vives, L.; Patterson, K.E.; *et al.* The contribution of de novo coding mutations to autism spectrum disorder. *Nature* **2014**, *515*, 216–221.
213. Manolio, T.A.; Collins, F.S.; Cox, N.J.; Goldstein, D.B.; Hindorff, L.A.; Hunter, D.J.; McCarthy, M.I.; Ramos, E.M.; Cardon, L.R.; Chakravarti, A.; *et al.* Finding the missing heritability of complex diseases. *Nature* **2009**, *461*, 747–753.
214. Stoner, R.; Chow, M.L.; Boyle, M.P.; Sunkin, S.M.; Mouton, P.R.; Roy, S.; Wynshaw-Boris, A.; Colamarino, S.A.; Lein, E.S.; Courchesne, E. Patches of disorganization in the neocortex of children with autism. *N. Engl. J. Med.* **2014**, *370*, 1209–1219.

11

Open Access Data Sharing in Genomic Research

Stacey Pereira [1], Richard A. Gibbs [2] and Amy L. McGuire [1,*]

[1] Center for Medical Ethics and Health Policy, Baylor College of Medicine, Houston, TX 77030, USA; E-Mail: spereira@bcm.edu

[2] Human Genome Sequencing Center, Baylor College of Medicine, Houston, TX 77030, USA; E-Mail: agibbs@bcm.edu

* Author to whom correspondence should be addressed; E-Mail: amcguire@bcm.edu;

Abstract: The current emphasis on broad sharing of human genomic data generated in research in order to maximize utility and public benefit is a significant legacy of the Human Genome Project. Concerns about privacy and discrimination have led to policy responses that restrict access to genomic data as the means for protecting research participants. Our research and experience show, however, that a considerable number of research participants agree to open access sharing of their genomic data when given the choice. General policies that limit access to all genomic data fail to respect the autonomy of these participants and, at the same time, unnecessarily limit the utility of the data. We advocate instead a more balanced approach that allows for individual choice and encourages informed decision making, while protecting against the misuse of genomic data through enhanced legislation.

Keywords: data sharing; genomics; privacy; genetic discrimination; open access database

---

1. Introduction

Last year marked the 10th anniversary of the completion of the Human Genome Project (HGP) [1]. One of the many accomplishments of the HGP was the broad sharing of data generated by genomic research studies in order to maximize the utility of the data and the public benefit of such projects [2]. This helped to create a culture of openness in genomic research that was codified in a joint policy from the National Human Genome Research Institute (NHGRI) and the Department of Energy (DOE) in 1991 [3] that called for the rapid public release of data generated by the HGP and subsequent projects. Additional policies in the following years, both domestic and international, reaffirmed and expanded these recommendations for publicly sharing large-scale DNA sequence data [4–7].

Initially, the means for protecting participants' privacy when these data were shared in open access (publicly accessible) databases rested upon the "de-identification" of the data by stripping them of all recognizable annotation before sharing. DNA has a very high information content, however, and in 2004, Lin *et al.* showed that it is possible to identify single individuals with as few as 30–80 single nucleotide polymorphisms (SNPs) [8,9], prompting new privacy concerns. In 2006, the U.S. National Institutes of Health (NIH) established the Database of Genotypes and Phenotypes (dbGaP) [10], which is a controlled access database, meaning that individual level genetic data are accessible only with approval from a Data Access Committee. The current NIH data sharing policy requires researchers to obtain approval from their institution before sharing genomic data in dbGaP, and provides guidance to institutions on how to review studies to ensure compliance with the policy, particularly with regard to the adequacy of informed consent documents.

In 2008, Homer *et al.* revealed further complications by showing that it was possible to uniquely identify individuals in aggregated data sets [11]. This led to the implementation of additional protections by restricting access to some aggregated data elements in dbGaP and other databases internationally [12]. Further, in early 2013, Gymrek *et al.* demonstrated that it was possible to identify individuals in the open access database of the 1000 Genomes Project by analysis of Y-chromosome short tandem repeats. They compared these data to genetic information available on a recreational genealogy website, and then used that information to link to additional publicly accessible data, such as obituaries and the National Institute of General Medical Sciences (NIGMS) Human Genetic Cell Repository, which banks samples from one of the same populations that took part in the 1000 Genomes Project [13]. This paper was the first to show unequivocally that individuals could be uniquely identified without first obtaining a reference sample. In response, the NIH worked with the NIGMS to move age information, which was previously publicly accessible, into the controlled-access part of the database [14].

Each successive policy decision to further restrict access to genomic data has received some pushback, with critics arguing that each was an overreaction and would unnecessarily impede science [12]. Nonetheless, limiting access to increasing amounts of data continues to be the primary policy response to mounting privacy concerns. Arguments against restricted access and for more open data sharing policies must balance the social and scientific benefits of unrestricted access to and use of data, with adequate protection of the rights and interests of individuals who contribute biological specimens and information to research. The almost exclusive focus on restricting access to genomic data as a matter of policy, however, impedes research and fails to respect the autonomy of those who choose to share their information openly. It has been observed that data in controlled access databases are used less frequently than data in open access databases, and as Rodriguez *et al.* [14] remind us, researchers and other custodians have an ethical responsibility not only to minimize the risk of harm to participants, but also to maximize the utility of generated data. These considerations have led some groups to advocate for a more balanced approach that expands options for open access genomic data release [15,16]. Providing research participants the opportunity to allow their data to be shared more broadly is consistent with the principle of respect for autonomy [9], and as we show below, at least among certain populations, there are a considerable number of "information altruists" [17] who would agree, if given the choice.

2. Participant Perspective

Although studies suggest that there is significant public concern about genetic privacy [18,19], in at least one study, the majority (60%) of more than 4600 U.S. adults surveyed reported willingness to participate in genomic research [20]. Likewise, we have found that a substantial number of research participants are even willing to consent to open access release of their genomic data. In a randomized trial of consent with 323 genomic research participants, the majority (84%) agreed to open access data release. Even after being debriefed, educated about all of the consent options, including the option to consent only to the release of data into controlled access databases like dbGaP, or not at all, and surveyed about their perspectives and concerns, the majority (53%) chose to allow their data to be shared in open access databases [21].

We found a similar response from participants in the Texas Cancer Research Biobank (TCRB), which aimed to establish a fully functional open access database incorporating cancer genomes and other participant data. Controlled access data release was a condition of participation in the TCRB, but the informed consent process allowed participants to opt in to broader sharing of their genomic information via open access data release. Of the 194 participants who were offered this choice, 122 (63%) agreed to open access data release.

These studies present an encouraging picture of research participants' altruistic motivations and lend support to the argument that restrictive data sharing policies fail to respect autonomy of participants who would choose to make their data more broadly available. However, they also raise two major challenges that deserve careful consideration: (1) genomic data sharing is a complex concept that can be difficult for participants to understand; and (2) there is a diversity of perspectives about open access data sharing and certain groups may be less willing to share their data publicly.

2.1. Participant Understanding

Autonomous decision-making requires adequate understanding of the options presented. Yet, ensuring adequate understanding is a challenge in all research involving human subjects. Studies suggest that research participants generally have difficulty understanding and remembering basic information described in research-informed consent documents (e.g., the purpose and risks of the research, as well as general concepts related to study design, like randomization) [22,23]. Genomic research and data sharing are complex concepts, so it is not surprising that participants also have difficulty understanding the differences between data sharing options. For example, in the randomized trial of consent mentioned above, a majority (54%) of participants who were surveyed either could not initially recall with whom they had agreed to share their data or did not understand that by agreeing to open access data sharing it meant that their data could be accessed and used by anyone on the internet without restriction [24]. One possible solution is to try to improve understanding with targeted educational interventions, such as brochures or videos. However, studies have shown that efforts to improve understanding have had only limited success, with the most effective intervention being on-on-one education [25,26].

Another approach to ensure participant awareness is to only release data into open access databases when participants can directly exhibit adequate understanding. For example, the Personal Genome Project, which aims to create a publicly available database of genomic and health information with no expectations of privacy, requires participants to correctly answer all questions in an enrollment examination prior to being allowed to participate, although they may retake the examination multiple times [27]. Similarly, in the TCRB, mentioned above, a subset ($n = 37$) of participants who had agreed to open access data sharing took part in an education session that described the difference between controlled and open access data release and the risks and benefits of each in a question and answer format with visual aids as appropriate. After completing the education session, participants were asked to take a survey, one aim of which was to assess understanding. We found that 73% of survey participants demonstrated adequate understanding, which we defined as (1) knowing that they agreed to open access data sharing; (2) knowing who could access data in an open access database; and (3) understanding the risk of discrimination associated with open access data sharing. Only data from those who had demonstrated adequate understanding were eligible for open access data release. We also assessed participants' risk tolerance and decisional conflict. Fifty-four percent of participants reported high risk tolerance, meaning that they (1) were comfortable sharing their genetic and health information with the general public; and (2) would still participate even if they knew someone would identify their genetic data. Using an adapted version of the decisional conflict scale [28], we found that 68% demonstrated low decisional conflict, meaning that they answered all six questions in a manner indicating that they had low decision uncertainty, no pressure from others, and high perceived effective decision making. Fourteen participants (38%) changed their consent and refused open access data sharing at the completion of the education session. Of the 23 participants (62%) who still agreed to open access data release, 19 had adequate understanding and were therefore eligible for participation. Data from those with high risk tolerance and low decisional conflict were prioritized for public release.

There is considerable debate elsewhere concerning the definition of adequate understanding in research and how best to measure it [29]. Some have focused on developing educational interventions, such as those mentioned above, while others have proposed simplified consent documents as a way to improve understanding [30]. In the PGP and the TCRB, extensive measures were taken to assess understanding and to release data only from those who met a predefined threshold of comprehension. This is time consuming and resource intensive and may not be feasible in all genomic studies. Additional research is needed to identify methods of measuring and improving understanding that are not only effective, but are also efficient, especially in the context of genomic research involving open access data sharing. This is particularly important because participants' right to withdraw from the research is necessarily limited by the inability to retrieve data that has been shared publicly. As these studies suggest, however, there is a subset of participants who understand the implications of open access data release and voluntarily agree to it.

2.2. Diverse Viewpoints

It is important to note that the participants in both the randomized trial of consent and the TCRB were primarily quite ill (sometimes with terminal disease), very trusting of their physicians, and highly motivated to participate in research. Even among this group, however, there was diversity of perspectives about open access data sharing. In the randomized trial of consent, for example, Hispanic, unmarried, and more educated participants were all less likely to choose public data release, as were parents who were making decisions about the release of their child's data [21].

Other populations may exhibit even more variation in their perspectives on data sharing. For example, Lemke *et al.* [31] explored public and biobank participants' attitudes toward genomic research and data sharing via focus groups. While different levels of data sharing (*i.e.*, open *versus* controlled access) were not specifically examined in those studies, the investigators found more generally that there was wide variation in views on genomic data sharing, with some study participants more comfortable than others. Similarly, Trinidad *et al.* [32] conducted focus groups with research participants, surrogate decision-makers, and members of a health maintenance organization to investigate perspectives toward data sharing. They also found that perspectives varied, although they report that study participants were generally supportive of genomic data sharing for scientific benefit. In a commentary on conducting research with tribal communities in the U.S., Harding *et al.* [33] argue that special considerations that take into account the populations' perspectives are important when developing data sharing agreements with Native American tribes.

Our focus in this paper is data sharing in the context of the United States. Research participants in other parts of the world may feel differently about their genomic data and whether or not it should be shared for research purposes [34]. Thus, although generally reported as positive, participant perspectives on data sharing vary between populations, as well as among individuals, based on context, clinical circumstances, and personal values and beliefs.

3. Toward a More Balanced Approach

The variation in individual and group preferences for and understanding about genomic data sharing suggests that both mandatory public data release, as well as blanket restriction of access to genomic research data as a matter of policy, are misguided. Regulatory bodies in general tend to address this "heterogeneity problem" by taking the most restrictive and risk averse approach [35], which, in this case, inhibits choice by prohibiting the broader release of data from those who understand and are comfortable with open access sharing. It also reportedly impedes research [14], although studies quantifying the added benefit of open access *versus* controlled access data sharing are required. We advocate instead for a more balanced approach that allows for individual choice, but provides protection to participants by supporting adequate understanding as part of the informed consent process, and by strengthening accountability and protections against the misuse of available data.

Recent accounts demonstrate that some sophisticated patients are exercising their autonomy by sharing data themselves using existing platforms, such as social media, in order to facilitate discovery for rare and serious diseases [36]. If people are to share their own data, it is important that they are aware of the risk of identifiability and understand the challenge of obscuring segments of data in the context of public release [37]. For those whose data are shared within the research context, novel approaches have been suggested to give participants more control over decisions about who can access their data, as well as the ability to continue to manage such choices. For example, a relatively new platform called Reg4All [38] facilitates the sharing of health information in order to find relevant clinical trials, but also gives its users the ability to make finely-tuned choices about who can access their information or contact them. Others have introduced new approaches to consent that allow participants to be more nuanced in their choices, as well as change those choices over time [39], though, arguably, once data are released in an open access manner, there is no way to guarantee their removal from the public domain.

Increased participant engagement and open access data sharing could both be accomplished with modifications to the existing dbGaP model. As it is currently designed, all individual level genomic data in dbGaP is accessible only via controlled access [10]. The NIH could support more broad sharing by creating a publicly accessible segment of dbGaP that includes data from those who agree to open access data release. Participants could also be provided the option for open access data sharing in the informed consent document when agreeing to participate in NIH-funded genomic research. If a participant changes her consent over time, a request could be made to dbGaP to move the relevant data from the open access portion of the database to controlled access.

Regardless of mechanism, if genomic data are made publicly available, then the individuals from whom those data originate ought to be protected against the misuse of that information. One way of providing some protection for these participants could be the use of "click-through" data use agreements. In this model, the person accessing the data would have to read and agree to a list of conditions of use of the data, including agreeing to not attempt to identify the individuals from whom the data came. However, while this may require those accessing the data to recognize that attempting identification would be a violation of the use of the data, such click-through data use agreements are not enforceable, and as such, may not provide adequate protection.

There are existing laws in the United States that provide protection against misuse of genetic information. The vast majority of states have laws that govern the use of genetic information in health insurance and employment [40]. Likewise, the Genetic Information Nondiscrimination Act (GINA) [41], in effect as of 2009, makes it illegal for health insurers and employers with 15 or more employees to discriminate against people based on their genetic information. GINA has both corrective and monetary penalties that vary based on the intention and severity of the infraction. However, it does not protect against genetic discrimination in other types of insurance, such as long-term, disability, and life insurance, or any other realm outside of health insurance and employment. Additionally, some report not feeling fully protected by GINA, leading some to decline acceptance of DNA sequencing in both clinical and research-related contexts for fear of discrimination [42]. In contrast, the Human Tissue Act of the Parliament of the United Kingdom [43], which regulates activities with human bodies and tissues and also provides protection against the use of DNA without consent, is not limited to such contexts, and carries criminal penalties for violations that range from a fine to up to three years in prison. Though criminal law may not be the best approach to discourage the misuse of genetic data in the U.S., stricter penalties and broader protections against misuse of data by any third party may be needed to protect individuals who agree to share their data broadly for the public's benefit.

4. Conclusions

In the context of research, investigators have a professional obligation to be good stewards of the data with which research participants have entrusted them. In order to fulfill this obligation, we need policies that respect participant autonomy and maximize the utility of the data, alongside strengthened legislation that protects those participants from the misuse of their genomic information. The field has made great progress in the 10 years since the completion of the Human Genome Project. We must find ways to protect participants, yet avoid unneeded hindrances of researchers' access to genomic information.

Acknowledgments

This study was supported by the Cancer Prevention Research Institute of Texas (CPRIT) RP101353, and the National Human Genome Research Institute (NHGRI) HG004333 and HG003273.

Author Contributions

Stacey Pereira, Richard Gibbs, and Amy McGuire all contributed to writing, editing, and approving the final version of this paper.

References

1. Lander, E.S.; Linton, L.M.; Birren, B.; Nusbaum, C.; Zody, M.C.; Baldwin, J.; Devon, K.; Dewar, K.; Doyle, M.; FitzHugh, W.; *et al.* Initial sequencing and analysis of the human genome. *Nature* **2001**, *409*, 860–921.
2. Wellcome trust sanger institude. The human genome project. Available online: https://www.sanger.ac.uk/about/history/hgp/ (accessed on 15 July 2014).
3. National Institutes of Health, Department of Energy. NIH-DOE guidelines for access to mapping and sequencing data and material resources. Available online: http://www.genome.gov/10000925 (accessed on 15 July 2014).
4. National human genome research institute. Reaffirmation and extension of NHGRI rapid data release policies: Large-scale sequencing and other community resource projects. Available online: http://www.genome.gov/10506537 (accessed on 3 July 2014).
5. National human genome research institute. NHGRI policy on release of human genomic sequence data. Available online: http://www.genome.gov/10000910 (accessed on 3 July 2014).
6. Wellcome trust sanger institude. Summary of principles agreed upon at the international strategy meeting on human genome sequencing, bermuda. Available online: http://www.ornl.gov/sci/techresources/Human_Genome/research/bermuda.shtml#1 (accessed on July 2014).
7. Wellcome Trust. Sharing Data from Large-Scale Biological Research Projects: A System of Tripartite Responsibility, Fort Lauderdale. Available online: http://www.genome.gov/Pages/Research/WellcomeReport0303.pdf (accessed on 3 July 2014).
8. Lin, Z.; Owen, A.B.; Altman, R.B. Genomic research and human subject privacy. *Science* **2004**, *305*, 183.
9. McGuire, A.L.; Gibbs, R.A. No longer de-identified. *Science* **2006**, *312*, 370–371.
10. Mailman, M.D.; Feolo, M.; Jin, Y.; Kimura, M.; Tryka, K.; Bagoutdinov, R.; Hao, L.; Kiang, A.; Paschall, J.; Phan, L.; *et al.* The NCBI dbGaP database of genotypes and phenotypes. *Nat. Genet.* **2007**, *39*, 1181–1186.
11. Homer, N.; Szelinger, S.; Redman, M.; Duggan, D.; Tembe, W.; Muehling, J.; Pearson, J.V.; Stephan, D.A.; Nelson, S.F.; Craig, D.W.; *et al.* Resolving individuals contributing trace amounts of DNA to highly complex mixtures using high-density SNP genotyping microarrays.

PLoS Genet. **2008,** *4*, e1000167.
12. Gilbert, N. Researchers criticize genetic data restrictions. Available online: http://www.nature.com/news/2008/080904/full/news.2008.1083.html (accessed on 3 July 2014).
13. Gymrek, M.; McGuire, A.L.; Golan, D.; Halperin, E.; Erlich, Y. Identifying personal genomes by surname inference. *Science* **2013**, *339*, 321–324.
14. Rodriguez, L.L.; Brooks, L.D.; Greenberg, J.H.; Green, E.D. The Complexities of genomic identifiability. *Science* **2013**, *339*, 275–276.
15. National human genome research institute. Establishing a central resource of data from genome sequencing projects. Available online: http://www.genome.gov/27549169 (accessed on 15 July 2014).
16. Presidential commission for the study of bioethical issues. Privacy and progress in whole genome sequencing. Available online: http://bioethics.gov/node/764 (accessed on 15 July 2014).
17. Kohane, I.S.; Altman, R.B. Health-information altruists—A potentially critical resource. *N. Engl. J. Med.* **2005**, *353*, 2074–2077.
18. Oliver, J.M.; Slashinski, M.J.; Wang, T.; Kelly, P.A.; Hilsenbeck, S.G.; McGuire, A.L. Balancing the risks and benefits of genomic data sharing: Genome research participants' perspectives. *Publ. Health Genet.* **2012**, *15*, 106–114.
19. Williams, S.; Scott, J.; Murphy, J.; Kaufman, D.; Borchelt, R.; Hudson, K. The genetic town hall: Public opinion about research on genes, environment, and health. Available online: http://www.dnapolicy.org/pub.reports.php?action=detail&report_id=27 (accessed on 1 July 2014).
20. Kaufman, D.; Murphy, J.; Scott, J.; Hudson, K. Subjects matter: A survey of public opinions about a large genetic cohort study. *Genet. Med.* **2008**, *10*, 831–839.
21. McGuire, A.L.; Oliver, J.M.; Slashinski, M.J.; Graves, J.L.; Wang, T.; Kelly, P.A.; Fisher, W.; Lau, C.C.; Goss, J.; Okcu, M.; *et al.* To share or not to share: A randomized trial of consent for data sharing in genome research. *Genet. Med.* **2011**, *13*, 948–955.
22. Bergler, J.H.; Pennington, A.C.; Metcalfe, M.; Freis, E.D. Informed consent: How much does the patient understand? *Clin. Pharmacol. Ther.* **1980**, *27*, 435–440.
23. Joffe, S.; Cook, E.F.; Cleary, P.D.; Clark, J.W.; Weeks, J.C. Quality of informed consent in cancer clinical trials: A cross-sectional survey. *Lancet* **2001**, *358*, 1772–1777.
24. Robinson, J.O.; Slashinski, M.J.; Wang, T.; Hilsenbeck, S.G.; McGuire, A.L. Participants' recall and understanding of genomic research and large-scale data sharing. *J. Empir. Res. Hum. Res. Ethics* **2013**, *8*, 42–52.
25. Flory, J.; Emanuel, E. Interventions to improve research participants' understanding in informed consent for research: A systematic review. *JAMA* **2004**, *292*, 1593–1601.
26. Tamariz, L.; Palacio, A.; Robert, M.; Marcus, E.N. Improving the informed consent process for research subjects with low literacy: A systematic review. *J. Gen. Intern. Med.* **2013**, *28*, 121–126.
27. Ball, M.P.; Bobe, J.R.; Chou, M.F.; Clegg, T.; Estep, P.W.; Lunshof, J.E.; Vandewege, W.; Zaranek, A.; Church, G.M. Harvard personal genome project: Lessons from participatory public research. *Genome Med.* **2014**, *6*, 10.
28. O'Connor, A.M. Validation of a decisional conflict scale. *Med. Decis. Mak.* **1995**, *15*, 25–30.
29. Joffe, S.; Cook, E.F.; Cleary, P.D.; Clark, J.W.; Weeks, J.C. Quality of informed consent: A new measure of understanding among research subjects. *J. Natl. Cancer Inst.* **2001**, *93*, 139–147.
30. Beskow, L.M.; Friedman, J.Y.; Hardy, N.C.; Lin, L.; Weinfurt, K.P. Developing a simplified

consent form for biobanking. *PLoS One* **2010**, *5*, e13302.

31. Lemke, A.A.; Wolf, W.A.; Hebert-Beirne, J.; Smith, M.E. Public and biobank participant attitudes toward genetic research participation and data sharing. *Publ. Health Genet.* **2010**, *13*, 368–377.
32. Trinidad, S.B.; Fullerton, S.M.; Bares, J.M.; Jarvik, G.P.; Larson, E.B.; Burke, W. Genomic research and wide data sharing: Views of prospective participants. *Genet. Med.* **2010**, *12*, 486–495.
33. Harding, A.; Harper, B.; Stone, D.; O'Neill, C.; Berger, P.; Harris, S.; Donatuto, J. Conducting research with tribal communities: Sovereignty, ethics, and data-sharing issues. *Environ. Health Perspect.* **2012**, *120*, 6–10.
34. Hart, S.; Muenke, M. Genetics and genomic medicine around the world. *Mol. Genet. Genomic Med.* **2014**, *2*, 1–2.
35. Meyer, M.N. Regulating the production of knowledge: Research risk-benefit analysis and the heterogeneity problem. *Adm. Law Rev.* **2013**, *65*, 237–298.
36. Mnookin, S. One of a kind. Available online: http://www.newyorker.com/magazine/2014/07/21/one-of-a-kind-2 (accessed on 29 July 2014).
37. Nyholt, D.R.; Yu, C.E.; Visscher, P.M. On Jim Watson's APOE status: Genetic information is hard to hide. *Eur. J. Hum. Genet.* **2009**, *17*, 147–149.
38. Reg4ALL^{Beta}. Learn more. Available online: https://www.reg4all.org/more.php (accessed on 29 July 2014).
39. Kaye, J.; Whitley, E.A.; Lund, D.; Morrison, M.; Teare, H.; Melham, K. Dynamic consent: A patient interface for twenty-first century research networks. *Eur. J. Hum. Genet.* **2014**, doi:10.1038/ejhg.2014.71.
40. National conference of state legislatures. Genetics and health insurance state anti-discrimination laws. Available online: http://www.ncsl.org/research/health/genetic-nondiscrimination-in-health-insurance-laws.aspx (accessed on 15 July 2014).
41. Genetic information nondiscrimination act. Available online: https://www.govtrack.us/congress/bills/110/hr493/text# (accessed on 15 July 2014).
42. Peikoff, K. Fearing punishment for bad genes. Available online: http://www.nytimes.com/2014/04/08/science/fearing-punishment-for-bad-genes.html?hp?smid=fbnytimes&WT.z_sma=SC_FPF_20140408&bicmp=AD&bicmlukp=WT.mc_id&bicmst=1388552400000&bicmet=1420088400000&_r=4 (accessed on 8 July 2014).
43. Human Tissue Act 2004. Available online: http://www.legislation.gov.uk/ukpga/2004/30/contents (accessed on 15 July 2014).

12

Epigenetic Control of the Genome—Lessons from Genomic Imprinting

Bjorn T. Adalsteinsson and Anne C. Ferguson-Smith *

Department of Genetics, University of Cambridge, Cambridge CB2 3EH, UK;
E-Mail: bta26@cam.ac.uk

* Author to whom correspondence should be addressed; E-Mail: afsmith@gen.cam.ac.uk;

Abstract: Epigenetic mechanisms modulate genome function by writing, reading and erasing chromatin structural features. These have an impact on gene expression, contributing to the establishment, maintenance and dynamic changes in cellular properties in normal and abnormal situations. Great effort has recently been undertaken to catalogue the genome-wide patterns of epigenetic marks—creating reference epigenomes—which will deepen our understanding of their contributions to genome regulation and function with the promise of revealing further insights into disease etiology. The foundation for these global studies is the smaller scale experimentally-derived observations and questions that have arisen through the study of epigenetic mechanisms in model systems. One such system is genomic imprinting, a process causing the mono-allelic expression of genes in a parental-origin specific manner controlled by a hierarchy of epigenetic events that have taught us much about the dynamic interplay between key regulators of epigenetic control. Here, we summarize some of the most noteworthy lessons that studies on imprinting have revealed about epigenetic control on a wider scale. Specifically, we will consider what these studies have revealed about: the variety of relationships between DNA methylation and transcriptional control; the regulation of important protein-DNA interactions by DNA methylation; the interplay between DNA methylation and histone modifications; and the regulation and functions of long non-coding RNAs.

Keywords: Epigenetics; imprinting; gene expression; gene regulation; CTCF; long non-coding RNA; histone modifications; DNA methylation

---

1. A Primer on Epigenetics, DNA Methylation and Histone Modifications

Epigenetic modifications perform three main functions in mammalian cells: they contribute to the control of chromosome architecture ensuring stability and appropriate segregation of chromosomes during mitosis; they contribute to regulation of the silencing and inaccessibility of repetitive elements and endogenous retroelements; and they can initiate and maintain the activity and repression of individual genes or clusters of genes. Here we focus on the role of epigenetic modifications in the control of mammalian transcription and the contribution of genomic imprinting studies to our understanding of epigenetic mechanisms.

In mammals, the different cells that make up an organism generally contain the same DNA yet their cellular morphology and function can vary greatly. This is largely a result of differential gene expression, which is developmentally regulated and can then be maintained after repeated cell divisions. The maintenance of expression states/levels requires heritable information to be passed through cell division to ensure propagation in each daughter cell, and it is this information that has been termed epigenetic. Further, cells are subject to dynamic changes in gene expression, dependent, for example, on intrinsic and extrinsic cues, which can be mediated through epigenetic processes. Epigenetic mechanisms include DNA methylation and post translational modifications to core histones. Other related components have been proposed as epigenetic such as non-coding RNAs (ncRNAs) and nucleosomal positioning, however these might also be considered mediators and/or facilitators of epigenetic states. The characterization and mapping of genome-wide epigenetic modifications represent an ever increasing field of research. These studies are revealing genome-wide patterns of epigenetic regulation that not only have confirmed many of the conclusions suggested from more traditional experimental approaches in model systems but also allow for the generation of new hypotheses that await experimental testing. One model system that contributed a foundation for these studies is the process of genomic imprinting.

DNA methylation is a process whereby a methyl (CH_3) group is added most commonly to a cytosine in DNA. In mammals it is generally found at CpG dinucleotides and can be correlated with gene repression in a variety of ways (discussed in more detail below). CpG sites are generally depleted in the genome, apart from stretches of DNA called CpG islands where CpG density is high. CpG islands can be concentrated at gene promoters and are generally unmethylated. CpG sites outside CpG islands are generally methylated (reviewed in [1,2])—resulting in a genome-wide methylation pattern that can be described as roughly bimodal. Acquisition of DNA methylation is catalyzed by a family of DNA methyltransferases (DNMTs, reviewed in [3]). DNMT1 has affinity for hemi-methylated DNA and is responsible for maintaining methylation after DNA replication and DNMT3A and DNMT3B catalyze *de novo* DNA methylation while the DNA methyltransferase homologue, DNMT3L acts as a cofactor and has no methyltransferase activity.

Waves of DNA methylation loss and acquisition are orchestrated during embryonic development. After fertilization the two parental genomes are mostly stripped of their epigenetic marks, a process that presumably "resets" the genome to a naive state applicable for pluripotency (DNA methylation at certain sequences in imprinted loci are among few genomic regions to "escape" this demethylation, see details in Section 2). Around blastocyst implantation *de novo* methylation then occurs and, to our knowledge, no further genome wide erasure/acquisition waves occur in somatic cells. Another wave of genome-wide reprogramming occurs in primordial germ cells (this time DNA methylation at imprinted loci is also lost, see details in Section 2); erasure of DNA methylation commences in the embryonic germline after embryonic day 7.5 (E7.5) in the mouse and progressive *de novo* methylation follows at E12.5 in prospermatogonia of male embryos, but occurs after birth in oocytes of female embryos (reviewed in [4]). This germline epigenetic reprogramming is required for generating functional germ cells and failure to do this appropriately usually results in infertility or developmentally abnormal embryos that die during gestation [5,6].

Covalent post-translational modifications to core histones (histone modifications henceforth) can impact the conformation of the nucleosome-nucleosome architecture within chromatin and influence its function such that some modifications are associated with an active chromatin state and others with a repressive state (for extensive review refer to [7]). The full repertoire of histone modifications is unknown, but is complex, with some specific amino acid residues influencing the ability of others to be modified, and some sites having the potential to be modified in multiple different ways. It is currently unclear whether many of the modifications truly are epigenetically heritable in a replication-dependent manner, like DNA methylation. Lysine methylation and lysine acetylation are among the best characterized histone modifications whose correlations with gene activity and repression have been extensively studied. Furthermore, enzymes involved in "writing" and "erasing" these epigenetic marks have been identified and characterized; histone lysine methyltransferases deposit methyl groups to lysine, and histone lysine demethylases remove them. Histone acetyltransferases (KATs) and histone deacetylases (HDACs) deposit and remove acetyl groups, respectively. Generally, regions with acetylated histones are associated with gene activity and regions devoid of acetylated histones are repressed, while associations between histone methylation and gene transcription are more site specific; histone 3 lysine 4 (H3K4) and H3K36 methylation are for example found on expressed genes while H3K9 methylation is associated with repressed genes. Their distribution in the genome can be associated with certain genomic motifs, e.g., gene regions such as promoters or open reading frames (ORFs), or intergenic regions such as repeats. For example, H3K4me3 (me3 denotes tri-methylated) is found at the promoters of active genes, whilst H3K4me1 is associated with enhancers, H3K20me3 is found at repressed repeat regions, and H3K9me3 at promoters of repressed genes, retroelements, imprinted loci and at pericentromeric repeat regions.

2. Genomic Imprinting and Targeting DNA Methylation

Genomic imprinting is a process causing the mono-allelic expression of a specific subset of mammalian genes in a parental origin specific manner (reviewed in [8,9])—*i.e.*, genes that are expressed either from the paternally inherited chromosome or from the maternally inherited chromosome (paternal allele and maternal allele henceforth) are imprinted. The non-equivalence of parental genomes in mammals was discovered in 1984 [10,11], and individual imprinted genes were first discovered in 1991 (reviewed in [8]). Today, over 100 imprinted genes have been identified, most of which are organized in clusters and are regulated in a coordinated manner by a single imprinting control region (ICR) [9]. Most clusters contain at least one non coding gene and multiple protein coding genes, whose functions regulate embryonic development, placentation and a range of post-natal processes.

Epigenetic mechanisms allow the transcriptional machinery of the cell to distinguish the two parental chromosomes at imprinted loci and hence provide an important paradigm for understanding epigenetic control of gene activity and repression. Specifically, the discovery of differences in DNA methylation in the same place on the two parental chromosomes suggested the importance of epigenetic mechanisms in regulating imprinting [12,13] and the potential for epigenetic control in a wider context. The identification of imprinting control regions and their validation genetically as functional elements essential for the imprinting of multiple genes in *cis*, elucidated imprinting control. The loss of imprinting after targeted deletion of DNMT1 proved that DNA methylation was required for imprinting [14]. Importantly, in the absence of DNMT1, some imprinted genes were activated but others became repressed, an indication that methylation could impact activity as well as repression.

The acquisition of methylation at ICRs occurs in the germ line *de novo* by DNMT3A and DNMT3L with a small number of ICRs becoming methylated in sperm cells, and the majority acquiring methylation in oocytes—paternal and maternal ICRs, respectively. It is of interest that paternal ICRs are always located in intergenic regions while maternal ICRs are located at promoter sequences. Importantly, erasure of imprints occurs in the wave of demethylation that occurs in the primordial germ cells. However, in order to retain the memory of the parental origin that is subsequently established after that reprogramming, imprints must be retained during the post-fertilization epigenetic reprogramming phase [4]. Interestingly, other regions of the genome seem refractory to zygotic reprogramming [15] though these are not necessarily parent-specific or retained like imprints during development. The relationship, if any, of these regions to ICRs remains unclear. In addition to the ICR, other differentially methylated regions (DMRs) are located at some imprinted clusters, but a notable difference between ICRs and these DMRs is that differential methylation of the latter is not germline established, but rather is acquired post-fertilization. In all cases, these so-called secondary DMRs—to distinguish them from regions such as ICRs that acquire differential methylation in the germline—require the ICR for their establishment. The mechanisms through which ICRs control gene expression in their respective clusters are diverse and remain the subject of active research, including analysis of regulation by ncRNAs and of the relationships between DNA methylation and histone and non-histone proteins.

Both in imprinted and non-imprinted contexts, little is known about why certain DNA sequences become methylated and not others, or how this may change dynamically within a sequence such as a particular CpG island at a gene promoter. Most likely, it is a process that must be targeted in some manner. Targeting of the DNA methylation machinery has received much attention and efforts made to identify intrinsic sequence specificities of DNMTs and their cofactors. It has thus generally been assumed that the acquisition of methylation represents the "active" process in establishing differential methylation. However, recent studies on DMRs in the germlines and their propagation after fertilization suggest it might also be protection from DNA methylation and maintenance at methylated regions that determine differential methylation (Figure 1A, reviewed in [16]): Rather than appearing as discrete methylated sequences in otherwise unmethylated regions, maternal ICRs (which represent the vast majority of ICRs) are surrounded by methylation at both flanks. In contrast, these ICRs are unmethylated in sperm but are also flanked by methylation at surrounding sequences, suggesting that DNA methylation may be the "default" state and that it is protection from methylation at the ICRs, and perhaps other non-imprinted sequences as well, that establishes their differential methylation. Furthermore, in the germline, far more sequences are differentially methylated between oocytes and sperm than the ICRs; recent genome-wide studies suggest they are in the counts of thousands in oocytes and hundreds in sperm [15,17,18]. In contrast to ICRs these sequences generally lose methylation after fertilization, suggesting targeted maintenance of DNA methylation at specific sequences is essential for the germline-derived differential methylation of imprinted loci. Hence perhaps, loss of maintenance, in addition to active removal of DNA methylation at non-imprinted loci, contributes to the mechanism through which demethylation occurs in somatic cells. KRAB zinc finger proteins (ZFP) represent a family of over 350 tetrapod-specific genes whose functions remain poorly understood. They bind DNA and have previously been shown to recruit the repressive chromatin machinery in a site-specific manner. One of these KRAB-ZFPs, ZFP57, has been shown to be required to maintain the DNA methylation memory at imprints during post-fertilization reprogramming when the bulk of the genome is changing its epigenetic state [19]. ZFP57 binds methylated DNA and is thought to recruit methyltransferases to imprinting control regions hence preventing them from loss of their imprints.

3. DNA Methylation and Gene Repression—The Chicken or the Egg?

3.1. DNA Methylation Correlates with Repression

The correlation between DNA methylation and gene repression was noted in several experiments assaying viral and endogenous gene expression in mammalian, frog and sea urchin cells in the late 1970s and early 80s [20–30]. Experiments were conducted to determine whether the observed relationship was purely correlational, or whether DNA methylation functionally regulated gene expression. This was, however, challenging, but the strong evidence in many different contexts, showing that hypomethylated regions were associated with activity and hypermethylated regions refractory to transcription, suggested that absence of DNA methylation may be necessary though not sufficient for transcription. Vardimon *et al.* injected bacterial plasmids containing *in vitro* methylated or unmethylated DNA encoding a viral gene into frog oocyte nuclei [31]. They observed maintenance of the respective methylation states over a 24 h period, and expression of the gene in oocytes that were injected with unmethylated DNA but not in those that were injected with methylated DNA [31]. In a similar experiment, Stein *et al.* transfected *in vitro* methylated or unmethylated plasmids containing the *Aprt* (adenine phosphoribosiltransferase) gene into cultured *Aprt* null mouse cells. They observed maintenance of the respective *Aprt* methylation states after integration into the endogenous genome over several cell divisions for both unmethylated and methylated plasmids, and that integration of the unmethylated but not the methylated gene rescued the *Aprt* null phenotype, suggesting methylation of the gene was associated with inhibition of its transcription [32].

Correlations between gene expression and DNA methylation have been assessed at CpG sites across whole chromosomes or the whole genome. Consistent with the earlier studies, DNA methylation of promoter sequences, though rare at CpG island promoters, was observed to correlate with gene repression [33–35]. The functional role of DNA methylation in repressing gene expression is further suggested by results from studies in which the genes encoding the DNA methyltransferases are deleted conditionally in various cell lineages. Generally, the loss of DNMTs results in dysregulation of multiple genes, with a trend towards gene activation rather than silencing, again suggesting that DNA methylation represses gene expression (reviewed in [36]). Furthermore, treatment of cells *in vivo* with the DNA methyltransferase inhibitor 5-Azacytidine was shown to result in gene activation in several experiments in the 1980s, with concomitant loss of DNA methylation (reviewed in [37]). Together all these findings have led to the general assumption that loss and acquisition of DNA methylation at a gene promoter results in gene activation and silencing, respectively, but none actually proved that the acquisition of DNA methylation itself causes the gene silencing in all contexts.

3.2. DNA Methylation as a Consequence of Transcriptional Silencing

Studies of the temporal onset of mono-allelic expression of imprinted genes and the acquisition of differential methylation at secondary DMRs during mouse development indicate that DNA methylation can be acquired *after* gene repression (Figure 1B). The imprinted genes *Gtl2*, *Cdkn1C*, *H19* and *Igf2r* each contain a secondary DMR in their promoters, which become differentially methylated days *after* their mono-allelic expression is observed (summarized in [38]). Generally, mono-allelic expression of these genes is initiated around the morula or blastocyst stage (E3.5-4.5), while differential methylation of the respective secondary DMR occurs after E6.5 [13,38–42]. In the most extreme case, *Igf2r* is mono-allelically expressed from the maternal allele from E6.5 onward but the silent paternal allele only becomes methylated at or after E15.5 [13,42]. It is reasonable to assume that this temporal relationship, where methylation is acquired as a consequence of gene repression, also applies to non-imprinted genes (Figure 1B). In particular, is has recently been shown that DNA methylation levels are secondary to the binding of transcription factors; Stadler *et al.* [43] identified multiple clusters of CpG sites that have low to intermediate levels of methylation, 10%–50%, in mouse embryonic stem (ES) cells. These low methylated regions (LMRs) are likely distal regulatory regions, and are bound by various transcription factors. Scrambling binding sites for the insulator protein CTCF or knocking out the transcription factor REST led to increased methylation at the LMRs. Furthermore, reintroduction of REST into the *REST*$^{-/-}$ cells reverted the methylation status of the LMRs to the normal low levels [43]. These findings suggest DNA methylation may not have a direct role in silencing gene expression in all situations. In such cases DNA methylation might rather be acquired after gene silencing to maintain the repressed state or as a secondary readout of other mechanisms of genome control. Nonetheless, there are situations where acquisition of DNA methylation unquestionably does regulate gene expression, notably at the germline DMRs of imprinted genes [1,8,9,14–16].

4. How Does DNA Methylation Confer Effects on Gene Expression?

4.1. Proteins Attracted and Repelled

In situations where DNA methylation does indeed direct gene repression there are currently two model mechanisms that are generally acknowledged [1,44]: First, DNA methylation can attract proteins that bring about gene repression through recruitment of chromatin modifiers. A group of proteins, collectively referred to as methyl binding proteins (MBPs) have been characterized and shown to specifically bind to methylated, but not unmethylated, DNA [44–49]. MBPs are known to interact with histone modifiers such as HDACs, e.g., in forming complexes, such as the nucleosome remodeling deacetylase (NuRD) complex, which through their histone deacetylase activity and subsequent chromatin condensation bring about gene repression [50–55]. Secondly, certain proteins may interact with DNA in a methylation dependent manner. Here, DNA methylation may be refractory to the binding of proteins, such as transcription factors or other regulatory proteins [56–58], that are necessary for gene expression (Figure 1C). For this latter model, the best characterized example is the regulation of CTCF binding at the imprinted H19/Igf2 cluster *via* differential DNA methylation on the two parental alleles (reviewed in [1]).

4.2. Regulation of CTCF Binding at the H19/Igf2 Imprinted Cluster; the Insulator Mechanism

In the H19/Igf2 imprinted cluster, the protein coding gene *Igf2* is expressed from the paternally inherited allele [59]. This expression pattern is dependent on the regional ICR [60], on its differential methylation [12,14,61] and on the insulator protein CTCF binding to the ICR. On the unmethylated maternal allele, CTCF can bind, while its binding is inhibited on the methylated paternally inherited chromosome [62–65]—thus CTCF binding to DNA is methylation-sensitive (Figure 1C). *Igf2* and a downstream non-coding RNA gene, *H19,* share enhancers that are located at the 3' end of *H19* [66,67] and the parental specific expression of *Igf2* and *H19* are ultimately determined by interaction with these sequences; on the paternally inherited chromosome, *Igf2*-enchancer interaction is possible and the gene is expressed. On the maternally inherited chromosome this contact is blocked by CTCF binding to the ICR and this facilitates enhancer interaction with a now active *H19* instead, and also results in *Igf2* repression.

What is the mechanism of CTCF's enhancer blocking activity? The current model (reviewed in [68]) suggests that in the H19/Igf2 cluster, chromatin loop formation on the maternal allele spatially inhibits enhancer interaction with *Igf2.* The process appears to depend on three elements; dimerization, CTCF binding to more than one region and physical contact between these neighboring sites via CTCF interaction [69–74]. The model suggests that on the unmethylated maternally inherited chromosome, CTCF binds to the ICR and also to an upstream somatic DMR located 5' of *Igf2*. Binding does not occur at the paternal allele where methylation inhibits the binding. On the maternal allele ICR-DMR contact is made possible by CTCF dimerization bringing together the two distinct loci, and because they flank *Igf2*, the gene is 'looped out' (Figure 1D). Further chromatin contacts within the cluster, some facilitated by CTCF, then result in physical separation between the *Igf2* loop and the enhancers. Recently cohesins have been shown to bind to over half of CTCF binding sites in the genome, including in the H19/Igf2 cluster [75]. Given the ability of cohesins to tether DNA strands (*i.e.*, sister chromatids after cell's S-phase) it is possible that cohesins contribute mechanistically to these chromatin contacts on the maternal H19/Igf2 locus. On the paternal allele, where CTCF cannot bind, long-range chromatin interactions are not observed within the cluster, suggesting a state that allows interaction between the 90 kb distant enhancers and *Igf2* (Figure 1D) [73]. Similar interactions involving CTCF have been noted at other loci (Figure 1D).

5. Relationship between DNA Methylation and Histone Modifications

Similar to DNA methylation, correlation between multiple histone modifications in various genomic elements, including promoters, have been associated with gene activity and repression, and early studies illustrating this indeed investigated the relationship in the context of imprinted loci [76–82]. A functional relationship may therefore potentially exist between DNA methylation and histone modifications whereby the acquisition of one may be dependent on, or mutually exclusive, with the other. Indeed, as noted above, MBPs can recruit histone modification enzymes. Well-defined examples of histone modifications that regulate *de novo* DNA methylation are however scarce [83,84]. One very compelling example again comes from the study of genomic imprinting, as discussed below.

DNMT3L lacks a DNA methyltransferase activity, but it is necessary for methylation of DNA in certain situations [85,86] because it forms a complex with DNMT3A and DNMT3B, impacts their activity and contributes to their structural interaction with chromatin [87–90]. The ability is likely a result of a recently discovered affinity of DNMT3L to histone H3 [90] and this interaction is dependent on the methylation state of H3 at lysine K4—the binding only occurs when the histone is unmethylated hence H3 methylation might shield from DNA methylation [90]. A functional role for H3K4 methylation in modulating DNA methylation came from an imprinting study where Ciccone *et al.* showed that this interaction has important regulatory implications. The group generated mice deficient for a H3K4 demethylase enzyme, KDM1B, which resulted in increased H3K4 methylation in oocytes, where KDM1B is almost exclusively expressed. Consistent with inhibition of the DNMT3L-DNMT3A complex binding to methylated histone H3, DMRs at four imprinted regions that normally acquire DNA methylation in the female germ line were unmethylated in the *Kdm1b* null oocytes and imprinted expression of the corresponding genes was lost in embryos from *Kdm1b* null females (Figure 1E) [91]. These results strongly suggest a functional link between loss of H3K4 methylation and acquisition of DNA methylation, at least at imprinted regions (Figure 1E).

Cedar and Bergman take this further proposing a model of how the bimodal methylation pattern of mammalian genomes may be dependent on this same relationship. They suggest that *de novo* DNA methylation at the blastocyst stage is prevented at particular loci by deposition of H3K4 methylation. They further suggest that H3K4 methyltransferases may be targeted to CpG islands by RNA polymerase II and as a consequence, the DNA methyltransferase machinery containing DNMT3L, cannot access CpG sites in regulatory regions that are CpG islands [84].

H3K9 di- and trimethylation is associated with repressive DNA. DNA methylation is often found at such regions. Furthermore, DNA is globally hypomethylated in mouse ES cells carrying deletion of a H3K9 methyltransferase, G9a [92]. In this case the loss of DNA methylation is not a result of the aberrantly low levels of histone methylation, but rather due to loss of the histone methyltransferase enzyme itself; the DNA methyltransferase machinery interacts with G9a, and this interaction is mediated through a protein domain that is independent of the histone methyltransferase catalytic activity by a SET protein domain. Therefore, in *G9a*$^{-/-}$ mouse ES cells carrying *G9a* transgenes that lack histone methyltransferase activity, e.g., due to a point mutation in the SET domain, DNA methylation levels are partially rescued [93,94]. Regulation of DNA methylation through interaction of the DNMTs with histone modifiers, rather than with the histone modifications themselves, seems to be common and is observed for multiple mammalian histone methyltransferases [95–97], as well as in plant [98] and fungal systems [99]. Interestingly, in *G9a*$^{-/-}$ ES cells DNA methylation is lost at some imprinted loci [94,100], but where tested this is not observed in embryos [100,101]. This behaviour at imprints may suggest that ES cell culture is not a faithful model for assessing a requirement for histone modifying enzymes in DNA methylation, but equally might also reflect different properties of imprint-specific maintenance in ES cells compared to *in vivo*.

Figure 1. Regulatory epigenetic phenomena at imprinted loci. On the left are examples of various epigenetic mechanisms as observed in imprinted loci, and on the right models are presented of how those principles may apply more generally.

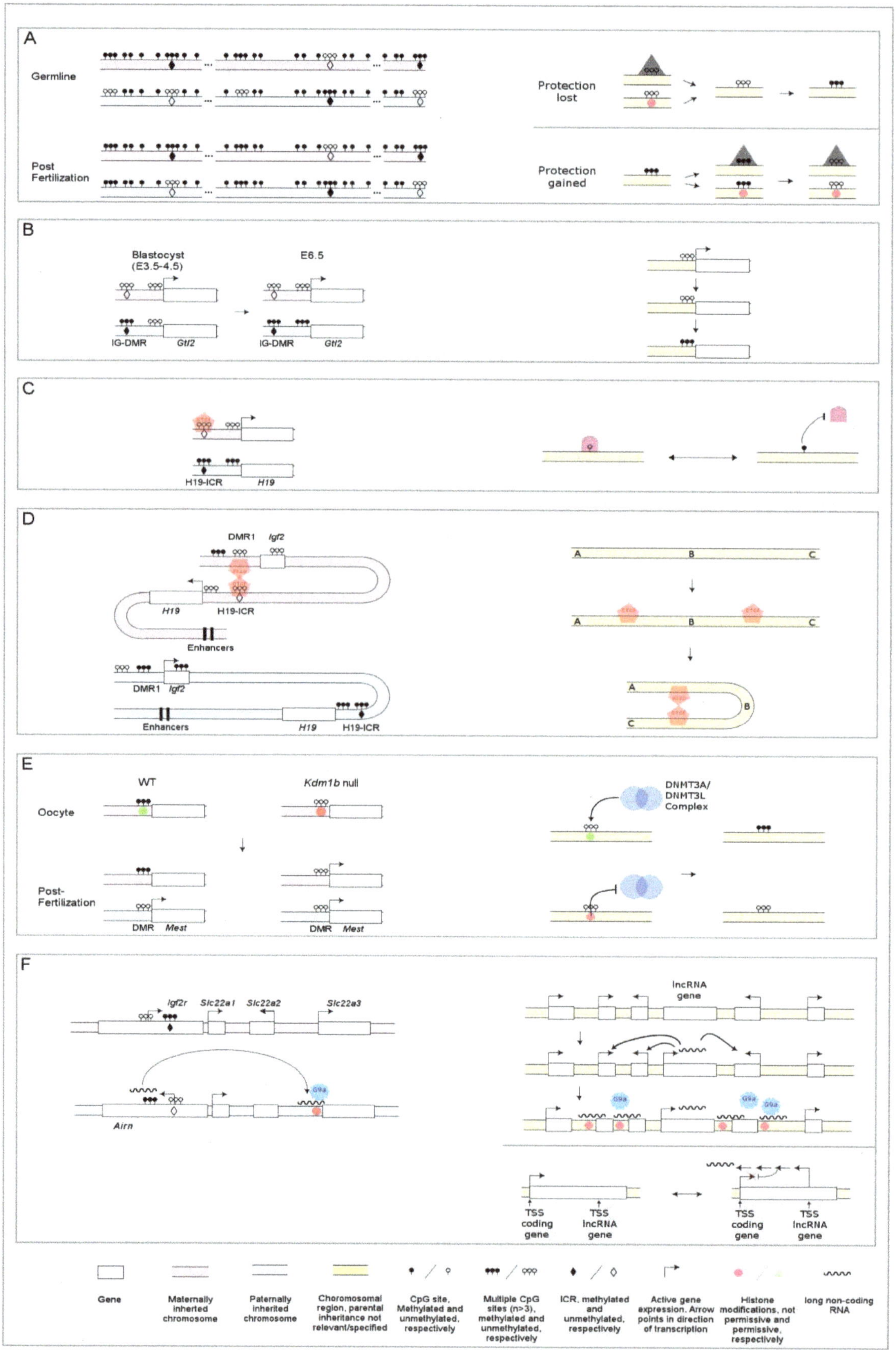

(**A**) left: In the male germline (sperm), CpG dense regions are generally unmethylated and less dense regions are methylated. In the female germline (oocytes) CpG rich regions are more frequently methylated. This results in multiple differentially methylated regions between the male and female germlines. After fertilization only a small subset of these regions retain differential methylation. Retention of differential methylation at imprinting control regions (diamonds) post-fertilization may therefore be a targeted protection from either demethylation or *de novo* methylation; right: Model; Changes in DNA methylation may be mediated through loss and gain of such protection—when protection is lost (e.g., upper—as a result of factor (black triangle) binding or a histone modification that is non-permissive (red circle) for DNMT binding) CpGs become methylated by the methyltransferase machinery. If protection is gained (lower) the machinery cannot access the CpG sites to maintain methylation and after cell divisions methylation is therefore lost; (**B**) left: The secondary DMR located in the promoter region of the imprinted gene *Gtl2* becomes methylated on the paternal allele after expression is silenced; right: Model; DNA methylation at CpG sites in promoter regions of non-imprinted genes may therefore, at least in some cases, occur after gene silencing; (**C**) left: In the H19/Igf2 imprinted locus, CTCF (red pentagon) binds the H19-ICR, on the unmethylated maternal allele, not the methylated paternal allele; right: Model; Methylation of CpG sites can inhibit protein binding (purple) to DNA; (**D**) left: In the H19-Igf2 imprinted locus CTCF (red pentagon) binds to regions flanking *Igf2* and dimerizes, looping the gene and physically inhibiting its interaction with distal enhancers. On the paternally inherited allele, CTCF does not bind and enhancers are in contact with *Igf2* and the gene is expressed; right: Model; Looping of DNA sequences through the action of CTCF (red pentagon) can separate regions or bring them into contact; (**E**) left: In *Kdm1b*$^{-/-}$ (histone methyltransferase) mouse oocytes, imprints are not established at multiple ICRs due to the inhibitory effect of H3K4 methylation on DNMT3L. Histone modification states in WT and *Kdm1b*$^{-/-}$ mice are depicted as green and red circles to signify permissive and non-permissiveness to *de novo* DNA methylation, respectively. In embryos from *Kdm1b*$^{-/-}$ mothers, imprinted expression is lost, and genes are biallelically expressed (*Mest,* depicted) or repressed; right: Model; Histone modifications (red and green circles) can regulate DNA methylation; (**F**) left: In the Igf2r imprinted locus *Igf2r* expression is inhibited by transcriptional interference from the *Airn* lncRNA transcript on the paternal allele. The lncRNA recruits histone modifiers such as G9a (blue) to proximal imprinted genes that contribute to silencing of the imprinted *Slc22a3* in a lineage specific manner, e.g., through deposition of histone marks that are non-permissive for transcription (red circle); upper right: Model; lncRNAs may exert their effects in *trans* at proximal genes. As illustrated, a lncRNA is expressed and silences proximal genes, but not the more distal genes; lower right: Model; In the example provided a lncRNA and a coding gene are expressed from within the same ORF. Transcription of the lncRNA inhibits expression of the coding gene.

6. lncRNAs

6.1. lncRNAs, Definition, Characterization and Potential Functions

In recent years the roles of long noncoding RNAs (lncRNAs) in regulating genome function have received considerable attention, and are now emerging as a large group of genes with potential functions of fundamental importance for cell biology (for review see [2,102–104]). lncRNAs are defined as noncoding RNA transcripts of >200 bp [104]. Transcription of lncRNAs resembles that of mRNA genes; they are transcribed by the same transcriptional machinery and by RNA polymerase II, the transcripts are 5' capped and can be spliced and shuttled to the cytoplasm [102]. The lack of an open reading frame and their size are therefore the only criteria that currently define lncRNAs as a group [102,104]. On basis of high-throughput RNA sequencing experiments, the numbers of lncRNA transcripts have been suggested to range between 5000–15,000 [105,106]. With higher sensitivity, targeted capture experiments have identified lncRNAs that are undetectable by high-throughput technology, suggesting that this range is an underestimate [107]. However, as a result of their loosely defined criteria, lncRNAs as a group may be very heterogeneous. Therefore, the functional roles discussed below may only apply to a subset of their estimated numbers.

Despite the current excitement surrounding "new" roles for lncRNAs, they were shown to regulate genomic imprinting over a decade ago. Multiple potential functions of lncRNAs have been proposed whereby lncRNAs either exert their effects by acting in *trans* or by the act of their transcription in *cis* (transcriptional interference). Both effects have been shown to act at imprinted loci. Some *trans* acting lncRNAs, such as HOTAIR, have been suggested to exert their effects throughout the genome [108], while others, including most imprinted lncRNAs defined to date, act over a limited area surrounding or close to their transcriptional origin. Some lncRNAs may utilize both *cis* and *trans* acting mechanisms. An example is the imprinted Airn lncRNA whose transcription on the paternal chromosome represses *Igf2r* expression in *cis* by transcriptional interference [109], while the Airn RNA molecule itself is also necessary for regulating other genes in the cluster in a *trans*-targeted manner (see below, and Figure 1F [110]). Transcriptional interference is proposed to occur as a result of a collision between the transcriptional machineries of two adjacent or overlapping transcripts which might result in termination of one or both transcriptional events. Alternatively it may occur by promoter occlusion via inhibition of formation of a transcriptional initiation complex due to existing transcription of one transcript through the promoter of another [111].

Functions of *trans* acting lncRNAs have been proposed to fall into the following categories [112]: (1) Decoys: lncRNAs that bind to DNA binding proteins and prevent their interaction with DNA; (2) Scaffolds: lncRNA that function to join two or more proteins into an lncRNA-RNP (ribonucleoprotein) complex; (3) Guides: lncRNAs that bind proteins to guide them to certain genomic locations, e.g., by lending them specificity and/or binding capacity to certain DNA sequences or chromatin states.

6.2. lncRNAs in the Epigenetic Control of Genome Function—Lessons from Imprinting

Every cluster of imprinted genes contains at least one lncRNA and these lncRNAs are regulated by DNA methylation. This was demonstrated in experiments where the genes encoding DNA methyltransferases were deleted in mice to gauge effects on imprinting regulation. Promoters for the *Airn*, *Nespas/Gnasxl*, *Snrpn* and *Kcnq1ot1* lncRNA genes lie within the ICR for their respective region and are differentially methylated on the two parental chromosomes. Upon loss of DNMT1, the maintenance methyltransferase, methylation is lost at these ICRs in E10.5 embryos (the genetic manipulation is lethal at later embryonic stages) and *Airn, Nespas/Gnasxl*, *Snrpn* and *Kcnq1ot1* are biallelically expressed, with effects on neighboring imprinted protein coding genes, some of which may lose imprinting as a result of the lncRNA dysregulation [14,113,114]. *Kcnq1ot1* and *Airn* promoters are located in the ICRs, exhibit differential methylation, and, importantly, are located within genes running antisense to them, hence these provide examples of critical regulatory DNA methylation at genomic regions considered by some to have little or no consequence, *i.e.*, intragenic. The existence of other epigenetically regulated elements within genes and acting in this way to potentially regulate lncRNAs, may have very widespread effects on genome function.

Furthermore, imprinted lncRNAs have been demonstrated to be necessary for epigenetic control of genome function, to guide chromatin modifying enzymes in *trans* to specific sites in the genome. This is thought to mediate changes in histone modifications and be associated with changes in transcriptional activity. Although challenging to address experimentally, this function for lncRNAs is currently the topic of much attention. It was studies on imprinted gene regulation at the Igf2r/Airn and Kcnq1/Kcnq1ot1 imprinted clusters that provided examples of this type of regulation [104]. *Kcnq1ot1* and *Airn* are estimated as greater than 100 kb lncRNA transcripts, transcribed in an antisense orientation from within protein coding genes; *Kcnq1ot1* from *Kcnq1* in a 1 Mb imprinted cluster that contains eight maternally expressed protein coding genes, and *Airn* from *Igf2r* in a 400 kb long imprinted cluster that contains three maternally expressed protein coding genes. Both transcripts generate unspliced lncRNAs that are localized in the nucleus [103,115,116]. The ICRs of both genes are methylated on the maternal, but not paternally inherited chromosome, and determine monoallelic expression of the lncRNAs from the paternal allele. In mouse genetic mutants, where the promoters of *Airn* and *Kcnq1ot1* are deleted or their transcripts truncated by insertion of premature polyA sequence into the endogenous genes, biallelic expression of the imprinted protein coding genes occurs within their respective clusters [110,117–119]. These results suggested that the lncRNAs or the act of their transcription is necessary for silencing of genes in *cis* (Figure 1F). In addition, several lines of evidence further indicate that lncRNAs guide chromatin modifying enzymes in *trans* to establish repressive histone marks and gene silencing on the paternal allele (Figure 1F): In the Kcnq1 imprinted cluster *Osbpl5*, *Cd81*, *Ascl2* and *Tscc4* are imprinted exclusively in the placenta [40] and so are *Slc22a2* and *Slc22a3* in the Igf2r/Airn cluster [120]. The paternal chromosomes are bound by the histone methyltransferases G9a and/or Ezh-Eed2 in the extraembryonic lineage [40,110], and both Airn and Kcnq1ot1 lncRNAs associate with G9a histone methyltransferase in a lineage specific manner—in placenta but not embryo [110,115]. These results showed that lncRNAs may be a contributing factor for targeting epigenetic marks (Figure 1F) with genetic models being used alongside biochemical approaches to generate a more tractable and comparable experimental paradigm for added robustness. These studies have paved the way for explorations of the roles of multiple other lncRNAs which are found in association with different chromatin modifying enzymes [121,122]. Most recently, the imprinted lncRNA Gtl2/Meg3 has been shown to function in *trans* to target polycomb regulatory complexes in mouse and human stem cells in culture [123].

7. Conclusions

The robust genetic approaches applied to the regulation of imprinting have allowed it to be an excellent hypothesis-driven model to investigate and understand the epigenetic control of genome regulation. One of its greatest strengths as a model is that it allows the comparison of differentially expressed alleles of the two inherited copies of a gene with identical sequence within the same cell. Because these two parentally inherited alleles have well-defined different epigenetic states the contributions of these to gene expression can be determined. Since imprinted clusters employ multiple different epigenetic mechanisms, acting through various different mediators (long non-coding RNA, CTCF, *etc.*), this has enabled investigators to explore their hierarchical interactions and relationships with one another. As evidenced by the examples presented here, imprinting has provided insight into some of the most fundamental aspects of a range of epigenetic phenomena and their mediators.

Nevertheless, many important aspects of imprinting and epigenetic control remain to be elucidated. These include: what allows epigenetic marks to be *de novo* targeted differently in the male and female germlines; whether they are modulated by extrinsic or intrinsic signals, for example in the context of development and disease; and how DNA methylation is actively removed during reprogramming and perhaps at other times in development. The mechanisms regulating some of these processes are beginning to emerge where the context of imprinting has contributed; the DNA binding proteins ZFP57 and PGC/Stella have been shown to target and maintain DNA methylation at imprinted clusters during postfertilisation epigenetic reprogramming [19,124] and selective loss of imprinting is necessary for stem cell regulation in the neurogenic niche of the developing mouse [125]. Whether we can apply more generally what we learn from these mechanisms—for example about the general targeting of epigenetic states or the dynamic changes in epigenetic state in specific cellular niches—remains to be determined. It is likely that future studies, addressing these and other similarly fundamental questions in the context of imprinting will continue to add new layers to our understanding of genome regulation and the epigenetic control of genome function more widely.

Acknowledgments

We thank previous and current members of the Ferguson-Smith lab for their contribution to the ideas presented here and to Carol Edwards and Lisa Neumann for critical reading of the manuscript. Bjorn Adalsteinsson is funded by the Peoples Programme (Marie Curie Actions) of the European Union's Seventh Framework Programme FP7/2007-2013/ under REA grant agreement No.: 290123.

Author Contributions

Bjorn Thor Adalsteinsson wrote the manuscript. Anne C. Ferguson-Smith conceived, wrote and edited the manuscript.

References

1. Miranda, T.B.; Jones, P.A. DNA methylation: The nuts and bolts of repression. *J. Cell. Physiol.* **2007**, *213*, 384–390.
2. Gibney, E.R.; Nolan, C.M. Epigenetics and gene expression. *Heredity (Edinb).* **2010**, *105*, 4–13.
3. Goll, M.G.; Bestor, T.H. Eukaryotic cytosine methyltransferases. *Annu. Rev. Biochem.* **2005**, *74*, 481–514.
4. Smallwood, S.A.; Kelsey, G. *De novo* DNA methylation: A germ cell perspective. *Trends Genet.* **2012**, *28*, 33–42.
5. Kaneda, M.; Okano, M.; Hata, K.; Sado, T.; Tsujimoto, N.; Li, E.; Sasaki, H. Essential role for *de novo* DNA methyltransferase Dnmt3a in paternal and maternal imprinting. *Nature* **2004**, *429*, 900–903.
6. Gu, T.-P.; Guo, F.; Yang, H.; Wu, H.-P.; Xu, G.-F.; Liu, W.; Xie, Z.-G.; Shi, L.; He, X.; Jin, S.; *et al*. The role of Tet3 DNA dioxygenase in epigenetic reprogramming by oocytes. *Nature* **2011**, *477*, 606–610.

7. Berger, S.L. The complex language of chromatin regulation during transcription. *Nature* **2007**, *447*, 407–412.
8. Ferguson-Smith, A.C. Genomic imprinting: The emergence of an epigenetic paradigm. *Nat. Rev. Genet.* **2011**, *12*, 565–575.
9. Edwards, C.A.; Ferguson-Smith, A.C. Mechanisms regulating imprinted genes in clusters. *Curr. Opin. Cell Biol.* **2007**, *19*, 281–289.
10. McGrath, J.; Solter, D. Completion of mouse embryogenesis requires both the maternal and paternal genomes. *Cell* **1984**, *37*, 179–183.
11. Surani, M.A.H.; Barton, S.C.; Norris, M.L. Development of reconstituted mouse eggs suggests imprinting of the genome during gametogenesis. *Nature* **1984**, *308*, 548–550.
12. Ferguson-Smith, A.C.; Sasaki, H.; Cattanach, B.M.; Surani, M.A. Parental-origin-specific epigenetic modification of the mouse *H19* gene. *Nature* **1993**, *362*, 751–755.
13. Stöger, R.; Kubicka, P.; Liu, C.-G.; Kafri, T.; Razin, A.; Cedar, H.; Barlow, D.P. Maternal-specific methylation of the imprinted mouse *Igf2r* locus identifies the expressed locus as carrying the imprinting signal. *Cell* **1993**, *73*, 61–71.
14. Li, E.; Beard, C.; Jaenisch, R. Role for DNA methylation in genomic imprinting. *Nature* **1993**, *366*, 362–365.
15. Smallwood, S.A.; Tomizawa, S.; Krueger, F.; Ruf, N.; Carli, N.; Segonds-Pichon, A.; Sato, S.; Hata, K.; Andrews, S.R.; Kelsey, G. Dynamic CpG island methylation landscape in oocytes and preimplantation embryos. *Nat. Genet.* **2011**, *43*, 811–814.
16. Kelsey, G.; Feil, R. New insights into establishment and maintenance of DNA methylation imprints in mammals. *Phil. Trans. R. Soc. B* **2013**, *368*, doi:10.1098/rstb.2011.0336.
17. Kobayashi, H.; Sakurai, T.; Imai, M.; Takahashi, N.; Fukuda, A.; Yayoi, O.; Sato, S.; Nakabayashi, K.; Hata, K.; Sotomaru, Y.; *et al.* Contribution of intragenic DNA methylation in mouse gametic DNA methylomes to establish oocyte-specific heritable marks. *PLoS Genet.* **2012**, *8*, e1002440.
18. Smith, Z.D.; Chan, M.M.; Mikkelsen, T.S.; Gu, H.; Gnirke, A.; Regev, A.; Meissner, A. A unique regulatory phase of DNA methylation in the early mammalian embryo. *Nature* **2012**, *484*, 339–344.
19. Li, X.; Ito, M.; Zhou, F.; Youngson, N.; Zuo, X.; Leder, P.; Ferguson-Smith, A.C. A maternal-zygotic effect gene, *Zfp57*, maintains both maternal and paternal imprints. *Dev. Cell* **2008**, *15*, 547–557.
20. Groudine, M.; Eisenman, R.; Weintraub, H. Chromatin structure of endogenous retroviral genes and activation by an inhibitor of DNA methylation. *Nature* **1981**, *292*, 311–317.
21. Sutter, D.; Doerfler, W. Methylation of integrated adenovirus type 12 DNA sequences in transformed cells is inversely correlated with viral gene expression. *Proc. Natl. Acad. Sci. USA* **1980**, *77*, 253–256.
22. Desrosiers, R.C.; Mulder, C.; Fleckenstein, B. Methylation of *Herpesvirus saimiri* DNA in lymphoid tumor cell lines. *Proc. Natl. Acad. Sci. USA* **1979**, *76*, 3839–3843.
23. Cohen, J.C. Methylation of milk-borne and genetically transmitted mouse mammary tumor virus proviral DNA. *Cell* **1980**, *19*, 653–662.
24. Guntaka, R.V.; Rao, P.Y.; Mitsialis, S.A.; Katz, R. Modification of avian sarcoma proviral DNA sequences in nonpermissive XC cells but not in permissive chicken cells. *J. Virol.* **1980**,

34, 569–572.

25. Van der Ploeg, L.H.T.; Flavell, R.A. DNA methylation in the human globin locus in erythroid and nonerythroid tissues. *Cell* **1980**, *19*, 947–958.
26. McGhee, J.D.; Ginder, G.D. Specific DNA methylation sites in the vicinity of the chicken beta-globin genes. *Nature* **1979**, *280*, 419–420.
27. Kuo, M.T.; Mandel, J.L.; Chambon, P. DNA methylation: Correlation with DNase I sensitivity of chicken ovalbumun and conalbumin chromatin. *Nucleic Acids Res.* **1979**, *7*, 2105–2113.
28. Mandel, J.L.; Chambon, P. DNA methylation: Organ specific variations in the methylation pattern within and around ovalbumin and other chicken genes. *Nucleic Acids Res.* **1979**, *7*, 2081–2103.
29. Bird, A.P.; Taggart, M.H.; Smith, B.A. Methylated and unmethylated DNA compartments in the sea urchin genome. *Cell* **1979**, *17*, 889–901.
30. Bird, A.; Taggart, M.; Macleod, D. Loss of rDNA methylation accompanies the onset of ribosomal gene activity in early development of *X. laevis*. *Cell* **1981**, *26*, 381–390.
31. Vardimon, L.; Kressmann, A.; Cedar, H.; Maechler, M.; Doerfler, W. Expression of a cloned adenovirus gene is inhibited by *in vitro* methylation. *Proc. Natl. Acad. Sci. USA* **1982**, *79*, 1073–1077.
32. Stein, R.; Razin, A.; Cedar, H. *In vitro* methylation of the hamster adenine phosphoribosyltransferase gene inhibits its expression in mouse L cells. *Proc. Natl. Acad. Sci. USA* **1982**, *79*, 3418–3422.
33. Ball, M.P.; Li, J.B.; Gao, Y.; Lee, J.-H.; LeProust, E.M.; Park, I.-H.; Xie, B.; Daley, G.Q.; Church, G.M. Targeted and genome-scale strategies reveal gene-body methylation signatures in human cells. *Nat. Biotechnol.* **2009**, *27*, 361–368.
34. Deng, J.; Shoemaker, R.; Xie, B.; Gore, A.; LeProust, E.M.; Antosiewicz-Bourget, J.; Egli, D.; Maherali, N.; Park, I.-H.; Yu, J.; *et al.* Targeted bisulfite sequencing reveals changes in DNA methylation associated with nuclear reprogramming. *Nat. Biotechnol.* **2009**, *27*, 353–360.
35. Rauch, T.A.; Wu, X.; Zhong, X.; Riggs, A.D.; Pfeifer, G.P. A human B cell methylome at 100-base pair resolution. *Proc. Natl. Acad. Sci. USA* **2009**, *106*, 671–678.
36. Trowbridge, J.J.; Orkin, S.H. DNA methylation in adult stem cells. *Epigenetics* **2010**, *5*, 189–193.
37. Razin, A.; Cedar, H. DNA methylation and gene expression. *Microbiol. Rev.* **1991**, *55*, 451–458.
38. Sato, S.; Yoshida, W.; Soejima, H.; Nakabayashi, K.; Hata, K. Methylation dynamics of IG-DMR and *Gtl2*-DMR during murine embryonic and placental development. *Genomics* **2011**, *98*, 120–127.
39. Bhogal, B.; Arnaudo, A.; Dymkowski, A.; Best, A.; Davis, T.L. Methylation at mouse *Cdkn1c* is acquired during postimplantation development and functions to maintain imprinted expression. *Genomics* **2004**, *84*, 961–970.
40. Umlauf, D.; Goto, Y.; Cao, R.; Cerqueira, F.; Wagschal, A.; Zhang, Y.; Feil, R. Imprinting along the *Kcnq1* domain on mouse chromosome 7 involves repressive histone methylation and recruitment of Polycomb group complexes. *Nat. Genet.* **2004**, *36*, 1296–1300.
41. Sasaki, H.; Ferguson-Smith, A.C.; Shum, A.S.W.; Barton, S.C.; Surani, M.A. Temporal and spatial regulation of H19 imprinting in normal and uniparental mouse embryos. *Development* **1995**, *121*, 4195–4202.

42. Lerchner, W.; Barlow, D.P. Paternal repression of the imprinted mouse *Igf2r* locus occurs during implantation and is stable in all tissues of the post-implantation mouse embryo. *Mech. Dev.* **1997**, *61*, 141–149.
43. Stadler, M.B.; Murr, R.; Burger, L.; Ivanek, R.; Lienert, F.; Schöler, A.; van Nimwegen, E.; Wirbelauer, C.; Oakeley, E.J.; Gaidatzis, D.; *et al.* DNA-binding factors shape the mouse methylome at distal regulatory regions. *Nature* **2011**, *480*, 490–495.
44. Klose, R.J.; Bird, A.P. Genomic DNA methylation: The mark and its mediators. *Trends Biochem. Sci.* **2006**, *31*, 89–97.
45. Meehan, R.R.; Lewis, J.D.; McKay, S.; Kleiner, E.L.; Bird, A.P. Identification of a mammalian protein that binds specifically to DNA containing methylated CpGs. *Cell* **1989**, *58*, 499–507.
46. Lewis, J.D.; Meehan, R.R.; Henzel, W.J.; Maurer-Fogy, I.; Jeppesen, P.; Klein, F.; Bird, A. Purification, sequence, and cellular localization of a novel chromosomal protein that binds to methylated DNA. *Cell* **1992**, *69*, 905–914.
47. Hendrich, B.; Bird, A. Identification and characterization of a family of mammalian methyl-CpG binding proteins. *Mol. Cell. Biol.* **1998**, *18*, 6538–6547.
48. Hendrich, B.; Tweedie, S. The methyl-CpG binding domain and the evolving role of DNA methylation in animals. *Trends Genet.* **2003**, *19*, 269–277.
49. Bird, A.P.; Wolffe, A.P. Methylation-induced repression—Belts, braces, and chromatin. *Cell* **1999**, *99*, 451–454.
50. Jones, P.L.; Veenstra, G.J.C.; Wade, P.A.; Vermaak, D.; Kass, S.U.; Landsberger, N.; Strouboulis, J.; Wolffe, A.P. Methylated DNA and MeCP2 recruit histone deacetylase to repress transcription. *Nat. Genet.* **1998**, *19*, 187–191.
51. Nan, X.; Ng, H.-H.; Johnson, C.A.; Laherty, C.D.; Turner, B.M.; Eisenman, R.N.; Bird, A. Transcriptional repression by the methyl-CpG-binding protein MeCP2 involves a histone deacetylase complex. *Nature* **1998**, *393*, 386–389.
52. Ng, H.-H.; Zhang, Y.; Hendrich, B.; Johnson, C.A.; Turner, B.M.; Erdjument-Bromage, H.; Tempst, P.; Reinberg, D.; Bird, A. MBD2 is a transcriptional repressor belonging to the MeCP1 histone deacetylase complex. *Nat. Genet.* **1999**, *23*, 58–61.
53. Wade, P.A.; Gegonne, A.; Jones, P.L.; Ballestar, E.; Aubry, F.; Wolffe, A.P. Mi-2 complex couples DNA methylation to chromatin remodelling and histone deacetylation. *Nat. Genet.* **1999**, *23*, 62–66.
54. Zhang, Y.; Ng, H.-H.; Erdjument-Bromage, H.; Tempst, P.; Bird, A.; Reinberg, D. Analysis of the NuRD subunits reveals a histone deacetylase core complex and a connection with DNA methylation. *Genes Dev.* **1999**, *13*, 1924–1935.
55. Sarraf, S.A.; Stancheva, I. Methyl-CpG binding protein MBD1 couples histone H3 methylation at lysine 9 by SETDB1 to DNA replication and chromatin assembly. *Mol. Cell* **2004**, *15*, 595–605.
56. Prendergast, G.C.; Lawe, D.; Ziff, E.B. Association of Myn, the murine homolog of Max, with c-Myc stimulates methylation-sensitive DNA binding and Ras cotransformation. *Cell* **1991**, *65*, 395–407.
57. Watt, F.; Molloy, P.L. Cytosine methylation prevents binding to DNA of a HeLa cell transcription

factor required for optimal expression of the adenovirus major late promoter. *Genes Dev.* **1988**, *2*, 1136–1143.

58. Comb, M.; Goodman, H.M. CpG methylation inhibits proenkephalin gene expression and binding of the transcription factor AP-2. *Nucleic Acids Res.* **1990**, *18*, 3975–3982.
59. DeChiara, T.M.; Robertson, E.J.; Efstratiadis, A. Parental imprinting of the mouse insulin-like growth factor II gene. *Cell* **1991**, *64*, 849–859.
60. Thorvaldsen, J.L.; Duran, K.L.; Bartolomei, M.S. Deletion of the *H19* differentially methylated domain results in loss of imprinted expression of *H19* and *Igf2*. *Genes Dev.* **1998**, *12*, 3693–3702.
61. Tremblay, K.D.; Duran, K.L.; Bartolomei, M.S. A 5' 2-kilobase-pair region of the imprinted mouse *H19* gene exhibits exclusive paternal methylation throughout development. *Mol. Cell. Biol.* **1997**, *17*, 4322–4329.
62. Szabó, P.E.; Tang, S.-H.E.; Rentsendorj, A.; Pfeifer, G.P.; Mann, J.R. Maternal-specific footprints at putative CTCF sites in the *H19* imprinting control region give evidence for insulator function. *Curr. Biol.* **2000**, *10*, 607–610.
63. Kanduri, C.; Pant, V.; Loukinov, D.; Pugacheva, E.; Qi, C.-F.; Wolffe, A.; Ohlsson, R.; Lobanenkov, V.V. Functional association of CTCF with the insulator upstream of the *H19* gene is parent of origin-specific and methylation-sensitive. *Curr. Biol.* **2000**, *10*, 853–856.
64. Hark, A.T.; Schoenherr, C.J.; Katz, D.J.; Ingram, R.S.; Levorse, J.M.; Tilghman, S.M. CTCF mediates methylation-sensitive enhancer-blocking activity at the *H19/Igf2* locus. *Nature* **2000**, *405*, 486–489.
65. Bell, A.C.; Felsenfeld, G. Methylation of a CTCF-dependent boundary controls imprinted expression of the *Igf2* gene. *Nature* **2000**, *405*, 482–485.
66. Webber, A.L.; Ingram, R.S.; Levorse, J.M.; Tilghman, S.M. Location of enhancers is essential for the imprinting of *H19* and *Igf2* genes. *Nature* **1998**, *391*, 711–715.
67. Leighton, P.A.; Saam, J.R.; Ingram, R.S.; Stewart, C.L.; Tilghman, S.M. An enhancer deletion affects both *H19* and *Igf2* expression. *Genes Dev.* **1995**, *9*, 2079–2089.
68. Phillips, J.E.; Corces, V.G. CTCF: Master weaver of the genome. *Cell* **2009**, *137*, 1194–1211.
69. Pant, V.; Kurukuti, S.; Pugacheva, E.; Shamsuddin, S.; Mariano, P.; Renkawitz, R.; Klenova, E.; Lobanenkov, V.; Ohlsson, R. Mutation of a single CTCF target site within the *H19* imprinting control region leads to loss of *Igf2* imprinting and complex patterns of *de novo* methylation upon maternal inheritance. *Mol. Cell. Biol.* **2004**, *24*, 3497–3504.
70. Yusufzai, T.M.; Tagami, H.; Nakatani, Y.; Felsenfeld, G. CTCF tethers an insulator to subnuclear sites, suggesting shared insulator mechanisms across species. *Mol. Cell* **2004**, *13*, 291–298.
71. Yoon, Y.S.; Jeong, S.; Rong, Q.; Park, K.-Y.; Chung, J.H.; Pfeifer, K. Analysis of the *H19ICR* insulator. *Mol. Cell. Biol.* **2007**, *27*, 3499–3510.
72. Li, T.; Hu, J.-F.; Qiu, X.; Ling, J.; Chen, H.; Wang, S.; Hou, A.; Vu, T.H.; Hoffman, A.R. CTCF regulates allelic expression of *Igf2* by orchestrating a promoter-polycomb repressive complex 2 intrachromosomal loop. *Mol. Cell. Biol.* **2008**, *28*, 6473–6482.
73. Kurukuti, S.; Tiwari, V.K.; Tavoosidana, G.; Pugacheva, E.; Murrell, A.; Zhao, Z.; Lobanenkov, V.; Reik, W.; Ohlsson, R. CTCF binding at the *H19* imprinting control region mediates maternally inherited higher-order chromatin conformation to restrict enhancer access to

Igf2. *Proc. Natl. Acad. Sci. USA* **2006**, *103*, 10684–10689.

74. Murrell, A.; Heeson, S.; Reik, W. Interaction between differentially methylated regions partitions the imprinted genes *Igf2* and *H19* into parent-specific chromatin loops. *Nat. Genet.* **2004**, *36*, 889–893.
75. Wendt, K.S.; Yoshida, K.; Itoh, T.; Bando, M.; Koch, B.; Schirghuber, E.; Tsutsumi, S.; Nagae, G.; Ishihara, K.; Mishiro, T.; *et al.* Cohesin mediates transcriptional insulation by CCCTC-binding factor. *Nature* **2008**, *451*, 796–801.
76. Hon, G.C.; Hawkins, R.D.; Ren, B. Predictive chromatin signatures in the mammalian genome. *Hum. Mol. Genet.* **2009**, *18*, R195–R201.
77. Ernst, J.; Kheradpour, P.; Mikkelsen, T.S.; Shoresh, N.; Ward, L.D.; Epstein, C.B.; Zhang, X.; Wang, L.; Issner, R.; Coyne, M.; *et al.* Mapping and analysis of chromatin state dynamics in nine human cell types. *Nature* **2011**, *473*, 43–49.
78. Li, B.; Carey, M.; Workman, J.L. The role of chromatin during transcription. *Cell* **2007**, *128*, 707–719.
79. Hon, G.; Wang, W.; Ren, B. Discovery and annotation of functional chromatin signatures in the human genome. *PLoS Comput. Biol.* **2009**, *5*, e1000566.
80. Zhou, V.W.; Goren, A.; Bernstein, B.E. Charting histone modifications and the functional organization of mammalian genomes. *Nat. Rev. Genet.* **2011**, *12*, 7–18.
81. Grandjean, V.; O'Neill, L.; Sado, T.; Turner, B.; Ferguson-Smith, A. Relationship between DNA methylation, histone H4 acetylation and gene expression in the mouse imprinted *Igf2-H19* domain. *FEBS Lett.* **2001**, *488*, 165–169.
82. Pedone, P.V.; Pikaart, M.J.; Cerrato, F.; Vernucci, M.; Ungaro, P.; Bruni, C.B.; Riccio, A. Role of histone acetylation and DNA methylation in the maintenance of the imprinted expression of the *H19* and *Igf2* genes. *FEBS Lett.* **1999**, *458*, 45–50.
83. Chen, T. Mechanistic and functional links between histone methylation and DNA methylation. *Prog. Mol. Biol. Transl. Sci.* **2011**, *101*, 335–348.
84. Cedar, H.; Bergman, Y. Linking DNA methylation and histone modification: Patterns and paradigms. *Nat. Rev. Genet.* **2009**, *10*, 295–304.
85. Bourc'his, D.; Xu, G.-L.; Lin, C.-S.; Bollman, B.; Bestor, T.H. Dnmt3L and the establishment of maternal genomic imprints. *Science* **2001**, *294*, 2536–2539.
86. Bourc'his, D.; Bestor, T.H. Meiotic catastrophe and retrotransposon reactivation in male germ cells lacking Dnmt3L. *Nature* **2004**, *431*, 96–99.
87. Jia, D.; Jurkowska, R.Z.; Zhang, X.; Jeltsch, A.; Cheng, X. Structure of Dnmt3a bound to Dnmt3L suggests a model for *de novo* DNA methylation. *Nature* **2007**, *449*, 248–251.
88. Margot, J.B.; Ehrenhofer-Murray, A.E.; Leonhardt, H. Interactions within the mammalian DNA methyltransferase family. *BMC Mol. Biol.* **2003**, *4*, doi:10.1186/1471-2199-4-7.
89. Suetake, I.; Shinozaki, F.; Miyagawa, J.; Takeshima, H.; Tajima, S. DNMT3L stimulates the DNA methylation activity of Dnmt3a and Dnmt3b through a direct interaction. *J. Biol. Chem.* **2004**, *279*, 27816–27823.
90. Ooi, S.K.T.; Qiu, C.; Bernstein, E.; Li, K.; Jia, D.; Yang, Z.; Erdjument-Bromage, H.; Tempst, P.; Lin, S.-P.; Allis, C.D.; *et al.* DNMT3L connects unmethylated lysine 4 of histone H3 to *de novo*

methylation of DNA. *Nature* **2007**, *448*, 714–717.

91. Ciccone, D.N.; Su, H.; Hevi, S.; Gay, F.; Lei, H.; Bajko, J.; Xu, G.; Li, E.; Chen, T. KDM1B is a histone H3K4 demethylase required to establish maternal genomic imprints. *Nature* **2009**, *461*, 415–418.
92. Ikegami, K.; Iwatani, M.; Suzuki, M.; Tachibana, M.; Shinkai, Y.; Tanaka, S.; Greally, J.M.; Yagi, S.; Hattori, N.; Shiota, K. Genome-wide and locus-specific DNA hypomethylation in G9a deficient mouse embryonic stem cells. *Genes Cells* **2007**, *12*, 1–11.
93. Tachibana, M.; Matsumura, Y.; Fukuda, M.; Kimura, H.; Shinkai, Y. G9a/GLP complexes independently mediate H3K9 and DNA methylation to silence transcription. *EMBO J.* **2008**, *27*, 2681–2690.
94. Dong, K.B.; Maksakova, I.A.; Mohn, F.; Leung, D.; Appanah, R.; Lee, S.; Yang, H.W.; Lam, L.L.; Mager, D.L.; Schübeler, D.; *et al.* DNA methylation in ES cells requires the lysine methyltransferase G9a but not its catalytic activity. *EMBO J.* **2008**, *27*, 2691–2701.
95. Lehnertz, B.; Ueda, Y.; Derijck, A.A.H.A.; Braunschweig, U.; Perez-Burgos, L.; Kubicek, S.; Chen, T.; Li, E.; Jenuwein, T.; Peters, A.H.F.M. Suv39h-mediated histone H3 lysine 9 methylation directs DNA methylation to major satellite repeats at pericentric heterochromatin. *Curr. Biol.* **2003**, *13*, 1192–1200.
96. Viré, E.; Brenner, C.; Deplus, R.; Blanchon, L.; Fraga, M.; Didelot, C.; Morey, L.; van Eynde, A.; Bernard, D.; Vanderwinden, J.-M.; *et al.* The Polycomb group protein EZH2 directly controls DNA methylation. *Nature* **2006**, *439*, 871–874.
97. Li, H.; Rauch, T.; Chen, Z.-X.; Szabó, P.E.; Riggs, A.D.; Pfeifer, G.P. The histone methyltransferase SETDB1 and the DNA methyltransferase DNMT3A interact directly and localize to promoters silenced in cancer cells. *J. Biol. Chem.* **2006**, *281*, 19489–19500.
98. Jackson, J.P.; Lindroth, A.M.; Cao, X.; Jacobsen, S.E. Control of CpNpG DNA methylation by the KRYPTONITE histone H3 methyltransferase. *Nature* **2002**, *416*, 556–560.
99. Freitag, M.; Hickey, P.C.; Khlafallah, T.K.; Read, N.D.; Selker, E.U. HP1 is essential for DNA methylation in *Neurospora*. *Mol. Cell* **2004**, *13*, 427–434.
100. Xin, Z.; Tachibana, M.; Guggiari, M.; Heard, E.; Shinkai, Y.; Wagstaff, J. Role of histone methyltransferase G9a in CpG methylation of the Prader-Willi syndrome imprinting center. *J. Biol. Chem.* **2003**, *278*, 14996–15000.
101. Wagschal, A.; Sutherland, H.G.; Woodfine, K.; Henckel, A.; Chebli, K.; Schulz, R.; Oakey, R.J.; Bickmore, W.A.; Feil, R. G9a histone methyltransferase contributes to imprinting in the mouse placenta. *Mol. Cell. Biol.* **2008**, *28*, 1104–1113.
102. Mercer, T.R.; Mattick, J.S. Structure and function of long noncoding RNAs in epigenetic regulation. *Nat. Struct. Mol. Biol.* **2013**, *20*, 300–307.
103. Mohammad, F.; Mondal, T.; Kanduri, C. Epigenetics of imprinted long noncoding RNAs. *Epigenetics* **2009**, *4*, 277–286.
104. Ponting, C.P.; Oliver, P.L.; Reik, W. Evolution and functions of long noncoding RNAs. *Cell* **2009**, *136*, 629–641.
105. Derrien, T.; Johnson, R.; Bussotti, G.; Tanzer, A.; Djebali, S.; Tilgner, H.; Guernec, G.; Martin, D.; Merkel, A.; Knowles, D.G.; *et al.* The GENCODE v7 catalog of human long noncoding RNAs: Analysis of their gene structure, evolution, and expression. *Genome Res.*

2012, *22*, 1775–1789.

106. Cabili, M.N.; Trapnell, C.; Goff, L.; Koziol, M.; Tazon-Vega, B.; Regev, A.; Rinn, J.L. Integrative annotation of human large intergenic noncoding RNAs reveals global properties and specific subclasses. *Genes Dev.* **2011**, *25*, 1915–1927.
107. Mercer, T.R.; Gerhardt, D.J.; Dinger, M.E.; Crawford, J.; Trapnell, C.; Jeddeloh, J.A.; Mattick, J.S.; Rinn, J.L. Targeted RNA sequencing reveals the deep complexity of the human transcriptome. *Nat. Biotechnol.* **2012**, *30*, 99–104.
108. Chu, C.; Qu, K.; Zhong, F.L.; Artandi, S.E.; Chang, H.Y. Genomic maps of long noncoding RNA occupancy reveal principles of RNA-chromatin interactions. *Mol. Cell* **2011**, *44*, 667–678.
109. Latos, P.A.; Pauler, F.M.; Koerner, M.V; Senergin, H.B.; Hudson, Q.J.; Stocsits, R.R.; Allhoff, W.; Stricker, S.H.; Klement, R.M.; Warczok, K.E.; *et al. Airn* transcriptional overlap, but not its lncRNA products, induces imprinted *Igf2r* silencing. *Science* **2012**, *338*, 1469–1472.
110. Nagano, T.; Mitchell, J.A.; Sanz, L.A.; Pauler, F.M.; Ferguson-Smith, A.C.; Feil, R.; Frase, P. The Air noncoding RNA epigenetically silences transcription by targeting G9a to chromatin. *Science* **2008**, *322*, 1717–1720.
111. Osato, N.; Suzuki, Y.; Ikeo, K.; Gojobori, T. Transcriptional interferences in *cis* natural antisense transcripts of humans and mice. *Genetics* **2007**, *176*, 1299–1306.
112. Rinn, J.L.; Chang, H.Y. Genome regulation by long noncoding RNAs. *Annu. Rev. Biochem.* **2012**, *81*, 145–166.
113. Howell, C.Y.; Bestor, T.H.; Ding, F.; Latham, K.E.; Mertineit, C.; Trasler, J.M.; Chaillet, J.R. Genomic imprinting disrupted by a maternal effect mutation in the *Dnmt1* gene. *Cell* **2001**, *104*, 829–838.
114. Lewis, A.; Mitsuya, K.; Umlauf, D.; Smith, P.; Dean, W.; Walter, J.; Higgins, M.; Feil, R.; Reik, W. Imprinting on distal chromosome 7 in the placenta involves repressive histone methylation independent of DNA methylation. *Nat. Genet.* **2004**, *36*, 1291–1295.
115. Pandey, R.R.; Mondal, T.; Mohammad, F.; Enroth, S.; Redrup, L.; Komorowski, J.; Nagano, T.; Mancini-DiNardo, D.; Kanduri, C. *Kcnq1ot1* antisense noncoding RNA mediates lineage-specific transcriptional silencing through chromatin-level regulation. *Mol. Cell* **2008**, *32*, 232–246.
116. Seidl, C.I.M.; Stricker, S.H.; Barlow, D.P. The imprinted *Air* ncRNA is an atypical RNAPII transcript that evades splicing and escapes nuclear export. *EMBO J.* **2006**, *25*, 3565–3575.
117. Shin, J.-Y.; Fitzpatrick, G.V.; Higgins, M.J. Two distinct mechanisms of silencing by the KvDMR1 imprinting control region. *EMBO J.* **2008**, *27*, 168–178.
118. Mancini-DiNardo, D.; Steele, S.J.S.; Levorse, J.M.; Ingram, R.S.; Tilghman, S.M. Elongation of the *Kcnq1ot1* transcript is required for genomic imprinting of neighboring genes. *Genes Dev.* **2006**, *20*, 1268–1282.
119. Sleutels, F.; Zwart, R.; Barlow, D.P. The non-coding *Air* RNA is required for silencing autosomal imprinted genes. *Nature* **2002**, *415*, 810–813.
120. Zwart, R.; Sleutels, F.; Wutz, A.; Schinkel, A.H.; Barlow, D.P. Bidirectional action of the *Igf2r* imprint control element on upstream and downstream imprinted genes. *Genes Dev.* **2001**, *15*, 2361–2366.

121. Khalil, A.M.; Guttman, M.; Huarte, M.; Garber, M.; Raj, A.; Morales, R.D.; Thomas, K.; Presser, A.; Bernstein, B.E.; van Oudenaarden, A.; *et al.* Many human large intergenic noncoding RNAs associate with chromatin-modifying complexes and affect gene expression. *Proc. Natl. Acad. Sci. USA* **2009**, *106*, 11667–11672.

122. Guttman, M.; Donaghey, J.; Carey, B.W.; Garber, M.; Grenier, J.K.; Munson, G.; Young, G.; Lucas, A.B.; Ach, R.; Bruhn, L.; *et al.* lincRNAs act in the circuitry controlling pluripotency and differentiation. *Nature* **2011**, *477*, 295–300.

123. Kaneko, S.; Bonasio, R.; Saldaña-Meyer, R.; Yoshida, T.; Son, J.; Nishino, K.; Umezawa, A.; Reinberg, D. Interactions between JARID2 and noncoding RNAs regulate PRC2 recruitment to chromatin. *Mol. Cell* **2014**, *53*, 1–11.

124. Nakamura, T.; Arai, Y.; Umehara, H.; Masuhara, M.; Kimura, T.; Taniguchi, H.; Sekimoto, T.; Ikawa, M.; Yoneda, Y.; Okabe, M.; *et al.* PGC7/Stella protects against DNA demethylation in early embryogenesis. *Nat. Cell Biol.* **2007**, *9*, 64–71.

125. Ferrón, S.R.; Charalambous, M.; Radford, E.; McEwen, K.; Wildner, H.; Hind, E.; Morante-Redolat, J.M.; Laborda, J.; Guillemot, F.; Bauer, S.R.; *et al.* Postnatal loss of *Dlk1* imprinting in stem cells and niche astrocytes regulates neurogenesis. *Nature* **2011**, *475*, 381–385.

Permissions

All chapters in this book were first published in MDPI; hereby published with permission under the Creative Commons Attribution License or equivalent. Every chapter published in this book has been scrutinized by our experts. Their significance has been extensively debated. The topics covered herein carry significant findings which will fuel the growth of the discipline. They may even be implemented as practical applications or may be referred to as a beginning point for another development.

The contributors of this book come from diverse backgrounds, making this book a truly international effort. This book will bring forth new frontiers with its revolutionizing research information and detailed analysis of the nascent developments around the world.

We would like to thank all the contributing authors for lending their expertise to make the book truly unique. They have played a crucial role in the development of this book. Without their invaluable contributions this book wouldn't have been possible. They have made vital efforts to compile up to date information on the varied aspects of this subject to make this book a valuable addition to the collection of many professionals and students.

This book was conceptualized with the vision of imparting up-to-date information and advanced data in this field. To ensure the same, a matchless editorial board was set up. Every individual on the board went through rigorous rounds of assessment to prove their worth. After which they invested a large part of their time researching and compiling the most relevant data for our readers.

The editorial board has been involved in producing this book since its inception. They have spent rigorous hours researching and exploring the diverse topics which have resulted in the successful publishing of this book. They have passed on their knowledge of decades through this book. To expedite this challenging task, the publisher supported the team at every step. A small team of assistant editors was also appointed to further simplify the editing procedure and attain best results for the readers.

Apart from the editorial board, the designing team has also invested a significant amount of their time in understanding the subject and creating the most relevant covers. They scrutinized every image to scout for the most suitable representation of the subject and create an appropriate cover for the book.

The publishing team has been an ardent support to the editorial, designing and production team. Their endless efforts to recruit the best for this project, has resulted in the accomplishment of this book. They are a veteran in the field of academics and their pool of knowledge is as vast as their experience in printing. Their expertise and guidance has proved useful at every step. Their uncompromising quality standards have made this book an exceptional effort. Their encouragement from time to time has been an inspiration for everyone.

The publisher and the editorial board hope that this book will prove to be a valuable piece of knowledge for researchers, students, practitioners and scholars across the globe.

List of Contributors

Navneet K. Matharu
Department of Bioengineering and Therapeutic Sciences, Institute for Human Genetics, University of California San Francisco, San Francisco, CA 94143, USA

Sajad H. Ahanger
Department of Ophthalmology, Lab for Retinal Cell Biology, University of Zurich, Wagistrasse 14, Zurich 8952, Switzerland

Matthew Brown
University of Queensland Diamantina Institute, Translational Research Institute, Princess Alexandra Hospital, 37 Kent Road, Woolloongabba, Brisbane 4102, Queensland, Australia

Emma Duncan
University of Queensland Diamantina Institute, Translational Research Institute, Princess Alexandra Hospital, 37 Kent Road, Woolloongabba, Brisbane 4102, Queensland, Australia
Department of Endocrinology and Diabetes, Royal Brisbane and Women's Hospital, Herston 4029, Queensland, Australia

Eileen M. Shore
Center for Research in FOP and Related Disorders, Department of Orthopaedic Surgery, Perelman School of Medicine, University of Pennsylvania, 3450 Hamilton Walk, Philadelphia, PA 19104, USA
Department of Genetics, Perelman School of Medicine, University of Pennsylvania, Curie Boulevard, Philadelphia, PA 19104, USA

Jessica N. Cooke Bailey and Jonathan L. Haines
Department of Epidemiology and Biostatistics, Case Western Reserve University Medical Center, Cleveland, OH 44106, USA

Margaret A. Pericak-Vance
Hussman Institute for Human Genomics, Miller School of Medicine, University of Miami, Miami, FL 33136, USA

Anneke Lucassen
Clinical Ethics and Law Unit, Wessex Clinical Genetics Service, The Princess Anne Hospital, Southampton, SO16 5YA, UK

Richard S. Houlston
Division of Genetics and Epidemiology, The Institute of Cancer Research, Sutton, Surrey SM2 5NG, UK

Elena Kypri
NIPD Genetics Ltd., Neas Engomis 31, Engomi, Nicosia 2409, Cyprus

Elisavet A. Papageorgiou and George Koumbaris
NIPD Genetics Ltd., Neas Engomis 31, Engomi, Nicosia 2409, Cyprus
The Cyprus Institute of Neurology and Genetics, 6 International Airport Avenue, Ayios Dometios, Nicosia 2370, Cyprus

Michael Hadjidaniel and Philippos C. Patsalis
The Cyprus Institute of Neurology and Genetics, 6 International Airport Avenue, Ayios Dometios, Nicosia 2370, Cyprus

Anke R. Hammerschlag and Tinca J. C. Polderman
Department of Complex Trait Genetics, Center for Neurogenomics and Cognitive Research, Neuroscience Campus Amsterdam, VU University Amsterdam, De Boelelaan 1085, 1081 HV Amsterdam, The Netherlands

Christiaan de Leeuw
Department of Complex Trait Genetics, Center for Neurogenomics and Cognitive Research, Neuroscience Campus Amsterdam, VU University Amsterdam, De Boelelaan 1085, 1081 HV Amsterdam, The Netherlands
Institute for Computing and Information Sciences, Radboud University Nijmegen, 6500 GL Nijmegen, The Netherlands

Henning Tiemeier and Tonya White
Department of Child and Adolescent Psychiatry, Erasmus University Medical Center and Sophia Children's Hospital, 3000 CB Rotterdam, The Netherlands

August B. Smit
Department of Molecular and Cellular Neurobiology, Center for Neurogenomics and Cognitive Research, Neuroscience Campus Amsterdam, VU University Amsterdam, De Boelelaan 1085, 1081 HV Amsterdam, The Netherlands

Matthijs Verhage
Department of Functional Genomics, Center for Neurogenomics and Cognitive Research, Neuroscience Campus Amsterdam, VU University Amsterdam, De Boelelaan 1085, 1081 HV Amsterdam, The Netherlands
Department of Clinical Genetics, VU University Medical Center, 1007 MB Amsterdam, The Netherlands

Danielle Posthuma
Department of Complex Trait Genetics, Center for Neurogenomics and Cognitive Research, Neuroscience Campus Amsterdam, VU University Amsterdam, De Boelelaan 1085, 1081 HV Amsterdam, The Netherlands
Department of Child and Adolescent Psychiatry, Erasmus University Medical Center and Sophia Children's Hospital, 3000 CB Rotterdam, The Netherlands
Department of Clinical Genetics, VU University Medical Center, 1007 MB Amsterdam, The Netherlands

Grégory Caignard
Department of Human Genetics, McGill University, Montréal, QC H3G 0B1, Canada
Complex Traits Group, McGill University, Montréal, QC H3G 0B1, Canada

Megan M. Eva and Danielle Malo
Department of Human Genetics, McGill University, Montréal, QC H3G 0B1, Canada
Complex Traits Group, McGill University, Montréal, QC H3G 0B1, Canada
Department of Medicine, McGill University, Montréal, QC, H3G 0B1, Canada

Philippe Gros and Rebekah van Bruggen
Complex Traits Group, McGill University, Montréal, QC H3G 0B1, Canada
Department of Biochemistry, McGill University, Montréal, QC, H3G 0B1, Canada

Robert Eveleigh and Guillaume Bourque
McGill University and Genome Quebec Innovation Center, Montréal, QC, H3A 1A4, Canada

Silvia M. Vidal
Department of Human Genetics, McGill University, Montréal, QC H3G 0B1, Canada
Complex Traits Group, McGill University, Montréal, QC H3G 0B1, Canada
McGill Life Sciences Complex, Bellini Building, 3649 Sir William Osler Promenade, Room 367, Montreal, QC, H3G 0B1, Canada

Yunxin Fu
Division of Biostatistics and Human Genetics Center, The University of Texas Health Science Center at Houston, 1200 Herman Pressler, Houston, TX 77025, USA
Laboratory for Conservation and Utilization of Bio-Resources, Yunnan University, Kunming 650091, China

Radoslaw K. Ejsmont and Bassem A. Hassan
VIB Center for the Biology of Disease, VIB, 3000 Leuven, Belgium
Center for Human Genetics, University of Leuven School of Medicine, 3000 Leuven, Belgium

Donald Freed
Program in Biochemistry, Cellular and Molecular Biology, Johns Hopkins School of Medicine, Baltimore, MD 21205, USA
Department of Neurology, Kennedy Krieger Institute, 707 N. Broadway, Baltimore, MD 21205, USA

Eric L. Stevens
Department of Neurology, Kennedy Krieger Institute, 707 N. Broadway, Baltimore, MD 21205, USA
Department of Psychiatry and Behavioral Sciences, Johns Hopkins School of Medicine, Baltimore, MD 21205, USA

Jonathan Pevsner
Program in Biochemistry, Cellular and Molecular Biology, Johns Hopkins School of Medicine, Baltimore, MD 21205, USA
Department of Neurology, Kennedy Krieger Institute, 707 N. Broadway, Baltimore, MD 21205, USA
Department of Psychiatry and Behavioral Sciences, Johns Hopkins School of Medicine, Baltimore, MD 21205, USA

Stacey Pereira and Amy L. McGuire
Center for Medical Ethics and Health Policy, Baylor College of Medicine, Houston, TX 77030, USA

Richard A. Gibbs
Human Genome Sequencing Center, Baylor College of Medicine, Houston, TX 77030, USA

Bjorn T. Adalsteinsson and Anne C. Ferguson-Smith
Department of Genetics, University of Cambridge, Cambridge CB2 3EH, UK

Index

www.ingramcontent.com/pod-product-compliance
Lightning Source LLC
LaVergne TN
LVHW061513220926
848485LV00018B/901
9781641167970